Enhancing Research Output in Higher Education:

Research Proposals, Profiles, and Publishing

Andile Alfred Mdikana
University of South Africa, South Africa

IGI Global
Scientific Publishing
Publishing Tomorrow's Research Today

Published in the United States of America by
IGI Global Scientific Publishing
701 East Chocolate Avenue
Hershey, PA, 17033, USA
Tel: 717-533-8845
Fax: 717-533-8661
E-mail: cust@igi-global.com
Website: https://www.igi-global.com

Library of Congress Cataloging-in-Publication Data

CIP Pending
ISBN: 979-8-3373-0806-7
EISBN: 979-8-3373-0808-1

British Cataloguing in Publication Data
A Cataloguing in Publication record for this book is available from the British Library.

Table of Contents

Detailed Table of Contents

This chapter serves as a comprehensive guide for emerging academics and postgraduate students navigating higher education research. Its primary objective is to equip readers with skills, knowledge, and strategies to excel in research, publish effectively, and contribute to knowledge production. The chapter empowers researchers by delving into best practices, methodologies, and trends. It provides guidance on effective writing and publication strategies, enabling researchers to disseminate their findings and contribute to scholarly discourse. Embracing diversity, the chapter encourages a multifaceted approach, exploring frameworks and marginalized voices. This challenges paradigms and fosters a more representative body of knowledge. The work supports research supervision, providing strategies to enhance postgraduate student guidance, recognizing supervision's crucial role in shaping emerging scholars' development. Ultimately, this work aspires to make a significant contribution to knowledge production in higher education.

This research explores strategies to boost research output among academicians in Tamil Nadu's self-financing colleges. Despite higher education's research emphasis, these professionals face challenges in conducting and disseminating research. Identifying effective empowerment strategies is crucial for enhancing the research culture. The study aims to link academician empowerment to improved research performance by examining personal, professional, technological, institutional, and community factors. It highlights the balancing act between teaching, administrative duties, and research responsibilities, compounded by limited resources like time and infrastructure. Using a quantitative approach with 197 participants, the research employs analysis, exploring relationships Findings underscore the positive impact of empowerment on research performance, with implications for policymakers and administrators to enhance environments and academic quality in these institutions, despite limitations in sample representativeness and the study's focus on self-financing colleges within Tamil Nadu.

Any researcher needs to develop a sound research profile to boost his or her standing in the scholarly and research community. Developing a research profile starts with an idea that is brainstormed to develop a research focus. It also includes defining a research niche and stating the rationale and the aims and objectives of the research niche (University of Birmingham). It is to be hoped that emerging academics would, through this input, be in a better position to develop appealing, well thought out, well-organized and professional research profiles.

This chapter explores the concept of research supervision in higher education, focusing on the roles and responsibilities of supervisors, the supervisory relationship, and the key factors that impact the effectiveness of research supervision. The chapter starts by defining research supervision and its significance for supervisors and supervisees. It then delves into the complex dynamics of the supervisory relationship, examining the different roles and responsibilities of supervisors, the importance of communication and feedback, and the challenges that can arise in this relationship. Finally, the chapter provides practical guidance and strategies for enhancing the effectiveness of research supervision, drawing on current research and best practices in the field.

This chapter delves into the challenges and transformative potential of research supervision within the African higher education landscape. It provides practical insights into effective supervision strategies, mentoring frameworks, and innovative methodologies aimed at fostering high-quality research proposals and dissertations among postgraduate students. By exploring culturally relevant supervision frameworks through the lens of the Africanisation concept, the chapter highlights the importance of integrating indigenous knowledge systems into the academic discourse. It addresses key issues such as the gap between traditional and modern supervision models, limited mentorship programs, and cross-cultural challenges. The chapter offers solutions including structured mentorship programs, supervisor training, and enhanced research infrastructure, aiming to empower supervisors and postgraduate students alike, ultimately contributing to the development of a more inclusive and supportive research environment.

Writing a research proposal is a requirement for obtaining an undergraduate or postgraduate degree at institutions of higher education globally, however many students struggle to write such a proposal. This is especially true for students who are writing a proposal for the first time, despite research methodology modules that introduce them to the process. This often leads to their studies being delayed, which can result in them not graduating on time. In certain extreme cases, students withdraw from their studies because they have not completed the dissertation required to obtain their intended degree, even when they have passed their coursework. Higher education institutions could use the content of this chapter to develop appropriate targeted interventions and ongoing strategies for use by students and study supervisors to address the challenges that students face when writing their research proposals.

Today's academicians and academic fraternity are very much dependent on research. A research proposal, if accepted, provides the funds for smooth performance of the same. Hence, a proper and concise proposal is required to be drafted for easy acceptance. This chapter offers a detailed roadmap for academics navigating the intricate process of research proposal writing. It also provides a thorough examination of developing a compelling research proposal. Beginning with an exploration of the fundamental principles underlying research proposal construction, the chapter progresses to issues such as formulating research objectives, conducting a comprehensive literature review, and designing a robust methodology. Ethical considerations and budgetary planning are also addressed, ensuring researchers are equipped to navigate potential challenges ethically and financially. By synthesizing theoretical frameworks with practical guidance, this chapter serves as an invaluable resource for academicians seeking to enhance their research proposal writing skills.

The chapter provides an overview of the essential components of a research proposal for academicians in higher education. It includes guidance on developing a research question, identifying key sources and literature, and outlining the research methodology. Additionally, it covers the importance of effective abstracts and keywords in a research proposal. The chapter provides tips for creating a clear and concise proposal that effectively communicates the purpose and goals of the research. This chapter discusses the key elements of a research proposal for academicians in higher education, including the research question, literature review, methodology, and outcomes. It also highlights the role of effective keywords in helping potential readers find and understand the research proposal. The research proposal is the skeleton or backbone of the research as it is structured in such a way that indicates what the writer intends to do. The mission of having a chapter on writing a research proposal is to provide higher education students and researchers with the knowledge and skills.

This chapter provides a comprehensive guide to writing an effective research proposal. It covers the fundamental components of a research proposal, including the title, abstract, introduction, literature review, research questions, methodology, timeline, budget, and references. The chapter emphasizes the importance of clear, concise, and well-organized writing and provides practical advice for developing a robust research design and preparing a realistic budget and timeline. Through detailed explanations, practical examples, and step-by-step instructions, readers will learn how to effectively communicate their research objectives, significance, methodology, and potential impact.

Preface

In higher education, research serves as the cornerstone of intellectual advancement, innovation, and societal impact. It is through the production of quality research that institutions contribute to solving complex global challenges, shape public policies, and establish themselves as beacons of academic excellence. Faculty, researchers, and scholars, as the driving forces behind this endeavor, bear the responsibility of not only generating new knowledge but also effectively communicating and disseminating their findings to a broader audience.

Enhancing Research Output in Higher Education: Research Proposals, Profiles, and Publishing aims to provide a comprehensive guide for academics seeking to strengthen their research impact. This volume offers practical insights into critical aspects of the research process, from crafting compelling research proposals to building a robust academic profile and navigating the intricate publishing landscape. By presenting best practices, real-world examples, and strategies for maximizing research visibility, this book equips readers with the tools necessary to elevate their scholarly contributions and achieve greater recognition for their work.

The chapters in this book address a wide range of topics, including academic authorship, doctoral research, proposal development, thesis publishing, and the role of research supervision in fostering quality output. Additionally, it highlights the implications of research on institutional reputation, funding opportunities, and the evolving expectations of higher education stakeholders. Together, these themes underscore the multifaceted nature of research in today's academic landscape.

CHAPTER OVERVIEW

In Chapter 1, this chapter offers an indispensable resource for emerging academics and postgraduate students embarking on the journey of higher education research. By equipping readers with the skills, methodologies, and strategies necessary for effective knowledge production and publication, it sets the stage for meaningful contributions to academic discourse. The chapter underscores the importance of diversity in research frameworks, integrating marginalized voices to challenge conventional paradigms and foster inclusivity. Furthermore, it highlights the critical role of research supervision, providing actionable strategies to support postgraduate students and advance their development into accomplished scholars.

In Chapter 2, this chapter explores the challenges and prospects of cultivating a research-oriented culture within Cambodian higher education. It delves into the role of limited resources and capacity in shaping the research landscape while emphasizing the importance of academic networks and collaborations. By examining the strategic role of the Ministry of Education, Youth and Sport, the chapter outlines initiatives to address existing barriers and foster long-term progress in research, contributing significantly to Cambodia's development.

In Chapter 3, this chapter investigates the challenges faced by academicians in Tamil Nadu's self-financing colleges in conducting and disseminating research. It emphasizes the need for empowerment strategies to enhance research performance and academic culture. Using quantitative analysis, the chapter links factors such as professional development, technological access, and institutional support to improved research outcomes. It also provides practical recommendations for policymakers and administrators to create environments conducive to academic excellence despite resource constraints.

In Chapter 4, this chapter focuses on the development of robust research profiles as a cornerstone for academic success. By guiding researchers through defining their niche, formulating objectives, and articulating a professional narrative, it empowers them to establish a credible presence in the scholarly community. The chapter offers practical advice aimed at enhancing visibility and impact in the academic landscape.

In Chapter 5, this chapter examines the pivotal role of research supervision in higher education. It explores the dynamics of the supervisor-supervisee relationship, including communication, feedback, and challenges, while offering best practices to enhance supervision effectiveness. With practical strategies informed by current research, this chapter provides valuable insights into fostering successful mentorships that nurture scholarly growth.

In Chapter 6, this chapter addresses the transformative potential of culturally relevant research supervision in African higher education. By integrating indigenous knowledge systems and embracing the Africanisation concept, it proposes innovative strategies for bridging traditional and modern supervision models. Structured mentorship programs and supervisor training are presented as solutions to overcome mentorship gaps and infrastructure challenges, ultimately fostering an inclusive and supportive academic environment.

In Chapter 7, this chapter emphasizes the critical importance of well-crafted problem statements in guiding research within the African context. By framing challenges through the lens of Africanization, it highlights the potential for indigenous perspectives to enrich educational practices and scholarship. Through a comprehensive literature review, this chapter underscores the role of problem statements in driving meaningful and contextually relevant research.

In Chapter 8, this chapter advocates for an applied approach to research in higher education, focusing on selecting real-world problems that yield actionable insights. By bridging the gap between academia and industry, it provides a practical process and checklist to transform real-world challenges into researchable questions. This methodology emphasizes stakeholder impact and relevance, fostering innovation and evidence-based solutions.

In Chapter 9, this chapter addresses the challenges faced by students in writing research proposals, particularly those new to the process. It explores the barriers that lead to delays and dropout rates while proposing targeted interventions and strategies for supervisors and institutions. By providing practical solutions, the chapter aims to equip students with the tools needed to successfully navigate the research proposal process.

In Chapter 10, this chapter serves as a detailed roadmap for academics tackling the complexities of research proposal writing. Covering fundamental principles, methodology design, ethical considerations, and budgeting, it offers practical guidance for crafting compelling and concise proposals. By integrating theoretical and practical insights, this chapter supports researchers in securing funding and advancing their academic pursuits.

In Chapter 11, this chapter outlines the essential components of a research proposal, from formulating research questions to creating effective abstracts and keywords. With a focus on clear and concise communication, the chapter provides practical tips to help researchers articulate their objectives and methodology. It emphasizes the research proposal as a foundational element of scholarly work, equipping academics with the skills to craft impactful submissions.

In Chapter 12, this chapter provides a step-by-step guide to writing effective research proposals. Covering key components such as literature review, methodology, budgeting, and timelines, it emphasizes the importance of organization and clarity. Practical examples and actionable advice ensure researchers are equipped to design robust proposals that communicate their objectives and significance effectively.

In Chapter 13, this chapter equips scholars with strategies for converting theses into published works. It covers the publishing landscape, content refinement, and journal selection, alongside practical solutions for navigating peer reviews and collaborative authorship. Ethical considerations and networking strategies are also addressed, empowering researchers to contribute meaningfully to their fields and advance their academic careers.

In Chapter 14, this chapter guides scholars in identifying promising research areas that align with current trends and literature gaps. By offering practical methodologies and examples, it emphasizes the importance of selecting topics that enhance knowledge and academic reputation. The chapter fosters a culture of scholarly excellence and effective dissemination.

In Chapter 15, this chapter provides a comprehensive guide to crafting strong research proposals. It demystifies key elements such as research questions, literature reviews, and methodological design while emphasizing the importance of communication and impact. By equipping researchers with practical tools and knowledge, the chapter lays a foundation for successful and impactful research projects.

CONCLUSION

This book would not have been possible without the dedication and expertise of our contributors, whose work reflects a deep commitment to enhancing the quality and visibility of academic research. We are also grateful to the institutions and colleagues who supported this effort, creating an environment that encourages collaboration and knowledge sharing. Their encouragement and resources have been instrumental in bringing this work to fruition.

We hope that this book serves as an invaluable resource for academicians, faculty, researchers, and higher education administrators, inspiring them to pursue excellence in research and embrace innovative approaches to scholarship. By fostering critical discussions and offering actionable guidance, it is our aspiration that *Enhancing Research Output in Higher Education* contributes meaningfully to the growth and development of academic communities worldwide.

As global challenges grow increasingly complex, the role of rigorous, impactful research in shaping solutions becomes more critical than ever. It is our belief that the insights and strategies presented in this book will empower readers to not only achieve their individual academic goals but also to drive meaningful changes in their institutions and beyond. By equipping current and future scholars with the tools and perspectives needed to navigate the evolving research landscape, this book seeks to inspire a culture of inquiry, innovation, and sustained academic excellence.

Andile Alfred Mdikana

Chapter 1
Rethinking Research in Higher Education: A Scholarly Exploration

Mohsen Mahmoudi-Dehaki
https://orcid.org/0000-0002-4962-6827
English Department, Islamic Azad University, Najafabad, Iran

Nasim Nasr-Esfahani
https://orcid.org/0000-0002-2388-6524
Department of English, Islamic Azad University, Isfahan, Iran

ABSTRACT

This chapter serves as a comprehensive guide for emerging academics and postgraduate students navigating higher education research. Its primary objective is to equip readers with skills, knowledge, and strategies to excel in research, publish effectively, and contribute to knowledge production. The chapter empowers researchers by delving into best practices, methodologies, and trends. It provides guidance on effective writing and publication strategies, enabling researchers to disseminate their findings and contribute to scholarly discourse. Embracing diversity, the chapter encourages a multifaceted approach, exploring frameworks and marginalized voices. This challenges paradigms and fosters a more representative body of knowledge. The work supports research supervision, providing strategies to enhance postgraduate student guidance, recognizing supervision's crucial role in shaping emerging scholars' development. Ultimately, this work aspires to make a significant contribution to knowledge production in higher education.

DOI: 10.4018/979-8-3373-0806-7.ch001

INTRODUCTION

This scholarly work serves as a comprehensive guide for emerging academics and postgraduate students navigating the realm of research in higher education. Its primary objective is to equip readers with the necessary skills, knowledge, and strategies to excel in their research endeavors, publish effectively, and contribute meaningfully to knowledge production. The chapter recognizes the ordinary struggles academics face in publishing and aims to assist them in navigating the intricate pathways of publishing and making informed decisions in their research pursuits. For postgraduate students, it promises to unlock the secrets to crafting sound research proposals and producing high-quality dissertations. By delving into the inner workings of multiple perspectives, this scholarly work aspires to broaden horizons, foster empathy, and deepen understanding of the world around us.

One of the primary aims of this chapter is to equip academics, postgraduate students, and researchers with the necessary skills and knowledge to excel in their research endeavors. By delving into best practices, methodologies, and emerging trends in higher education research, this chapter empowers its readers to design and execute high-quality studies that push the boundaries of knowledge. Additionally, it provides valuable guidance on effective writing and publication strategies, enabling researchers to disseminate their findings through reputable academic channels and contribute to the scholarly discourse in their respective fields. This chapter encourages a multifaceted approach to higher education research to recognize the importance of diverse perspectives in shaping research practices. Exploring various theoretical frameworks, cultural contexts, and marginalized voices aims to challenge existing paradigms and foster a more inclusive and representative body of knowledge. By embracing diversity and inclusivity, this chapter empowers researchers to produce research that is relevant, impactful, and responsive to the needs of diverse stakeholders.

Another key aim of this chapter is to support academics involved in research supervision and mentorship. Providing practical strategies and evidence-based best practices aims to enhance the quality of supervision for postgraduate students. This chapter recognizes the crucial role that adequate supervision plays in shaping emerging scholars' research skills, academic writing abilities, and professional development. By equipping supervisors with the tools and resources necessary to guide their students through the research process, this chapter ultimately contributes to the success and well-being of postgraduate students and the broader academic community. Ultimately, this chapter aspires to make a significant contribution to knowledge production in the field of higher education. By inspiring and empowering academics, postgraduate students, and researchers to engage in rigorous, innovative, and socially relevant research, it aims to advance knowledge and understanding in the higher education sector. Through its emphasis on diverse perspectives and impactful research, this chapter seeks to inspire a new generation of scholars committed to addressing the complex challenges facing higher education institutions and contributing to the betterment of society.

1. EMPOWERING ACADEMICS AND POSTGRADUATE STUDENTS

In the dynamic landscape of higher education, the need to empower academics and postgraduate students has become increasingly paramount. This multifaceted endeavor aims to cultivate a thriving research culture by focusing on several key areas: enhancing research skills and publication outcomes, fostering diverse and inclusive research perspectives, strengthening research supervision and mentorship, and advancing knowledge production within the higher education ecosystem. Equipping academics and postgraduate students with robust research skills is crucial for their success. Institutions can provide comprehensive training, mentorship, and resources to help scholars navigate the complexities of the research process, from literature review to data analysis and effective dissemination of findings. This approach can increase publication success, elevating the institution's research profile and contributing to the broader advancement of knowledge.

In an increasingly globalized and interconnected world, embracing diverse research perspectives is paramount. By actively promoting inclusivity and representation within the research community, institutions can unlock innovative ideas, challenge existing paradigms, and foster a more comprehensive understanding of complex societal issues. This approach not only enriches the research landscape but also contributes to the development of a more equitable and socially conscious academic environment. Adequate research supervision and mentorship are pivotal in nurturing the growth and development of academics and postgraduate students. Through robust mentorship programs, institutions can provide their scholars with personalized guidance, constructive feedback, and opportunities for professional development. This holistic approach not only enhances research outcomes but also fosters a supportive and collaborative research culture, ultimately contributing to the long-term success of the institution's scholarly community.

At the heart of this multifaceted approach lies the overarching goal of advancing knowledge production within the higher education sector. By empowering academics and postgraduate students, institutions can catalyze the generation of groundbreaking research, disseminate innovative ideas, and cultivate a vibrant intellectual discourse. This, in turn, can lead to expanding disciplinary boundaries, resolving complex societal challenges, and enriching the global knowledge ecosystem. Empowering academics and postgraduate students is a strategic imperative for higher education institutions seeking to enhance research capabilities, foster diverse perspectives, strengthen research supervision, and advance knowledge production. By embracing a holistic and multifaceted approach, institutions can unlock the full potential of their scholarly community and position themselves as leaders in the pursuit of academic excellence and societal impact.

1.1. Enhancing Research Skills and Publication Outcomes

As early career academics navigate the competitive landscape of research and publication, developing robust research skills is paramount for success. A multi-faceted approach, encompassing collaborative learning, skill diversification, and critical thinking development, can significantly enhance research productivity and publication outcomes. Engaging in research collaboration activities provides early career academics with invaluable learning opportunities from experienced researchers. By working alongside seasoned professionals, young academics can acquire essential skills such as scientific writing, proposal writing, and data analysis. The feedback and guidance received during collaborative projects help build early career academics' research capabilities and self-confidence, enabling them to tackle complex research challenges with greater ease and effectiveness.

Early career academics must go beyond the basics and explore diverse learning resources to become versatile and impactful researchers. This issue includes show-casing their skills in different scenarios, such as interdisciplinary collaborations or international research projects, and incorporating research into their career development plans. By expanding their skill set, early career academics can adapt to changing research landscapes and contribute to various projects, ultimately enhancing their research impact and visibility within the academic community. Developing critical thinking skills is essential for early career academics to analyze information objectively and draw well-founded conclusions (Gruwell, 2022). By cultivating a mindset that questions assumptions, considers multiple perspectives, and seeks evidence-based solutions, early-career academics can enhance the quality and rigor of their research. This skill set not only improves their ability to conduct high-quality research but also strengthens their capacity to engage in constructive peer review and contribute to the advancement of their respective fields.

The definition of "research skills" should be broadened to include transferable competencies such as critical appraisal, information synthesis, and effective com-munication. By incorporating these skills into academic curricula, undergraduate and master's programs can better prepare students for the demands of research and publication. Explicitly addressing research skill development in academic programs is critical to enhancing students' research competence and ensuring that they are equipped with the necessary tools to succeed in their academic careers. Enhancing research skills for early career academics requires a multifaceted approach that combines collaborative learning, skill diversification, critical thinking development, and the incorporation of transferable competencies into academic curricula. By embracing this holistic approach, early career academics can significantly boost their research productivity, publication outcomes, and overall impact within the academic community.

1.2. Fostering Diverse and Inclusive Research Perspectives

In the ever-evolving landscape of higher education and research, embracing diverse perspectives and voices has become crucial in driving innovation, excellence, and social impact. By creating inclusive environments that celebrate and harness the richness of diverse experiences and knowledge systems, institutions can unlock the full potential of their academic communities and achieve remarkable breakthroughs in research and learning. One of the primary steps in fostering diverse perspectives is establishing safe spaces where individuals from various backgrounds can express their views freely without fear of judgment or recrimination. Experienced facilitators play a vital role in this process, as they can skillfully elicit a wide range of perspectives and ensure that diverse voices are heard and valued. The "Vision 2020" project in Jammu and Kashmir is a testament to the power of such initiatives, demonstrating how open and inclusive dialogues can lead to transformative change.

Encouraging teachers and researchers to adopt a more facilitative role is another critical strategy for promoting diverse perspectives. By stepping back and allowing for open discussion and reflection rather than simply transferring their knowledge and values, educators and researchers can create an environment that is conducive to diverse conversations. This approach fosters critical thinking and problem-solving skills and enables students and researchers to learn from one another's unique perspectives and experiences. Implementing proactive measures to address systemic barriers in research team recruitment and composition is crucial for promoting a diversity of perspectives (Sergeant, 2021). By using best practices to encourage diverse applicant pools and avoid disadvantaging underrepresented candidates, institutions can create research teams that reflect the communities they serve and the issues they aim to address. This commitment to inclusive recruitment not only enhances the diversity of perspectives but also contributes to the overall excellence and impact of the research.

Considering identity factors, accessibility needs, and potential unintended impacts throughout the research process is essential for capturing the nuances and complexities of diverse experiences. By applying an intersectional lens, researchers can reveal significant differences between groups and ensure that their work is relevant, accessible, and beneficial to all stakeholders. This approach also helps to mitigate the risk of perpetuating harmful stereotypes or reinforcing existing inequities. Engaging diverse sources, authors, and knowledge systems in the research process and involving relevant communities in co-developing objectives helps to ensure that a range of perspectives are represented. By drawing upon a diverse array of scholarly works and incorporating community members' lived experiences and insights, researchers can create a more holistic and impactful understanding of the issues they are studying. This collaborative approach enhances the research quality and fosters trust and mutual understanding between academia and the communities it serves.

Prioritizing effective knowledge mobilization strategies to share research findings with the groups who can use and benefit from them is crucial for maximizing the impact of diverse perspectives. By considering language and accessibility barriers, researchers can ensure that their work is disseminated in a meaningful and actionable way for diverse audiences. This commitment to knowledge mobilization amplifies the voices of underrepresented groups and contributes to creating a more equitable and just society. Fostering diverse perspectives and voices is essential for enhancing student learning experiences and achieving research excellence. By creating safe spaces for open dialogue, adopting facilitative approaches in teaching and research, addressing systemic barriers, applying an intersectional lens, engaging diverse sources and knowledge systems, and prioritizing effective knowledge mobilization, institutions can unlock the benefits of varied perspectives and experiences. This commitment to diversity and inclusion is a moral and strategic necessity for institutions seeking to make a meaningful impact in an increasingly complex and interconnected world.

1.3. Enhancing Research Supervision and Mentorship

Adequate research supervision and mentorship are crucial for the success and professional development of PhD students and early-career researchers. This extended academic discussion will explore the fundamental principles and best practices that can elevate these critical support systems. Supervisors are pivotal in setting the stage for a productive and rewarding PhD journey. By collaborating with students to establish clear objectives, timelines, and milestones for the research project and professional development, supervisors provide a structured framework that helps students stay focused and on track. The shared understanding of expectations ensures that the research goals are well-defined and enables students to develop a sense of ownership and accountability. Frequent meetings between supervisors and students are essential for monitoring progress, offering constructive feedback, and providing guidance and encouragement. This ongoing dialogue allows supervisors to identify and address any challenges or roadblocks early on, fostering a supportive environment where students feel empowered to share their ideas, concerns, and learning experiences. By balancing providing direction and nurturing independence, supervisors can help students develop the critical thinking and problem-solving skills necessary for future success and stability in their work.

The supervisor-student relationship should be built on mutual respect and open communication, where both parties feel comfortable exchanging ideas and addressing challenges collaboratively. Techniques such as joint problem-solving and co-authoring publications can help cultivate a collaborative dynamic where students actively engage in the research process and feel their contributions are valued. This collaborative approach enhances the research quality and prepares students for the collaborative nature of academic and professional environments. Adequate supervision extends beyond the immediate research project, as supervisors should also provide career advice, networking opportunities, and guidance in developing essential professional skills. By taking a holistic approach to mentorship, supervisors can help students navigate the complexities of academic and research careers, preparing them for the challenges and opportunities. This comprehensive support system equips students with the necessary tools and resources to thrive in their future endeavors.

Fostering the habit of self-reflection is a crucial component of adequate supervision and mentorship (Hewson & Carroll, 2016). Supervisors should guide students in regularly reflecting on their progress, learning from setbacks, and continuously improving their research and professional skills. This process of self-evaluation and introspection promotes autonomy and lifelong learning and helps students develop the resilience and adaptability needed to navigate the dynamic landscape of academia and research. By implementing these principles of effective research supervision and mentorship, supervisors can enhance the overall experience for PhD students and early-career researchers, leading to more successful, fulfilling, and impactful academic and research journeys.

1.4. Advancing Knowledge Production in Higher Education

Advancing knowledge production in higher education is crucial for addressing the complex global challenges and fostering a more sustainable future. Universities, as centers of learning and research, play a pivotal role in generating and disseminating knowledge, but they also face various opportunities and challenges in this pursuit. One essential aspect of advancing knowledge production in higher education is the need for universities to integrate sustainability principles into their core functions, including education, research, and campus operations (Scoffham & Rawlinson, 2022). This issue requires a whole-systems approach that involves breaking down the barriers to change and institutionalizing sustainable development across the institution. Universities can model and promote sustainable development by aligning their educational, research, and operational practices with sustainability principles, setting an example for their students, communities, and the broader society.

Another critical area in advancing knowledge production for sustainable development is the diffusion of sustainable development knowledge through university curricula. It can be achieved by incorporating sustainability topics into regular courses and programs and adopting interdisciplinary and problem-based learning approaches. By integrating sustainability-related content and pedagogical methods, universities can empower students to develop the knowledge, skills, and mindsets necessary to tackle complex sustainability challenges. Collaboration and partnerships are also essential for advancing knowledge production in higher education. Universities should work closely with research institutes, private firms, government agencies, and civil society organizations to tackle complex sustainability challenges and co-create solutions. By fostering these collaborative relationships, universities can leverage diverse expertise, resources, and perspectives, leading to more comprehensive and innovative approaches to sustainable development.

Finally, universities must critically reflect on their role and purpose in the 21st century. This issue includes rethinking traditional knowledge production and dissemination models and embracing new, more inclusive, and participatory approaches that are responsive to the evolving needs of society. By engaging in this critical self-reflection, universities can redefine their mission and strategies to better align with the pressing challenges of our time, ensuring that their knowledge production efforts contribute to a more sustainable and equitable future. Advancing knowledge production in higher education for sustainable development requires a multifaceted approach that combines curricular reform, institutional change, collaborative partnerships, and critical reflection. By embracing these strategies, universities can position themselves as key drivers of sustainable development, contributing to creating a more resilient and just world. This transformation will require universities to foster a culture of innovation and adaptability, empowering faculty, staff, and students to explore new ideas and approaches. Additionally, universities should strengthen their engagement with local and global communities, leveraging their expertise and resources to address real-world challenges. Finally, universities must prioritize developing interdisciplinary and transdisciplinary programs that bridge traditional academic boundaries and foster holistic, systems-level thinking.

2. DEVELOPING OWN RESEARCH PROFILE

As early career academics navigate the competitive academic landscape, developing a strong research profile and portfolio is crucial for career advancement and recognition. This comprehensive approach outlines critical strategies to help emerging scholars establish a unique research identity and build an impactful research portfolio.

2.1. Crafting a Research Identity

A distinctive research identity is essential for standing out in the academic community (Savitha & Venkatachalam, 2024). This identity involves the following key elements:

- **Identifying a Specific Research Niche:** Narrowing the focus to a specific area of expertise allows for meaningful contributions and developing a coherent research narrative.
- **Establishing an Online Presence**: Creating profiles on academic social networks, such as ResearchGate, LinkedIn, and institutional websites, enables researchers to showcase their work, connect with peers, and share their research.
- **Strategic Self-Branding:** Actively engaging in activities like attending relevant conferences, delivering presentations, and publishing in high-impact journals can help build a reputation as an expert in the chosen research area.

2.2. Building a Research Portfolio

Curating an impactful research portfolio is vital for demonstrating research capabilities and achievements. Effective strategies include:

- **Targeted Journal Publications:** Carefully selecting high-quality, relevant journals that align with the research focus can maximize the visibility and impact of the work.
- **Collaborative Partnerships:** Collaborating with established researchers to co-author publications can provide valuable opportunities for skill development and exposure.
- **Multimedia Showcasing:** Incorporating multimedia elements like videos into online profiles can make the research portfolio more engaging and accessible.

- **Social Media Engagement:** Actively participating in social media platforms, like LinkedIn, to share research, interact with peers, and build a broader network can contribute to the growth of the research profile.
- **Integrating Teaching and Research:** Incorporating research into course content and supervising students in the area can demonstrate the work's real-world impact and practical applications.

Developing a robust research identity and portfolio is a continuous process that requires strategic planning, dedication, and a willingness to collaborate and leverage online tools (Joireman & Van Lange, 2015). By focusing their efforts, early career academics can accelerate the growth of their research profile and advance their academic careers.

3. RESEARCH APPROACHES

Quantitative, qualitative, and mixed methods research are the three main research approaches used by researchers to investigate various phenomena. Quantitative research is a systematic investigative approach that emphasizes numerical data collection and statistical analysis. This methodology is well-suited for studies to test hypotheses and identify generalizable relationships. Quantitative researchers seek to objectively measure and analyze variables to uncover patterns, trends, and causal associations. In contrast, qualitative research focuses on understanding the meanings, experiences, and contexts surrounding a phenomenon. This approach relies on non-numerical data, such as interviews, observations, and textual analysis, to gain in-depth insights into the complexities of human behavior and social interactions. Qualitative researchers often adopt an interpretive stance, seeking to explore and interpret the subjective perspectives of research participants.

The mixed methods approach combines quantitative and qualitative research techniques within a single study. This integrated methodology allows researchers to leverage the strengths of each approach to provide a more comprehensive understanding of the research problem. Mixed methods designs can take various forms, including convergent (parallel data collection and analysis), explanatory sequential (quantitative data followed by qualitative), and exploratory sequential (qualitative data followed by quantitative). The choice of research approach is contingent upon the nature of the research question, the researcher's philosophical assumptions, and the intended outcomes of the study. Quantitative research is well-suited for theory testing and making predictions, while qualitative research is more appropriate for exploring new phenomena and understanding context-specific experiences. The mixed methods approach can offer the most robust and well-rounded insights by combining the strengths of both quantitative and qualitative methodologies.

3.1. Quantitative Research

Quantitative research is a methodological paradigm emphasizing the systematic collection and statistical analysis of numerical data to enhance understanding of concepts, opinions, and experiences (Williams et al., 2022). This objective approach centers on measuring and quantifying phenomena, employing a range of distinct characteristics defining their nature and application. The quantitative research process often commences with formulating a hypothesis or articulating a theoretical framework, which then guides the subsequent data collection efforts. This deductive reasoning approach allows researchers to establish a clear and logical connection between the theoretical underpinnings and the empirical evidence, facilitating the validation or refutation of the proposed assumptions. The primary focus of quantitative inquiry is gathering numerical data, which can then be subjected to rigorous statistical analysis. It enables researchers to identify and examine relationships, trends, and correlations within the data, ultimately providing objective and generalizable findings. The emphasis on numerical data and statistical methods is a defining characteristic of the quantitative research paradigm.

Quantitative studies often aim to collect data from large, representative, and randomly selected samples to ensure the generalizability of the research findings. Standardized instruments, such as surveys and experiments, are employed to maintain consistency and reliability in the data collection process. This systematic approach to sampling and instrumentation is crucial for enhancing the validity and replicability of the research. Quantitative research is particularly well-suited for addressing research questions that seek to understand a phenomenon's "what" and "how much" aspects. This approach is practical in establishing cause-and-effect relationships and testing hypotheses, providing insights that can be applied across diverse contexts. The systematic and standardized nature of quantitative research allows for the replication of studies, ensuring the objectivity and reliability of the findings. This issue, in turn, contributes to the broader scientific understanding of the investigated phenomena and enhances the credibility of the research.

3.2. Qualitative Research

Qualitative research is a distinctive approach that delves into non-numerical data, focusing on understanding concepts, opinions, and experiences. This subjective methodology emphasizes exploring the underlying meaning behind human behavior, offering a unique perspective that complements the more quantitative approaches

(Flick, 2022). One of the key characteristics of qualitative research is its reliance on inductive reasoning. Rather than starting with predetermined hypotheses, qualitative researchers begin with observations and patterns, using these as a foundation to develop theories and concepts. This approach allows for a more open-ended and exploratory investigation, as researchers are not constrained by pre-existing assumptions. Additionally, qualitative research gathers data in the form of words, images, and behaviors, providing a rich tapestry of textual and visual information to analyze. This type of data can capture the nuances and complexities of human experiences that may be difficult to quantify.

Qualitative research often employs small sample sizes, as the primary goal is to gain an in-depth understanding of a particular phenomenon rather than to generalize findings. This approach allows researchers to delve deeply into the lived experiences of individuals or small groups, uncovering the unique perspectives and contextual factors that shape their worldviews. Qualitative research utilizes flexible and open-ended methods, such as interviews, focus groups, and observations, allowing for a more nuanced exploration of the research topic. These methods allow participants to share their stories, experiences, and insights in their own words, enabling researchers to gain a more authentic understanding of the phenomenon under study. Qualitative data analysis involves interpretive analysis, where researchers employ thematic analysis to identify patterns and extract meaning from the collected information. This approach requires researchers to immerse themselves in the data, carefully examining the language, emotions, and cultural contexts that shape the participants' responses. By engaging in this interpretive process, researchers can uncover the deeper meanings and underlying assumptions that may not be immediately apparent in the surface-level data. This approach provides rich, contextual insights that can uncover unexpected findings, making qualitative research well-suited for answering "why" and "how" questions, exploring complex social phenomena, and developing new theories.

3.3. Mixed Methods Research

The integration of quantitative and qualitative approaches, known as mixed methods research, offers several critical advantages for higher education researchers. This methodological approach allows researchers to leverage the unique strengths of both quantitative and qualitative methods, leading to a more comprehensive and nuanced understanding of the research topic (Watkins, 2022). One of the primary benefits of mixed methods research is the ability to triangulate data from multiple sources and methods. By combining quantitative survey data with qualitative insights from interviews or focus groups, researchers can validate and corroborate their findings, increasing the overall reliability and credibility of the study. This data triangulation helps overcome the limitations of relying on a single method, providing a more robust and trustworthy set of conclusions.

Mixed methods research also enables researchers to use qualitative and quantitative data in a complementary manner, allowing them to elaborate on, illustrate, and clarify the results. For example, a large-scale survey on student engagement might be followed by focus groups that delve deeper into the underlying reasons and experiences behind the quantitative findings. This combination of methods can lead to a more nuanced and comprehensive understanding of the phenomenon under study. In cases where the quantitative and qualitative findings are contradictory or divergent, mixed methods research can serve as a catalyst for initiating new research questions and perspectives. These contradictions can challenge existing assumptions and theories, leading to innovative approaches and the exploration of previously overlooked aspects of the research topic. This " initiation " process can significantly advance the field of higher education research and foster new avenues of inquiry.

By employing different methods for different components of a study, mixed methods research can expand the breadth and range of inquiry. For instance, a researcher might use a quantitative survey to measure engagement levels across a large student population, followed by in-depth qualitative interviews to explore the nuances of the student experience. This combination of methods allows the researcher to gain a more comprehensive understanding of the research topic, addressing the phenomenon's breadth and depth. The choice between quantitative, qualitative, or mixed methods research in higher education should be guided by the specific research questions, the nature of the phenomenon under study, and the desired scope and depth of understanding. By embracing the strengths of both quantitative and qualitative approaches, mixed methods research has the potential to significantly advance our understanding of complex issues in higher education and inform policies and interventions that positively impact student success and institutional effectiveness.

4. RESEARCH PARADIGMS: PHILOSOPHICAL PERSPECTIVES

4.1. Positivism vs. Constructivism

Positivism and constructivism are two prominent research paradigms that offer contrasting perspectives on the nature of reality, knowledge, and the role of values in research. The positivist paradigm assumes that there is an objective reality that exists independently of human perception and can be measured and understood through neutral, unbiased scientific inquiry. Positivist researchers favor quantitative methods like experiments, surveys, and statistical analysis to test hypotheses and uncover universal laws and causal relationships. The aim of positivist research is explanation, prediction, and control. In contrast, the constructivist paradigm views reality as socially constructed, subjective, and relative. Constructivists believe that knowledge is actively built through individual and collective processes of interpretation and meaning-making, which are inevitably shaped by social, cultural, and historical contexts. Constructivist researchers use qualitative methods like interviews, observations, and interpretive analysis to develop contextualized, nuanced understandings of phenomena. Constructivist research aims to understand and reconstruct the multiple constructions that people hold.

4.1.1. Positivism

Positivism, a research paradigm closely aligned with the natural sciences, emphasizes the importance of objective reality, empirical observation, quantifiable data, and generalizable laws (Pawar et al., 2023). Positivists believe an objective reality exists independently of human perception and can be measured and understood through neutral and unbiased scientific inquiry. This research approach has been widely adopted in higher education, particularly in studies focused on the effectiveness of teaching methods, student learning outcomes, and the impact of institutional policies. According to the positivist research paradigm, reality is considered objective and can be studied independently of the researcher. Knowledge is based on observable and measurable facts, and research should be conducted value-free and unbiased.

Positivist research in higher education strongly emphasizes quantitative methods, such as experiments, surveys, and statistical analysis, to measure and evaluate various aspects of the educational process. The deductive approach, which involves testing hypotheses and theories with empirical data, is a fundamental aspect of positivist research. Positivist research in higher education often aims to uncover universal truths and causal relationships that can be generalized. By employing quantitative methods, researchers seek to contribute to a better understanding of the factors that influence student learning and institutional effectiveness. This approach allows for testing hypotheses and theories based on empirical data, ultimately informing educational policies, improving teaching practices, and enhancing student learning outcomes.

4.1.2. Constructivism

Constructivism, a prominent research paradigm, posits that reality is inherently subjective and socially constructed, with multiple coexisting realities (Pawar et al., 2023). This epistemological stance holds that knowledge is actively built through individual and collective processes of interpretation and meaning-making, which are inevitably shaped by social, cultural, and historical contexts. Fundamental to constructivism is the belief that knowledge is subjective and context-dependent and that researchers are an integral part of the research process, influencing the findings. Constructivism posits that reality is not an objective, external entity but rather a subjective construct shaped by individual and collective interpretations. This stance challenges the notion of a single, universal truth and acknowledges the existence of multiple, equally valid realities. Constructivists view knowledge as an active process of interpretation and meaning-making rather than a passive acquisition of facts. Individuals and groups construct knowledge through their interactions with the world and their unique social, cultural, and historical contexts.

In constructivist research, the researcher is not an objective, detached observer but rather an active participant in the research process. Researchers' perspectives, experiences, and biases inevitably shape the research findings, and constructivists acknowledge this influence as an integral part of the research process. Constructivist research typically employs qualitative methods, such as interviews, observations, and interpretive analysis, to capture the nuances and complexities of social phenomena. The inductive approach, which involves generating theories from data, allows researchers to develop rich, contextualized understandings of the research topic. In higher education research, constructivism has been widely applied to explore topics such as student experiences, faculty perspectives, and the sociocultural dynamics of academic institutions. By adopting a constructivist lens, researchers can develop a deeper understanding of the complex interplay between individuals, institutions, and the broader societal context, ultimately contributing to advancing knowledge in higher education.

4.2. Critical Theory and Postmodernism

Critical theory and postmodernism are united in their overarching aim to interrogate and challenge dominant power structures and narratives within society. Both perspectives seek to amplify marginalized voices and disrupt the status quo. However, the epistemological foundations underpinning each approach exhibit notable distinctions. Postmodernism is more skeptical about universal truths, grand narratives, and objective knowledge. It embraces a pluralistic orientation, emphasizing the socially constructed nature of knowledge and the importance of localized, contextual understandings. In contrast, while rejecting positivist claims to value-neutrality, critical theory believes in certain universal moral principles, such as human emancipation, which should guide the critical process.

4.2.1. Critical Theory

Critical theory is a research approach that challenges prevailing power structures and advocates for social justice and emancipation (Strydom, 2011). Adherents of this paradigm believe that research should expose and transform oppressive social conditions. This perspective rests on several key principles. Critical theorists posit that reality is shaped by a complex interplay of social, political, cultural, economic, and ethnic values. These forces collectively construct the lived experiences of individuals and communities, rejecting the notion of an objective, value-free reality. Critical theory asserts that power relations and the interests of dominant groups within society inherently influence knowledge. Research, therefore, cannot be divorced from its social and political context, as it is imbued with the values and biases of those who produce it.

The aim of critical theory research is to critique the status quo and empower marginalized groups. This approach seeks to amplify the voices and experiences of those historically marginalized or silenced by exposing and challenging oppressive social conditions. Critical theorists emphasize the importance of participatory and action-oriented research methods. These approaches involve stakeholders as active participants in the research process and aim to generate knowledge that can be directly applied to address social inequities. The ultimate goal of critical theory research is to raise critical consciousness and catalyze social change. By exposing the underlying structures of oppression and empowering marginalized groups, this paradigm seeks to transform both individual mindsets and broader societal conditions

4.2.2. Postmodernism

Postmodernism is a research paradigm that rejects the notion of universal truth and challenges traditional conceptions of knowledge and power (Leslie, 2010). Adherents of this perspective believe that all knowledge claims are partial, situated, and contingent, shaped by the complex interplay of language, discourse, and social context. This paradigm views reality as fragmented, unstable, and open to multiple interpretations. Postmodernists reject the idea of a singular, objective truth, instead embracing the inherent complexity and fluidity of the social world. From a postmodern perspective, knowledge is local, provisional, and shaped by the language and discourses that constitute it. Knowledge claims are not seen as neutral or value-free but rather as products of power relations and cultural contexts.

Postmodern research emphasizes the importance of deconstructing dominant narratives and amplifying the voices and perspectives of those who have been marginalized or silenced. This approach seeks to challenge taken-for-granted assumptions and open up new ways of thinking about the complexities of the social world. Postmodernists are highly skeptical of grand theories and metanarratives that claim to offer universal explanations. Instead, they focus on the particular, the contextual, and the contingent, rejecting the idea of a single, overarching framework for understanding social phenomena. Postmodern research places a strong emphasis on reflexivity, acknowledging the role of the researcher and their positionality in the research process. Researchers are encouraged to critically examine their biases, assumptions, and power dynamics that shape knowledge production.

5. DEVELOPING A SOUND RATIONALE AND PROBLEM STATEMENT

5.1. Articulating Research Rationale

The justification for conducting research is a critical component of any research proposal or study, as it establishes the significance and importance of the chosen topic. A well-crafted rationale should clearly explain the investigation's reasons and elucidate its contributions to the existing body of knowledge or its capacity to address a pressing issue. When developing this rationale, several key elements should be incorporated. Offering a brief overview of the relevant literature and current knowledge in the field is essential. It allows the researcher to identify gaps, inconsistencies, or limitations in existing research that the study aims to address, thereby highlighting the need for the proposed investigation.

Explicating the importance of the research topic and its relation to broader theoretical, practical, or societal concerns is crucial. The researcher should emphasize the potential implications and impact of the study, underscoring its relevance and significance. Discussing how the research question or approach differs from or builds upon previous studies is important. The researcher should emphasize the unique contribution the study will make to the field, demonstrating its originality and innovative nature. If applicable, the researcher should discuss the practical applications and real-world relevance of the research. This involves explaining how the findings could inform policies, practices, or interventions in higher education or other relevant domains. Briefly mentioning personal interest, experience, or expertise related to the research topic can help establish the researcher's credibility and investment in the study, further strengthening the rationale.

5.2. Crafting a Problem Statement

The problem statement is a crucial component of any research endeavor, as it clearly delineates the research problem or issue the study aims to address. A well-formulated problem statement provides a clear direction for the research and guides the development of research questions and hypotheses. There are several key elements of an effective problem statement. The problem statement should be specific, avoiding vague or overly broad language. It should clearly define the scope and boundaries of the research problem. The problem statement should be relevant to the field of study and address a significant theoretical, practical, or societal issue in the context of higher education. The problem statement should be researchable, meaning that the problem can be investigated using appropriate research methods and data collection techniques. The problem statement should be written in clear, concise language easily understood by experts and non-experts in the field. The problem statement should align with and be supported by the research rationale. It should logically follow from the background information and significance of the study. When crafting a problem statement, researchers may consider using a template such as: "The problem is [describe the problem], as evidenced by [provide supporting evidence], which affects [identify the population affected] in [specify the context]." This structure helps ensure the problem statement is focused, relevant, and researchable. By developing a well-crafted problem statement, researchers can establish a strong foundation for their study and increase the likelihood of producing significant and impactful research within the field of higher education.

6. RESEARCH DESIGNS AND THEORETICAL FRAMEWORKS

The research design outlines the overall strategy for addressing the research question(s) and achieving the study's objectives. It specifies the methods, procedures, and techniques that will be used to collect and analyze the data. The research design ensures the research is conducted in a systematic and rigorous manner. It determines the type of data to be collected (qualitative, quantitative, or mixed), the sampling approach, the data collection instruments, and the data analysis methods. The research design provides the structural blueprint for the entire research process. The theoretical framework is the foundation upon which the research is built. It consists of the existing theories, concepts, and models from the literature that inform the research problem and guide the study. The theoretical framework provides the lens through which the researcher views and interprets the phenomena under investigation. It helps situate the research within the broader academic discourse and demonstrates the researcher's understanding of the relevant theories.

The theoretical framework shapes the research questions, hypotheses, and overall data collection and analysis approach.

6.1. Experimental vs. Descriptive Designs

Experimental research is a scientific method that involves the manipulation of variables to establish cause-and-effect relationships. Researchers in this approach actively intervene in the research process by introducing an independent variable and observing its impact on a dependent variable. Descriptive research, on the other hand, aims to describe the characteristics, behaviors, or phenomena of a particular population or situation without manipulating variables. Researchers in this approach observe and measure variables as they naturally occur, without intervention. In summary, the main distinction between experimental and descriptive research lies in the level of control and manipulation of variables. Experimental research actively intervenes to establish cause-and-effect relationships, while descriptive research observes and measures variables in their natural state to describe characteristics, patterns, and relationships. Both approaches are valuable in advancing scientific knowledge, but they serve different research objectives and require distinct methodological considerations.

6.1.1. Experimental Designs

Experimental research designs are a powerful tool that allows researchers to delve into the complex relationships between variables and establish causal connections. By systematically manipulating one or more independent variables and measuring their impact on a dependent variable, researchers can gain valuable insights into the underlying mechanisms driving various phenomena. This approach enables them to move beyond mere correlations and uncover the true causes of observed effects, which is crucial for advancing scientific knowledge and developing effective interventions. The key features of experimental research designs ensure the rigor and validity of the research process. Randomly assigning participants to treatment and control groups helps control for potential confounding variables, ensuring that any observed differences can be attributed to the intervention itself. The researcher's active manipulation of the independent variable allows for a more direct assessment of its impact on the outcome, rather than relying on naturally occurring variations.

Measuring the dependent variable before and after the intervention provides a basis for comparison and helps establish a temporal relationship between the independent variable and the outcome. By comparing the changes in the dependent variable between the treatment and control groups, researchers can determine whether the intervention had a significant effect. Statistical analysis plays a crucial role in this process, quantifying the magnitude of the effect and assessing the likelihood that the observed differences are due to chance. The example of investigating the impact of a new teaching method on student learning outcomes illustrates the practical application of experimental research designs. By randomly assigning students to either a control group receiving traditional instruction or an experimental group exposed to the new teaching method, the researcher can isolate the effect of the intervention. The pre-test and post-test measurements allow for a direct comparison of learning gains between the two groups. Suppose the experimental group demonstrates significantly higher learning gains compared to the control group. In that case, the researcher can infer that the new teaching method has a causal effect on improving student learning outcomes.

6.1.2. Descriptive Designs

Descriptive research designs play a crucial role in scientific inquiry by providing a comprehensive understanding of phenomena as they naturally occur, without any manipulation by the researcher. These designs aim to describe the current state of variables or relationships, offering a snapshot of the reality being studied. Unlike experimental designs, which actively manipulate independent variables to determine their effect on dependent variables, descriptive designs focus on observing and measuring variables in their natural environment. The key features of descriptive research designs include the observation of variables without intervention, no active manipulation of independent variables by the researcher, measurement of variables at a single point in time, and the use of descriptive statistics to summarize and interpret the collected data. This approach allows researchers to gain insights into the variables' current state and their relationships. To illustrate the application of descriptive research design, let us consider an example where a researcher is interested in exploring the association between student engagement and academic performance. The researcher selects a sample of students, measures their engagement levels using a survey instrument, and collects data on the students' academic performance, typically represented by their grade point average.

The researcher then employs correlational analysis to determine the nature and strength of the relationship between student engagement and academic performance. If a positive correlation is found, it suggests that higher levels of student engagement are associated with better academic performance. However, it is crucial to note that a correlational relationship does not imply causation, as other factors may influence both variables. While descriptive research designs provide valuable insights into the current state of variables and relationships, they have certain limitations. Since no manipulation of variables occurs, descriptive designs cannot establish causal relationships. Additionally, the cross-sectional nature of data collection may not capture changes over time or account for potential confounding variables. Researchers must carefully consider the sampling methods, measurement techniques, and statistical analyses employed in descriptive research to ensure the validity and reliability of their findings. Appropriate sample size, representativeness, and control of potential biases are crucial for drawing meaningful conclusions from descriptive studies.

6.2. Theoretical Framework Development

A theoretical framework serves as a guiding structure for research by identifying key concepts, variables, and relationships to be examined. It is grounded in existing theory and research and provides a lens through which to investigate a phenomenon. The process of developing a theoretical framework involves several crucial steps. Conducting a thorough review of relevant theories and empirical studies related to the research topic is essential. This process involves identifying key concepts, variables, and relationships that have been previously investigated and discussed in the literature. Choosing one or more theories that provide a suitable framework for the study is a critical step. Factors such as the theory's explanatory power, scope, and alignment with the research questions and methodology should be considered when making this selection. Clearly defining the key concepts and variables from the chosen theory within the context of the study is crucial. Researchers must explain how these variables will be operationalized and measured in the study.

Based on the selected theory, researchers must specify the hypothesized relationships between the key variables. This step involves explaining how the theory predicts these relationships will unfold in the study. Aligning the research questions and hypotheses with the theoretical framework is essential. Researchers must ensure that the theory provides a clear rationale for investigating the specific research problem. Aligning the research questions and hypotheses with the theoretical framework is essential. Researchers must ensure that the theory provides a clear rationale for investigating the specific research problem. Finally, researchers must discuss how the theoretical framework will guide the data collection, analysis, and interpretation processes. They should explain how the findings will be used to support, refine, or extend the theory.

For instance, a researcher investigating the impact of a mentoring program on student retention in higher education might use Tinto's theory of student departure as a theoretical framework. The theory posits that student persistence is influenced by their academic and social integration into the university. The researcher would define and measure academic and social integration as key variables, hypothesize that the mentoring program will enhance integration and, thus, retention, and discuss how the findings will be used to refine the theory. In summary, constructing a well-grounded theoretical framework is crucial for guiding the research process and situating findings within the broader context of existing knowledge. By following these steps and critically engaging with the literature, researchers can develop a robust theoretical foundation for their studies and contribute to the advancement of knowledge in their field.

7. PUBLISHING FROM A THESIS AND RESEARCH SUPERVISION

7.1. Publishing from a Thesis

Transitioning a doctoral thesis into a series of publishable research articles is crucial in establishing oneself as an academic researcher. This process requires a strategic and methodical approach to ensure the research findings are effectively disseminated and contribute meaningfully to the scholarly discourse. Doctoral theses often encompass a broad range of topics and research questions. However, for successful journal publication, the focus should be on concise, self-contained studies that make a clear and impactful contribution to the literature. By carefully reviewing the thesis, researchers can identify discrete studies or experiments that could be developed into standalone articles, thereby maximizing the publication potential of their doctoral work. Theses are typically comprehensive and lengthy documents, whereas journal articles demand a more streamlined approach. In the process of transforming a thesis into publishable articles, researchers must condense the introduction, tighten the literature review, and hone the data analysis and discussion sections to emphasize the key findings and their significance. This strategic revision ensures that the manuscript adheres to the conciseness and focus required by academic journals.

Successful publication requires a thorough understanding of the target journals and their specific aims, formatting requirements, and style guidelines. By carefully researching the scope and preferences of potential publication outlets, researchers can align their manuscript to increase the chances of favorable peer review and acceptance. Involving co-authors, such as thesis supervisors or other researchers who contributed to the work, can strengthen the manuscript and expand its reach. Collaborative authorship not only demonstrates the ability to work effectively in a research team but also brings diverse perspectives and expertise to the publication process. Presenting the research at conferences, workshops, or departmental seminars can help to refine the ideas and build connections with potential co-authors or journal editors. This engagement with the academic community can also increase the visibility of the work, leading to valuable feedback and potential collaborations that can further enhance the publication prospects.By adopting these strategic approaches, doctoral researchers can effectively transform their thesis into a series of high-quality, impactful research articles that contribute to the advancement of their respective fields and establish their reputation as emerging academic scholars.

7.2. Effective Research Supervision

Providing high-quality research supervision is crucial to supporting postgraduate students in their academic development and research endeavors. Implementing strategic approaches can significantly contribute to postgraduate students' success and scholarly knowledge advancement. From the outset of the supervisory relationship, it is essential to work collaboratively with the student to set realistic goals, timelines, and milestones for the research project. Regularly reviewing and adjusting these parameters as needed can maintain momentum and ensure the student remains on track throughout the research process. Encouraging the student to take ownership of their research and develop their own ideas and approaches is crucial for their intellectual growth. While providing guidance and feedback, it is important to avoid micromanaging or stifling the student's critical thinking abilities, allowing them to flourish as independent researchers. Recognizing that each student has unique strengths, weaknesses, and learning styles, the supervisory approach should be adapted to best suit the individual. Providing targeted support and feedback can help the student reach their full potential and navigate the challenges of postgraduate research.

Discussing publication strategies with the student early in the research process can be beneficial. It may include exploring the potential for a thesis by publication or developing standalone journal articles from the research. Offering advice and support throughout publication can enhance the student's scholarly output and professional development. Encouraging the student to build connections with other researchers in their field, both within and outside the institution, can lead to valuable networking opportunities, potential co-authorship, and a stronger sense of belonging to the academic community. This collaborative approach can enrich the student's research experience and contribute to the broader scholarly discourse. By implementing these strategies, academics can make significant contributions to the development of the next generation of scholars, transforming theses into publishable research and providing adequate research supervision that supports the growth and success of postgraduate students.

8. ETHICAL CONSIDERATIONS AND ONTOLOGY

Researchers must adhere to a set of principles and guidelines to protect the rights and well-being of research participants, maintain scientific integrity, and ensure the credibility of their findings. Participants must freely choose to take part in the research and be fully informed about the study's purpose, procedures, and any potential risks or benefits. Researchers must protect the privacy of participants by not revealing their identities and keeping data confidential. The research design should avoid or minimize any potential physical, psychological, social, or economic harm to participants. Researchers must accurately represent their methods and findings and avoid deception or falsification of data. Ontology refers to the philosophical assumptions about the nature of reality that underpin a research study. Ontological considerations involve questions about whether social reality exists independently of human interpretation (objectivism) or is constructed through human interactions and perceptions (constructivism). These ontological assumptions shape the researcher's epistemological and methodological approach.

8.1. Ethical Considerations in Higher Education Research

Ethical research practices are of paramount importance in the realm of higher education studies, as they serve to safeguard the rights and well-being of participants, maintain the integrity of data, and uphold the credibility of the research endeavor. Adherence to rigorous ethical standards is crucial in ensuring the validity and reliability of scholarly investigations within the academic community. Several key ethical considerations must be addressed to ensure the ethical conduct of research in higher education. Researchers are ethically bound to obtain informed consent from all participants, ensuring that they have a comprehensive understanding of the purpose, procedures, and potential risks or benefits associated with the study. This is particularly imperative when working with vulnerable populations, such as students or junior faculty, who may feel coerced or pressured to participate due to power dynamics or perceived consequences of non-participation. Researchers must take proactive measures to mitigate these risks and ensure that participants' consent is freely given, informed, and voluntary.

Maintaining confidentiality and protecting the privacy of participants is a fundamental ethical obligation for researchers. Researchers must safeguard participants' personal information and ensure that it remains confidential throughout the research process. This may involve de-identifying data, securely storing records, and limiting access to sensitive information. Researchers must also be mindful of potential breaches of confidentiality and take appropriate measures to prevent unauthorized access or disclosure of participants' information. Researchers have an ethical duty to collect, analyze, and report data accurately and objectively, without falsification or fabrication. Upholding data integrity is essential in preventing research misconduct and preserving the trustworthiness of the research findings. Researchers must adhere to established protocols and guidelines for data collection, management, and analysis, ensuring that the data is handled in a manner that maintains its accuracy, completeness, and reliability. Falsifying or manipulating data not only undermines the credibility of the research but also has far-reaching consequences for the academic community and society as a whole.

8.2. Ontological Assumptions in Higher Education Research

Researchers' ontological perspectives and fundamental beliefs about the nature of reality can significantly influence the design, data collection, and interpretation of higher education studies. The researcher's ontological stance serves as a guiding framework that shapes the approach to investigating phenomena within higher education. Understanding the implications of different ontological perspectives is crucial for evaluating the validity and trustworthiness of research findings. Researchers with an objectivist ontological view may design quantitative studies to identify causal relationships and uncover universal truths about higher education phenomena. They may collect numerical data and employ statistical analysis to draw generalizable conclusions about the nature of educational processes, outcomes, and trends. This approach assumes that reality exists independently of the researcher and can be objectively measured and observed.

In contrast, researchers with a constructivist ontological perspective may conduct qualitative studies to explore how individuals or groups construct meaning and understanding within the higher education context. They may collect in-depth interview data and use interpretive analysis to develop contextual insights into the subjective experiences and perceptions of students, faculty, or administrators. This approach acknowledges that reality is socially constructed and that multiple, context-dependent realities may exist. Researchers with a pragmatic ontological stance may employ a mixed-methods approach, combining quantitative and qualitative techniques to gain a more comprehensive understanding of higher education issues. They may prioritize the research questions and choose methods best suited to address them rather than adhering to a single ontological position. This approach recognizes the value of multiple perspectives and seeks to leverage the strengths of different research paradigms to generate a more holistic understanding of the phenomenon under investigation.

Regardless of the ontological position, it is crucial for higher education researchers to be transparent about their philosophical assumptions and how these shape the research process and findings. This transparency allows readers to evaluate better the validity and trustworthiness of the study, as well as the extent to which the research aligns with the researcher's stated ontological and epistemological perspectives. By understanding the implications of different ontological stances, higher education researchers can make informed decisions about their methodological approaches, ensuring that their studies are designed and conducted in a manner that is consistent with their underlying beliefs about the nature of reality and the generation of knowledge. This awareness enhances the rigor and credibility of higher education research, ultimately contributing to the advancement of the field.

9. LIMITATIONS OF CONVENTIONAL RESEARCH APPROACHES IN HIGHER EDUCATION

The prevailing research paradigms employed within the realm of higher education often fall short of addressing the multifaceted challenges and demands of the contemporary world. While valuable in their own right, these conventional approaches exhibit several limitations that hinder their efficacy in fostering meaningful change and progress. One of the primary shortcomings of conventional research lies in its limited real-world applicability. The research questions and hypotheses are frequently derived from theoretical frameworks and existing literature rather than being grounded in the actual needs and concerns of organizations, communities, and society at large. Consequently, the findings generated through these studies may lack the necessary relevance and actionability required to drive tangible improvements and address pressing issues. Moreover, conventional research often follows a unidirectional and expert-driven approach, where the researcher assumes the role of an external authority figure. This approach can create a disconnect between the research process and the intended beneficiaries, as it fails to foster a sense of ownership and engagement among the stakeholders. In contrast, action research emphasizes a collaborative and participatory approach, where the researcher works in tandem with the organization or community to co-create knowledge and solutions.

The learning methodology employed in conventional research is predominantly discipline-oriented, focusing on the acquisition of specialized knowledge within specific academic fields. While this approach is valuable in building disciplinary expertise, it may overlook the importance of cross-disciplinary collaboration and the integration of diverse perspectives. Action research, on the other hand, promotes mutual learning between the researcher and the participants, enabling the generation of contextually relevant and holistic knowledge.Furthermore, conventional research often relies on testing theories and predicting outcomes in controlled settings like laboratories. This approach may not adequately capture the complexity and dynamism of real-world organizational and community challenges, which are influenced by a myriad of interconnected factors. Action research, in contrast, embraces the inherent complexity of these settings and seeks to develop context-specific solutions.Finally, the primary audience for conventional research is typically limited to the academic community, with limited engagement with external stakeholders. This narrow focus can hinder the dissemination and application of research findings beyond the confines of academia, ultimately reducing the potential impact of the research on societal progress and development.

9.1. Siloed Mentality and Disciplinary Boundaries

The research literature highlights the limitations of conventional research approaches in higher education that are characterized by a siloed mentality and disciplinary boundaries. Disciplinary silos are a common issue, where research tends to operate in isolation within specific academic fields, hindering cross-disciplinary collaboration and the development of holistic perspectives. This siloed mentality can prevent researchers from addressing complex, multifaceted problems that require the integration of diverse knowledge and viewpoints. The lack of collaboration across disciplines is a significant barrier, as it limits the ability to foster innovative solutions that draw upon complementary expertise and perspectives. Researchers often work within the confines of their fields, with limited incentives or opportunities to engage in truly interdisciplinary work. This disciplinary fragmentation can lead to duplication of efforts, gaps in the research landscape, and an overall disconnect between academic work and real-world challenges facing higher education and society. Overcoming these silos and promoting more collaborative, transdisciplinary approaches is crucial for enhancing the relevance and impact of research in the evolving higher education landscape.

9.2. Disconnect from Real-World Challenges

The literature has identified a critical limitation in the disconnect between research conducted in higher education institutions and the real-world challenges faced by various stakeholders. This disconnect manifests in several key areas. Much of the research undertaken in higher education settings is often focused on theoretical or academic pursuits, with limited consideration for practical applications and real-world impact. This disconnect can result in research being perceived as irrelevant or disconnected from the pressing issues and needs of students, faculty, and the broader community. Traditionally, research in higher education has been insular, with limited engagement with external stakeholders and the local/regional community. The lack of community involvement can hinder the ability to identify and address the actual needs and challenges faced by various stakeholders, reducing the relevance and applicability of the research.

Many of the challenges facing higher education and society are inherently complex, requiring the integration of diverse perspectives and the ability to tackle multifaceted issues. To enhance the relevance and impact of research in higher education, there is a growing need to integrate research with practical applications, foster more robust community engagement, and adopt a more holistic perspective that can effectively address the complex, real-world challenges facing the higher education landscape. By addressing these limitations and fostering a stronger connection between higher education research and practical applications, the research conducted in these institutions can become more relevant, impactful, and responsive to the evolving needs of students, faculty, and the broader community.

10. EMERGING OPPORTUNITIES FOR COLLABORATIVE RESEARCH

The landscape of research has undergone a transformative shift, with collaborative endeavors becoming increasingly pivotal in driving scientific and technological progress. This evolving paradigm presents a myriad of compelling opportunities that transcend traditional boundaries and catalyze innovative breakthroughs. One of the primary advantages of collaborative research lies in the ability to pool diverse resources and expertise. By uniting the specialized knowledge and capabilities of researchers from various backgrounds, complex problems can be addressed through a comprehensive and multifaceted approach. This synergistic integration of perspectives and skill sets enables researchers to tackle challenges that may have been beyond the scope of individual efforts. The digital revolution has ushered in a new era of global collaboration, where geographical barriers have become increasingly surmountable. Advancements in communication technologies and virtual collaboration platforms have facilitated worldwide seamless interactions and knowledge exchange among researchers. This has expanded the horizons of collaborative research, fostering a truly interconnected and dynamic research ecosystem. Collaborative research projects can leverage a suite of technology-enabled tools to streamline coordination, task allocation, and progress monitoring. These innovative solutions enhance the overall efficiency of collaborative endeavors and contribute to the successful execution of research initiatives.

The integration of Artificial Intelligence (AI) algorithms into the collaborative research process has opened up new avenues for data processing and analysis. AI-powered tools can automate the handling of large datasets, identify patterns, and generate preliminary insights, allowing researchers to focus on the more complex and conceptual aspects of their work. This synergistic integration of human expertise and machine capabilities can significantly accelerate the pace of scientific discovery. Initiatives like the Emerging Researchers' Group (ERG) have emerged as platforms for early-career and doctoral researchers to engage with the broader research community, exchange ideas, and build international connections. These efforts contribute to the diversification and strengthening of the collaborative research landscape, ensuring that a wide range of perspectives and voices are represented in the pursuit of scientific advancement. Funding organizations have increasingly recognized the value of collaborative research projects, acknowledging the potential benefits of pooling resources and expertise. This shift has created more opportunities for researchers to secure the necessary funding and support to undertake ambitious collaborative endeavors, further catalyzing the progress of scientific inquiry.

10.1. Bringing Together Scholars From Diverse Fields

The integration of diverse disciplinary expertise and perspectives has become increasingly crucial for addressing complex, multifaceted problems facing society. Interdisciplinary and transdisciplinary research approaches, which transcend traditional academic boundaries, offer innovative solutions by combining complementary knowledge and viewpoints from various fields. This cross-pollination of ideas and methodologies can lead to groundbreaking discoveries and impactful interventions that would be difficult to achieve within the confines of a single discipline. Advances in digital technologies have facilitated new avenues for researchers to collaborate across geographical barriers. Online platforms, videoconferencing, shared workspaces, and other collaboration tools enable real-time communication, data sharing, and joint project management, thereby enhancing the effectiveness of research collaborations. These digital tools have revolutionized the way researchers interact, exchange ideas, and work together, fostering a more connected and collaborative research ecosystem.

Collaborations between nearby colleges and universities can leverage shared resources, facilities, and expertise, allowing institutions to pool their strengths and pursue common initiatives through shared funding and joint projects. This local academic partnership model enables institutions to capitalize on their collective strengths and tackle complex challenges more effectively. Within individual universities, new opportunities are emerging for researchers to collaborate across departments and disciplines. By breaking down silos and fostering connections between researchers with complementary interests, intra-university collaborations can lead to novel research directions and synergies. These cross-disciplinary interactions can spark innovative ideas, facilitate knowledge transfer, and create new avenues for interdisciplinary research.

To encourage more collaborative research, universities need to align their incentive structures and recognition mechanisms. Adjusting promotion and tenure criteria to value collaborative work, providing dedicated funding for interdisciplinary projects, and celebrating team achievements can help shift the culture towards greater collaboration. By recognizing and rewarding the contributions of researchers who engage in collaborative endeavors, universities can foster an environment that nurtures and supports interdisciplinary and transdisciplinary research. By embracing these emerging opportunities and addressing the inherent challenges, higher education institutions can harness the collective expertise and resources needed to drive innovation, advance knowledge, and address pressing societal challenges through collaborative research. This collaborative approach to research holds the promise of unlocking new frontiers of understanding, fostering innovative solutions, and making a meaningful impact on the world.

10.2. Interdisciplinary and Transdisciplinary Approaches

Interdisciplinary and transdisciplinary approaches have emerged as innovative research paradigms that transcend traditional disciplinary boundaries and offer novel ways of addressing complex problems. These approaches emphasize the integration of diverse perspectives, methods, and knowledge domains to gain a more comprehensive understanding of multifaceted phenomena.

Interdisciplinary research involves the collaboration of scholars from various academic disciplines to tackle research questions that cannot be adequately addressed by a single field alone. By combining their unique expertise and insights, researchers from different disciplines work together to create new knowledge at the intersection of their fields. This collaborative process requires ongoing dialogue, negotiation, and synthesis of diverse disciplinary perspectives, while maintaining the core competencies and identities of each discipline involved. The outcome of interdisciplinary research is a more holistic understanding that integrates the strengths of multiple disciplines, leading to novel solutions and advancements in knowledge.

In contrast, transdisciplinary research goes beyond the boundaries of individual disciplines, aiming to transcend and transform them altogether. This approach involves the active participation of both academic researchers and non-academic stakeholders, such as policymakers, community members, and industry partners, in a collaborative effort to address complex real-world problems. The co-creation of knowledge between academics and societal partners is a key feature of transdisciplinary research, which is often iterative and reflexive, with ongoing feedback and adjustment based on practical needs and outcomes. The resulting knowledge is not confined to academic domains but is intended to be directly applicable and actionable, contributing to the resolution of complex societal challenges. Both interdisciplinary and transdisciplinary approaches have gained prominence in various fields, including climate change research, sustainable urban planning, healthcare innovation, and beyond. These research paradigms offer innovative ways of generating knowledge and fostering collaboration across disciplines and sectors, ultimately leading to more comprehensive and impactful solutions to our world's pressing issues today.

10.3. Emphasis on Community Engagement

The extant research literature underscores the burgeoning emphasis on community engagement in higher education and its paramount importance for fostering more salient and impactful research. Community engagement in higher education denotes the collaboration between universities and their larger communities (local, regional, national, global) for the mutually beneficial exchange of knowledge and resources in a context of partnership and reciprocity. Empirical studies have found that community engagement is associated with several salutary benefits for higher education institutions, including heightened retention rates for faculty and students, increased public appropriations and funding, enhanced capacity in research and teaching functions, and greater campus diversity. By engaging with external stakeholders and communities, universities can better align their research agendas with real-world needs, leading to more salient and impactful outcomes. There is a growing recognition that universities need to move beyond a unidirectional "outreach" model, where they disseminate their expertise to the community, to a bidirectional "engagement" framework that involves co-constructing solutions with external partners. This paradigmatic shift requires intentional planning and the development of institutional policies that enable and incentivize community engagement as a core function of higher education. To effectively implement community engagement, universities are actively defining it in strategic documents and academic policies to better enable and unlock its potential. By firmly situating community engagement at the heart of the academic mission, higher education institutions can enhance their societal impact and responsiveness to local, national, and global challenges.

11. TOWARD A HOLISTIC PERSPECTIVE ON RESEARCH IN HIGHER EDUCATION

To move toward a more holistic perspective on research in higher education, several key elements should be considered. Embracing interdisciplinary and transdisciplinary approaches is crucial for tackling complex, multifaceted problems. By bringing together scholars from various fields and engaging stakeholders beyond academia, research can gain a more comprehensive understanding of issues and foster innovative solutions. This approach allows for the cross-pollination of ideas, the integration of diverse methodologies, and the creation of a synergistic environment that transcends disciplinary boundaries. Enhancing the relevance and impact of research requires a stronger emphasis on community engagement. Universities should actively collaborate with local, regional, and global communities to co-create knowledge and develop solutions aligned with real-world challenges and priorities. This engagement fosters a reciprocal relationship between academia and society, ensuring that research addresses pressing societal issues and that the knowledge generated is accessible and applicable to those who can benefit from it. While research is a core function of higher education, it should be balanced with a focus on holistic student development. By integrating academic knowledge with other areas, such as mental well-being, cultural values, and community building, universities can cultivate students who are better equipped to apply their skills and thrive in the modern world. This approach recognizes the multifaceted nature of student growth and emphasizes the importance of nurturing well-rounded individuals who can navigate complex personal and professional landscapes.

Realizing a holistic approach to research in higher education necessitates a shift in institutional cultures, policies, and incentive structures. Universities should firmly situate community engagement and interdisciplinary collaboration at the heart of their academic missions, providing dedicated funding, adjusting promotion criteria, and celebrating team achievements. This transformation requires a paradigm shift that values not only individual excellence but also collective impact, fostering an environment that encourages and rewards cross-disciplinary collaboration and community-engaged scholarship. By embracing these elements and fostering a culture of collaboration, innovation, and social responsibility, higher education institutions can position themselves to drive progress, advance knowledge, and make a meaningful impact on the world through their research endeavors. This holistic approach to research in higher education has the potential to transform the way knowledge is created, shared, and applied, ultimately contributing to the betterment of society as a whole.

SUMMARY

This chapter is a comprehensive and innovative guide for academics, postgraduate students, and researchers in the field of higher education. It aimed to address the common struggles faced by academics in the realm of publishing and research supervision, while also providing valuable insights and strategies for postgraduate students in crafting sound research proposals and producing high-quality dissertations. The chapter emphasized the importance of embracing diverse perspectives in shaping research practices, challenging existing paradigms, and fostering a more inclusive and representative body of knowledge. It equipped its readers with the necessary skills and knowledge to excel in their research endeavors, including best practices, methodologies, and emerging trends in higher education research. Furthermore, the chapter recognized the crucial role of effective research supervision in shaping the success and well-being of postgraduate students. It provided practical strategies and evidence-based best practices to enhance the quality of supervision, ultimately contributing to the development of emerging scholars. This chapter was poised to make a significant contribution to knowledge production in the field of higher education. By inspiring and empowering academics, postgraduate students, and researchers to engage in rigorous, innovative, and socially relevant research, it aimed to drive the advancement of knowledge and understanding in the higher education sector.

REFERENCES

Flick, U. (2022). *The SAGE handbook of qualitative research design*. Sage. DOI: 10.4135/9781529770278

Gruwell, C. (2022). *Critical thinking in academic research*. Open Textbook Library.

Hewson, D., & Carroll, M. (2016). *Reflective practice in supervision: companion volume to the reflective supervision toolkit*. Moshpit Publishing.

Joireman, J., & Van Lange, P. A. (2015). The context of research in the behavioral sciences: Facing and embracing complexity. *The American Psychologist*, 70(3), 213–221. DOI: 10.1037/a0038999

Leslie, L. Z. (2010). *Communication research methods in postmodern culture*. Routledge.

Pawar, P. B., Verma, R., Daniel, C. O., & Sayyad, L. (2023). *Foundation of research methodology: A comprehensive guide*. Red Unicorn Publication.

Savitha, K., & Venkatachalam, J. (2024). Crafting and validating the essence of professional identity in higher secondary educators: *Educational Administration. Theory into Practice*, 30(5), 10565–10572. DOI: 10.53555/kuey.v30i5.3801

Scoffham, S., & Rawlinson, S. (2022). *Sustainability education: A classroom guide*. Bloomsbury Publishing. DOI: 10.5040/9781350262119

Sergeant, S. (2021). *Working together, learning together: Towards universal design for research*. Gompel and Svacina.

Strydom, P. (2011). *Contemporary critical theory and methodology*. Routledge. DOI: 10.4324/9780203875568

Watkins, D. C. (2022). Secondary data in mixed methods research. *Sage (Atlanta, Ga.)*.

Williams, M., Wiggins, R., Vogt, W. P., & Vogt, P. R. (2022). Beginning quantitative research. *Sage (Atlanta, Ga.)*.

ADDITIONAL READING

Heller, D. E., & Shaw, K. M. (Eds.). (2023). *State postsecondary education research: New methods to inform policy and practice.* Taylor & Francis. DOI: 10.4324/9781003447092

Jacob, M., Meek, V. L., & Kearney, M. L. (Eds.). (2022). *Higher education and research in the post-knowledge society: Scenarios for a future world.* Cambridge Scholars Publishing.

Overton, B., Pasque, P. A., & Burkhardt, J. C. (Eds.). (2023). *Engaged research and practice: Higher education and the pursuit of the public good.* Taylor & Francis.

Tight, M., & Huisman, J. (Eds.). (2015). *Theory and method in higher education research.* Emerald Group Publishing.

Tiwari, N. (2023). *Innovative approaches to research in higher education.* Academic Guru Publishing House.

Valcik, N. A., & Johnson, J. A. (Eds.). (2018). *Institutional research initiatives in higher education.* Routledge.

KEY TERMS AND DEFINITIONS

Constructivism: Constructivism is a philosophical approach to research that emphasizes the active role of the researcher and participants in constructing knowledge. Constructivist researchers seek to understand the subjective experiences and perspectives of their participants, rather than attempting to uncover objective, universal truths. This approach rejects the notion of a single, fixed reality, and instead views knowledge as being socially constructed through interactions and interpretations. Constructivist research often utilizes qualitative methods, such as interviews and observations, to gain rich, contextual insights into the phenomenon under study.

Critical Theory: Critical theory in research is an approach that aims to challenge existing power structures and promote social justice. Researchers working within this paradigm view knowledge as inherently political, and seek to uncover and address inequalities, oppression, and marginalization. Critical theorists may employ a range of qualitative and quantitative methods, but the underlying goal is to emancipate individuals and communities and bring about transformative change.

Descriptive Design: Descriptive research design is a type of study that aims to describe the characteristics, behaviors, or attitudes of a particular population or phenomenon. This approach does not typically involve manipulating variables or establishing causal relationships, but rather seeks to provide a detailed, accurate, and comprehensive account of the subject matter. Descriptive studies often utilize surveys, observations, and case studies to gather data and generate insights.

Experimental Design: Experimental design refers to the systematic process of planning and structuring research studies to investigate causal relationships between variables. This approach involves manipulating one or more independent variables and observing the effects on the dependent variable(s), while controlling for potential confounding factors. Experimental designs are commonly used in quantitative research to establish cause-and-effect relationships and test hypotheses.

Mentorship: Mentorship is a developmental relationship in which a more experienced or knowledgeable individual (the mentor) provides guidance, support, and advice to a less experienced individual (the mentee). Mentorship can take various forms, such as career development, personal growth, or academic support, and is often seen as a valuable tool for fostering learning, skill development, and professional advancement.

Mixed-Methods Research: Mixed-methods research is an approach that combines both qualitative and quantitative methods within a single study or series of studies. This approach allows researchers to leverage the strengths of both methodologies, providing a more comprehensive understanding of the research problem. Mixed-methods studies may involve collecting and analyzing different types of data, integrating the findings, and using the results to address complex research questions.

Ontology: Ontology is a branch of philosophy that deals with the nature of reality and the fundamental categories of being. In the context of research, ontology refers to the researcher's assumptions about the nature of the world and the nature of the phenomena being studied. Ontological considerations are closely linked to the researcher's epistemological and methodological choices, as they shape the way in which the research is conducted and the knowledge that is produced.

Positivism: Positivism is a philosophical approach to research that emphasizes the importance of empirical observation and the use of scientific methods to uncover objective, universal truths. Positivist researchers typically adopt a deductive approach, starting with a theory or hypothesis and then testing it through quantitative data collection and analysis. Positivism is often associated with the natural sciences, but it has also been applied in the social sciences.

Postmodernism: Postmodernism is a philosophical perspective that challenges the notion of a single, objective reality and the idea of universal, foundational truths. Postmodern researchers are often skeptical of grand narratives and metanarratives, and instead focus on the diversity of perspectives, the contextual nature of knowledge, and the role of power in shaping what is considered "truth". Postmodern research may employ a range of qualitative and interpretive methods to explore the multiplicity of meanings and experiences.

Qualitative Research: Qualitative research is an approach that emphasizes the exploration and understanding of the meanings, experiences, and perspectives of research participants. This type of research often utilizes methods such as interviews, focus groups, observations, and document analysis to gather rich, contextual data. Qualitative researchers are typically interested in understanding the "how" and "why" of a phenomenon, rather than simply measuring its prevalence or distribution.

Quantitative Research: Quantitative research is an approach that focuses on the measurement and analysis of numerical data to test hypotheses, establish relationships, and identify patterns. This type of research often employs experimental, correlational, or survey designs, and utilizes statistical methods to analyze the data. Quantitative researchers are typically interested in identifying causal relationships, generalizing findings to larger populations, and making predictions based on the data.

Research Skills: Research skills refer to the set of competencies and abilities that enable individuals to effectively conduct research, including critical thinking, problem-solving, information literacy, data analysis, and effective communication. These skills are essential for researchers across various disciplines, as they allow for the systematic and rigorous investigation of research questions, the interpretation of findings, and the dissemination of knowledge.

Chapter 2
Facilitating Scholarly Pursuits in Higher Education:
Cambodia's Endeavors Toward Research Advancement

Sereyrath Em
https://orcid.org/0000-0002-6803-195X
Suranaree University of Technology, Nakhon Ratchasima, Thailand

Buratin Khampirat
Suranaree University of Technology, Nakhon Ratchasima, Thailand

ABSTRACT

This chapter examines initiatives intended to cultivate a research-oriented advancement in Cambodian higher education and presents potential strategies for ensuring long-term progress in research. Cambodian higher institutions face significant challenges due to their limited capacity and resources. Still, they have aimed to boost research endeavors through various supports and make substantial contributions to the country's development. It is critical to cultivate academic networks and collaboration with different partners to advance research. There is a need to establish a research-oriented culture among novice researchers and educators by providing workshops and training, enabling them to recognize and address obstacles proficiently. The Ministry of Education, Youth and Sport is the most important body in helping higher education institutions enhance research capabilities. There are substantial prospects for expansion and improvement, with some obstacles, to research. However, Cambodia seems to be able to settle them with a crucial impact on driving research and social advancement.

DOI: 10.4018/979-8-3373-0806-7.ch002

BACKGROUND

Currently, the growing importance of research in Cambodia's higher education (HE) has become more apparent. The country is emphasizing improving its research capacities in HE to strengthen its position in the global knowledge economy (Chamnan & Ford, 2018). In addition to tackling urgent socioeconomic issues, fostering innovation, and increasing competitiveness, this goal aims to enhance academic endeavors (Heng, 2024). The Royal Government of Cambodia (RGC), the Ministry of Education, Youth and Sport (MoEYS), and other pertinent stakeholders are actively collaborating to support research projects (MoEYS, 2014). These institutions have tried to foster a culture of research productivity among students and faculty members, established financial support channels, and strengthened research infrastructure (Chea, 2020).

The primary goal is to raise the standard and relevance of research findings in Cambodia while ensuring that they are in line with both national development plans and international trends (Kim et al., 2020). In order to advance sustainable development, raise living standards, and address pressing social issues in areas like healthcare, agriculture, environmental sustainability, and technological convergence, Cambodia's vision calls for the establishment of a knowledge-driven economy through significant investments in research and innovation (World Bank, 2018). Taking a more global view, Cambodia's dedication to bolstering research inside its university system represents an amplified recognition of the significant impact that knowledge and creativity can have. Cambodia has aspired to become a center of academic research and technological innovation by developing a thriving research environment, which will benefit both its own development and the broader global knowledge economy (UNESCO, 2019).

Background of HEIs in Cambodia

As of 2023, Cambodia had a total of 189 higher education institutions (HEIs). Out of them, 79 were government and 110 were private institutions. These institutions are located in the provinces and Cambodia's capital city, Phnom Penh. 17 ministries and organizations govern these institutions. MoEYS supervises a total of 90 HEIs. These 90 institutions are located in Phnom Penh, as well as in 20 provinces. Out of these 90 institutions, there are 17 government and 73 private institutions. Both private and government-run institutions, under MoEYS, have a total of 40 campuses. Besides, there are a total of 51 institutions that provide master's degree programs, with 10 being government institutions and 41 being privately owned. Moreover, 25 institutions offer doctorate programs, with 7 being government-run and 18 being private-run (MoEYS, 2024a).

The HE has played a critical role in nation-building and socioeconomic development. This perspective places the HE role primarily at the micro level rather than the macro level. Students and educationalists are more likely to serve their self-interest or personal motives at this level, as it provides them with a higher social status, a higher income, knowledge, and power in society. However, HE also functions at a macro level. Supported by a strong knowledge base, the process of economic development and social advancement becomes more effective, which also suggests a higher pace of development (Marginson, 2011). On the whole, it can definitely assist the environment, health, art, social justice, peaceful coexistence, science, education, and technology (Salmi, 2009).

Prior to the mid-1990s, when the HE sectors began to emerge after two decades of war and political turmoil, the School of Higher Pedagogy, established in 1980, primarily provided HE. It enrolled academic staff and students from various institutions, such as the Buddhist Institute of Teacher Training, the Institute of Technology of Cambodia, and the medical staff of the military hospital (Dy, 2004).

Importance of Research Advancement in HEIs

The advancement of HE research has become a global concern. Many governments prioritize investing in and promoting research endeavors in HE due to their multitude of benefits, which extend beyond economic development to societal growth (Dy, 2004). Cambodia is also aiming to develop a knowledge-based society. The HE system equips a human resource pool with sound education to forge a knowledge-based society, particularly for those involved in academic research and development activities. Thus, in order to ensure the competitive capacity of a country in education, research in HE is essential (Heng et al., 2023). So, now is a good time to look into the current state of scholastic rights efforts in HE in Cambodia and figure out what the biggest problems are that need to be fixed in order to support and encourage more academic success in Cambodia and around the world.

HISTORICAL PERSPECTIVES ON RESEARCH IN CAMBODIA

Research in Cambodia was rooted in pre-colonial, colonial, and post-independence periods. However, any picture of knowledge-based thought was demolished during the Khmer Rouge Regime from 1975 to 1979 (Heng et al., 2023). In the early 1980s, the Khmer people, aided by others, liberated their country from the Khmer Rouge, leading to a period of relief from long-standing atrocities. At the time, Cambodia was well known to be one of Asia's poorest countries, with only scientists who were sympathetic to the Kampuchean revolutionary movement and kept themselves alive to plan and manage the country's reconstruction from time to time (Em et al., 2022). Cambodia then seriously needed human resource development and rehabilitation in various aspects, and the education sector, of course, was not an exception. Cambodia's bureaucracy suffered heavily, and its ability in all fields was very poor (Ayres, 2000). Without access to the country's old system of HE, Cambodia had to rely on foreigners for training, and Cambodia then sent many students abroad to study mainly in the Soviet Union and other socialist countries (Neau, 2003; Rany et al., 2012). At present, Cambodia has some returning students, which reflects how their experiences could be useful to the further development of research at higher learning institutions, particularly in areas that are beneficial for research advancement (Chen et al., 2007).

HEIs in Cambodia have faced numerous challenges, especially in the two to three decades following the ousting of the Khmer Rouge in the late 1970s (Clayton, 1998). April 17, 1975, marks the day when the Khmer Rouge regime took over control of Cambodia. The Khmer Rouge regime immediately closed universities and all other forms of education, viewing them as surplus and unnecessary (Kiernan, 2002). The few intellectuals who managed to evade execution by concealing their identities in the past or facing repercussions met their demise at the hands of youthful individuals who, despite lacking proper reasoning and judgment, believed they were protecting the soldiers' paradise they had hoped to establish. Undoubtedly, other intellectuals pretended to be different, as their unusual behavior would have exposed them otherwise (Kiernan, 2002).

CURRENT STATE OF RESEARCH IN CAMBODIAN HEIS

Currently, research in Cambodia has been extremely limited. Many students and academics hold misconceptions about the concept of research. A significant number of them got the terms printing and publication mixed up. In light of this, Cambodians must encourage studying and publishing (Em, 2024).

Due to a system that heavily prioritizes teaching over research, most universities are relatively new to managing, conducting, and promoting research. In general, former Universidad scholars lack experience applying for higher-level international funding, publishing in indexed international scientific journals, or collaborating with institutions from high-profile foreign universities (Heng et al., 2021). The lack of knowledge about research development is a major barrier for young scholars to advance further in their academic careers, and it can become an obstacle to improving the conditions for newly recruited talents who underwent their graduate studies in high-ranking universities (Harris, 2015).

Challenges Faced by Researchers

Resolving the challenges faced by Cambodia's growing number of researchers depends largely on how well the research environment fosters their progress in both qualitative and quantitative terms. The general problems faced by administrators, advisors, and research professionals in Cambodia when carrying out their respective roles are the same ones that face their counterparts in the developed and developing worlds alike (Eam, 2018). Young, developing research communities often face these threshold problems, which repeatedly impede the desired transition from teenage to mature research status. Consequently, the quality, quantity, and impact of research significantly increase, reflecting the potential and aspirations of Cambodia's HEIs (Busza, 2004). These threshold problems are: (1) ensuring that research time is available to an increasing number of poorly prepared researchers; and (2) the lack of experienced and professionally trained academic researchers, research administrators, and advisors in the area of postgraduate research education in every discipline. The section uses these threshold problems as the basic framework to present the challenges faced by the struggling Cambodian research environment and strategies available to assist in strengthening the country's research capacities. With HE growing in Cambodia, now may be the optimum time to ensure strong research training plays a large part in this growth. While experience is an ideal teacher, experience plus good research training offers the most to students of all ages who are ready to absorb (Heng, 2024).

HEIs hold a special place in Cambodia's life, given the country's rich learning tradition dating back to the 5th century (Rany et al., 2012). Many HEIs understand that a strong research base in the field informs and guides plans and choices for maximum effectiveness in teaching and learning. Academics and students are at the heart of every institution; academic staff have much to give, and students are willing to learn (Greenhow & Galvin, 2020). However, Cambodia, as a young democracy, is only just beginning to develop its scientific and research culture. Before the 1990s, many academics grew up in an environment that heavily emphasized propaganda or ideology in their research. Most academic research consists of small-scale educational research projects, as well as research aimed at simply publishing academic articles rather than undertaking projects with genuine academic or social impact. It is our belief that increasing the research capacity of Cambodia's HEIs through developing experienced and professional academic staff, particularly those already in teaching positions, and exposing students to research training is, in the long term, the most effective way to enhance Cambodia's HE sectors in general (Clayton, 2005).

Research Funding and Support Programs

As a matter of policy, there is a need to support, guide, promote, patronize, and lead various sectors to be deeply involved in strengthening the research system of the country. The policy is to invest in infrastructure, build and maintain research potential, human resources, research finance, and collaboration with foreign experts and other partners. Researchers and those involved in research activities aim to acquire the necessary tools needed to take part in research (Heng & Sol, 2021; Heng, 2024; Sok & Bunry, 2021). The government will pursue technical and vocational education as the central strategy for transferring the skills, competencies, and technologies that are required to address industry demand and enhance national productivity. The HE and research institutions aim to have an institutional cluster of scientific equipment resources that allow access to a wide range of technology and advanced equipment at a cost that our researchers can afford. These modern equipment resources support pioneering research to solve real-world problems and enhance students' learning experiences. The researchers obtain income or some of the salary, by enlarging lab facilities, buying equipment useful for research and instruction, and covering travel expenditures for scientific meetings and training course expenses in the form of support for scientific research (Indrawati & Kuncoro, 2021; Li and Pilz, 2023).

Cambodia's drive for research development, especially in HEIs, is part of the country's efforts to improve its capacity to perform, seek, facilitate, and disseminate research undertakings. Research endeavors in HEIs enable academicians and scholars from national and sub-national agencies, private institutions, and non-governmental organizations to take part in and support research for improving Cambodian prospects (Heng & Sol, 2021; Heng, 2024). An additional objective of our goals is to steadily inch Cambodia upward in the international pecking order of upper-middle-income countries in 2050 (Em et al., 2022). We are striving to provide our citizens with the opportunity to lead a better life and to avail themselves of broad education, usable skills, and an advanced standard of living. Research policies are being formulated systematically to correspond to the new phase of development and changing circumstances in research and development at the international, regional, and national levels. The success of continuously increasing our country's capacity in terms of output and capacity of the research system will, in part, reflect the effectiveness of the strategy, plan, and scheme initiated by the government in this sector (Heng et al., 2020; Heng et al., 2021).

GOVERNMENT INITIATIVES FOR RESEARCH ADVANCEMENT

Cambodia's research plan being developed by MOEYS in collaboration with the related stakeholders represents another important policy milestone to promote the advancement of national capabilities for research of social importance. This initiative is part of the MoEYS's goal to become the National Research Foundation. With assistance from the Asia Foundation, the ministry is collaborating with many researchers and stakeholders from Cambodian institutions, including universities and governmental departments, as well as international partner institutions, to define strategies and an action plan to build national research capacity (Em et al., 2022). The plan is intended to describe priority research areas for Cambodia's development and, more broadly, to support the strategic process of applying international standards in academic research in line with the institution's vision of capacity building in sciences and technology (Heng & Sol, 2021). Furthermore, it is consistent with terms of reference identified by the National Institute for HE and is intended to help establish a national research network, link researchers with international counterparts, and fulfill the RGC's plan to convert the Cambodian economy into a knowledge-based society (Heng, 2024).

RGC continues to make progress in addressing the country's needs and challenges for research advancement (Heng & Sol, 2021). Government statements and national academic planning documents signal that Cambodian leaders have come to recognize the value of research that addresses the country's most pressing social needs. The country's medium-term and long-term strategies for education and human resources development underscore the importance of building a knowledge-based economy and of applied teaching and learning to solve problems and promote innovation in society (Heng et al., 2022). The government also recognizes that Cambodia is sorely lacking in researchers and research institutions. A range of government-sponsored scholarships exist to train master's and doctoral students in Cambodia and abroad. Efforts to support and improve the research environment have involved a variety of actions, such as the development of institutional enabling environments, provision of start-up research funds, research awards, collaboration with international donor organizations, and promotion of local research partnerships with diaspora and foreign institutions (Heng, 2024).

National Policies and Strategies

Furthermore, participants of the regional workshop agreed that initiating research work capacity focus should rely on actual circumstances of global changes and local impacts from research activities. The participants also accentuated that the provision of mechanisms to promote the professional and academic development of young scholars and their quality research outputs, and the generation of research groups learning from the curriculum courses from outside universities, with monitored and valued self-evaluation system to measure quality, should also be in place (Alberio & Soubirou, 2022). Nevertheless, despite the degree of agreement and gravitas attributed to the guidelines, there appears to be a large question mark lingering on the hearts and minds of the academics as to whether recommendations will decisively be taken into account. A governance and scholarship framework must be set (Ntabe, 2020).

Presenting a narrative discussion. Attention toward enhancing the capacity of HE to undertake research is beginning to surface among policymakers (Hayes et al., 2021). A pilot study was conducted with academics in the two major public universities; the opinions and propositions garnered from these participants provided initial insights. These scholars believe that enhancement of research capacity goes hand in hand with outstanding yielding development results and vice-versa, that Cambodia's institutions of HE must first and foremost engage in legitimizing research as equally important as teaching and community services (Heng & Sol, 2021; Heng et al., 2023; Seng et al., 2020).

In HE, policies toward research have been focused on integrating research with teaching and knowledge transfer and recruiting more high-quality researchers (Evans et al., 2021). National policies of many countries now emphasize research in HE as an integral contribution to national development, promoting opportunities for linkages between universities and surrounding as well as international communities. Drafting Cambodia's Education Strategic Planning 2014-2018 without explicitly mentioning the importance of research could well be a missed milestone. However, it is not too late to take on board that research as a priority in HE could enhance the quality of the nation's human resources (M. Heng, 2022; Sok & Bunry, 2021).

Establishment of Research Centers

The establishment of research centers is in line with the assessment of the gap constraints between the demands and barriers that were one hundred percent rated for the foundation of all research types, centers, or sources (Petraki & Khat, 2022). They could be more effective with well-equipped and skilled staff and students. The structure of these centers is still modest. For achieving the most effective outcome, which can contribute toward the enhancement of sustainable development, the proper infrastructure, techniques, and expertise personnel are necessarily required (Heng & Sol, 2021). A large portion of funds from both local and international organizations should be engaged in these institutions. Concerning the reasons above, if these centers are operated with proper objectives focusing on carrying out research activities that reflect Cambodia’s vision plan by the students and academic staff, they can be asset-sharing for all educational bodies (Sun et al., 2021). They might be the dynamic engines in promoting and facilitating dedicated scholars, neutral policy recommendations, and the development of excellent students in this post-conflict society. In the end, it is hopeful that these research centers to be top-line institutions that lead to shaping the face of the whole educational context of Cambodia's society (Nhem, 2023).

The increasing demand for quality research was noticed and undertaken by rectors, professors, and lecturers in various fields of study. In struggling toward this realm, various Cambodian universities began to establish their research centers (Heng & Sol, 2021; Heng et al., 2023; Heng, 2024). Fields such as geography and development, archaeology, agriculture and food, health education, Buddhism, etc. are among the research centers that already exist and rapidly conduct their research activities. Moreover, the research center, Cambodian Development Research Institute (CDRI), on tourism research was also established and ran its research projects independently. Thus, these research centers are expected to be the channels that trigger capacity-building and sustainable research activities in Cambodia's universities (Heng, 2024).

PROMOTING A RESEARCH CULTURE IN CAMBODIAN HEIS

Many HEIs face constraints as they try to promote a research culture. The prevailing situation of overstretched academic staff and their low remuneration hardly encourages research activities. The limited access to current academic discourse is another factor, more appreciated as an obstacle by Khmer scholars than by expatriates. Access to state-of-the-art academic literature at research universities occurs partially through individual initiative. Resources for the preparation of serious collective research proposals also exceed the capacity of institutions (K. Heng, 2022; Heng, 2024). Seeing these difficulties and some possible solutions, MoEYS started providing professorship titles through research and publication (Em, 2023a, 2023b). Table 1 presents the requirements for professorship qualifications in the Cambodian education sector prior to 2024.

Table 1. Initial minimum criteria for the professorship title in the Cambodian education sector

No.	Criteria	Assistant Professor	Associate Professor	Professors
1.	Degree	Holds a PhD degree for at least two years after graduation or holds a master's degree for at least five years after graduation	Holds a PhD degree for at least five years after graduation or holds a master's degree for at least 10 years after graduation	Holds a PhD degree for at least 10 years after graduation
2.	Article publication	Publishes at least three articles in national or international journals or publishes at least one recognized textbook in the relevant field	Publishes at least five articles in national or international journals	Publishes at least seven articles in national or international journals
3.	Conference presentation	-	Has presented or served as a speaker in national or international conferences for at least five times	Has presented or served as a speaker in national or international conferences for at least seven times
4.	Textbook publication	-	Has published at least one recognized textbook in the relevant field	Has published at least two recognized textbooks in the relevant field
5.	Teaching and research experience	Has at least five years of experience in teaching and research recognized by the scientific council of the applicant's university	-	-
6.	Recognition	-	-	Is a person who is recognized in their own respective field
7.	Professional ethics	Has never violated professional codes of conduct or has never been convicted of a crime or misdemeanor	Has never violated professional codes of conduct or has never been convicted of a crime or misdemeanor	Has never violated professional codes of conduct or has never been convicted of a crime or misdemeanor

Source: Heng and Heng (2023, pp. 65-66)

Currently, the nomination process for professorships in Cambodia has undergone changes as the country strives to advance its research and publications. They have made the criteria for professorship nominations clearer and more challenging. Applicants are awarded 40 points for research and publication (articles and books). For teaching, presenting at conferences or public forums, recognizing good work, or supervising or advising students earns 15 points. The applicants receive 15 points for additional recognized work, as well as 15 points for qualifications such as an MA or a doctorate.

Table 2. New and updated criteria (rubric for professorship evaluation)

<table>
<tr><th>No.</th><th>Indicator</th><th>Point</th><th colspan="5">Grade and point</th></tr>
<tr><td rowspan="2">1.</td><td rowspan="2">Degree</td><td rowspan="2">15</td><td>-</td><td>5 points</td><td>7.5 points</td><td>10 points</td><td>25 points</td></tr>
<tr><td>-</td><td>Holds MA from 5 to 10 years</td><td>Holds an MA from 11 to 15 years
Holds PhD less than 5 years</td><td>Holds an MA for more than 15 years or a doctoral degree from or more than 5 years</td><td>Holds a doctoral degree for more than 10 years</td></tr>
<tr><td rowspan="2">2.</td><td rowspan="2">Article publication</td><td rowspan="2">20</td><td>4 points</td><td>8 points</td><td>12 points</td><td>16 points</td><td>20 points</td></tr>
<tr><td>1 article</td><td>2 articles</td><td>3 or 4 articles</td><td>5 or 6 articles</td><td>7 articles</td></tr>
<tr><td colspan="8">Note:
- The applicants who have published from 8 to 10 articles receive 5 extra points.
- The applicants who have published from 11 to 20 articles receive 10 extra points.
- The applicants who have published more than 20 articles receive 15 extra points.
- The applicants who have published from 5 to 10 articles in Scopus or Web of Science-indexed journals receive 5 extra points.
- The applicants who have published more than 10 articles in Scopus or Web of Science-indexed journals receive 10 extra points.
- The extra points for articles published in Scopus or Web of Sciences-indexed journals are the second additional points as incentives and the recognition of high-quality work accepted internationally.</td></tr>
<tr><td rowspan="2">3.</td><td rowspan="2">Book publication</td><td rowspan="2">20</td><td>-</td><td>5 points</td><td>10 points</td><td>15 points</td><td>20 points</td></tr>
<tr><td>-</td><td>- 1 international report book or
- 2 national report books</td><td>- 1 academic book or
- 2 international report books or
- Four national report books</td><td>- 1 academic book and 1 international report book or 2 international report books
- 3 international report books</td><td>- 2 academic books or 1 academic book and 2 international report books or 4 national report books</td></tr>
<tr><td colspan="8">Note:
If the applicants have published from 3 to 5 books, they receive 5 extra points.
- If the applicants have published from 6 to 10 books, they receive 10 extra points.
- If the applicants have published more than 10 books, they receive 15 extra points.
- If the applicants have published books or report books with international famous organizations (e.g., ADB, OECD, ILO, UNESCO, UNICEF, World Bank, etc.), from 4 to 5 books, they receive 5 extra points, and if they have published with these organizations of more than 5 books, they receive 10 extra points.
- If the applicants have published books or report books with international famous publishers (e.g., Springer, Routledge/ Taylor and Francis, John Wiley & Sons, Palgrave MacMillan, Pearson, Sage Publications, Cambridge University Press, Oxford University Press, etc.), from 2 to 3 books, they receive 5 extra points, and if they have published with these publishers of more than 3 books, they receive 10 extra points.</td></tr>
<tr><td rowspan="2">4.</td><td rowspan="2">Teaching experiences/ speakers or presenters in conferences/ speakers in public forums</td><td rowspan="2">15</td><td>-</td><td>-</td><td>5 points</td><td>10 points</td><td>15 points</td></tr>
<tr><td>-</td><td>-</td><td>- 5-7 years of teaching experience and holds a master's degree
- 2-4 years of teaching experience and holds a PhD or
- At least 2 years of teaching experience and being a speaker or a presenter in national or international conferences 3-4 times/being a speaker or a presenter in public forums 6-9 times</td><td>- 8-9 years of teaching experience and holds a master's degree
- 5-7 years of teaching experience and holds a doctoral degree
- At least 4 years of teaching experience and being a speaker or a presenter in national or international conferences 5-6 times/being a speaker or a presenter in the public forums 10-13 times</td><td>- At least 10 years of teaching experience and holds a master's degree
- At least 8 years of teaching experience and holds a doctoral degree or
- At least 6 years of teaching experience and being a speaker or a presenter in national or international conferences of at least 7 times/ being a speaker or a presenter in public forums of at least 14 times</td></tr>
</table>

continued on following page

Table 2. Continued

No.	Indicator	Point	Grade and point				
Note: - Attending national and international conferences or public forums as participants, not speakers/presenters are not counted.							
5.	Recognition of good teaching / supervising students	15	-	-	5 points	10 points	15 points
			-	-	- 1 recognized certification of good teaching experience or - Supervised 3-5 master's degree students/ supervised 1-2 doctoral students	- 2 recognized certifications of good teaching experience or -Supervised 6-10 master's degree students/ supervised 3-5 doctoral students	- 3 recognized certifications of good teaching experience or - Supervised more than 10 master's degree students/ supervised more than 5 doctoral students
Note: - The recognition of good teaching refers to those who receive teaching awards, on semester/yearly wards, outstanding evaluations from students, yearly outstanding evaluations from students or the institution where the applicants work, certificates of good teaching experience issued by the institution where the applicants work, etc. - Supervising or advising master's degree students refers to supervising or advising their master's thesis or master's research report. Other supervising or advising such as the term papers works in the courses are not counted. - Supervising or advising doctoral degree students refers to supervising or advising their doctoral thesis. Other supervising or advising such as the term papers works in the courses are not counted. Being the main supervisor or advisor and co-supervisor or co-advisor are counted, but being the doctoral committee is not counted. - Supervising or advising associate or bachelor's degree students is not counted.							
6.	Other additional recognized works	15	-	-	5 points	10 points	15 points
			-	-	- Has 1 or 2 additional works	- Has 2 or 4 additional works	- Has 5 additional works or more
Note: - Other additional recognized works encompass prototypes, research products, patents, software applications, formulation of theories and models, involvement in policy startups, curriculum design, production of instructional videos, execution of experiments, undertaking research endeavors, evaluation of research articles as an editor or expert reviewer, innovative teaching utilizing technology or novel pedagogical approaches, collaboration with universities, communities, or social development initiatives, participation in the advancement of teaching methodologies, and acknowledgment of exceptional individuals within the applicants' respective fields, among others.							
7.	Professional ethics	-	Requirements	Requirements	Requirements	Requirements	Requirements
Note: Requirements have no point, but they are the most essential need to apply for the three levels of professorships: assistant professor, associate professor, and full professor.							
Total points according to each indicator		Academic degree (15 points)	Article publication (20 points)	Book publication (20 points)	Teaching experience or being a speaker in conferences/ public forums (15 points)	Recognition of good teaching / supervising students (15 points)	Other additional recognized works (15 points)
Cross total points							

Source: Adapted and translated from the Khmer version of MoEYS (2024b, pp. 19-24)

A. Eligible criteria

Applicants applying for professorship titles in the education sector need to fulfill the following criteria:

1. Have teaching experience and be an instructor at an HE institution where he/she has been fully nominated in the public framework for 2 years for the applicants who have a doctoral degree but have no prior experience.
2. Have teaching experience and be an instructor at an HE institution where he/she has been fully nominated in the public framework for 5 years for the applicants who have a master's degree.
3. Have a doctoral degree and have published at least an academic book for applicants applying for a professor title.

   B. Points and results

      1. From 85 points and above (Receive professorship title)
      2. From 75 to 84 points (Receive associate professorship title)
      3. From 60 to 74 points (Receive assistant professorship title)
      4. Less than 60 points (Receive no title)

Importance of Research Ethics and Integrity

It is essential to underscore the veracity that the optima of higher-level research can only be best realized in a setting endowed with full awareness of the nuances of research ethics and with integrity uncompromisingly upheld (Gonzalez et al., 2022). As mentors, faculty members must inculcate in the minds of these neophyte researchers the values they must hold dear whenever they tread the road to scholarly pursuits. HEIs enforce and impose research integrity standards through encouragement, excitement, and rewards. It is the fundamental obligation of any institution involved in teaching and research undertakings to drive ethical conduct in research as an essential ingredient in scholarly work. Frequent exposure of students to research environments and engagements in scientific research has been shown to aid in the development of a strong sense of research integrity (Abbass et al., 2022). The known beneficial effects of mentorship are also apt conduits by which research integrity could effectively be driven. Such ethical prerequisites inculcated and demonstrated in academic exercises remain in the hearts and minds of these individual learners, magnifying the need to deal with distrust and betrayal. With a clear understanding of what is required, these young researchers set their moral compass specifically during research proposal development and are able to cruise freely through the rough seas of research conduct and responsibility (Chin et al., 2023).

Given the steady progress of both undergraduate and postgraduate programs at HEIs over recent years in Cambodia, sustained effort is also being made by various universities to similarly concentrate on knowledge generation, focusing on, among others, impactful research initiatives. The HE settings are likeliest to be where research endeavors are undertaken, usually in consort with the teaching function (Heng, 2024; Heng & Sol, 2021). In these organizations, research activities are heavily reliant on faculty members, students, and would-be scholars or researchers. Therefore, they must be provided ample exposure and orientation to enhance their skills in both research capabilities and integrity long before they are expected to conduct actual research activities on their own (Sovansophal & Shimizu, 2020).

ENCOURAGING INTERDISCIPLINARY RESEARCH

With the advent of interdisciplinarity as one of the latest trends in university education, the academic system stands to enhance the creation and dissemination of knowledge, a primary goal of tertiary education, while also developing the competencies that graduates need to tackle the 21st-century grand challenges (Peycam, 2020). However, in opposition to the potential for and advantages of the advancement of interdisciplinary research, international acknowledgment that while productivity is maximized at the interfaces of academic disciplines, such research is most difficult to undertake. Central to interdisciplinary research capacity and competencies across tertiary education levels and types facilitate social progress and critically explore the boundaries of emergent technologies (Ullah et al., 2022). The international association notes a huge unmet demand for legal experts on climate change, and a similar conclusion may be drawn from the number of joint degrees or research initiatives around land management, land tenure, or nutrition. Colleges and universities that explicitly implement curricula and invest in intellectual infrastructures that facilitate interdisciplinary teaching and open up research opportunities for scientists at the faculty level will be best placed to attract and retain experts who combine these competencies with scientific skills and social awareness (Żuk & Szulecki, 2020).

Interdisciplinary research is essential in meeting the challenges and requirements of sustainable development as it integrates knowledge and skills from several disciplines to address complex problems and identify solutions at the intersections of traditional academic disciplines (Heng, 2024). The international conference on HE is an effort to foster a structural dialogue among HE, science, and development stakeholders across the continents and a contribution to promoting strategic collaboration within and between academic and international research organizations and HEIs worldwide. Moreover, from a social perspective, interdisciplinary research is necessary to increase our understanding of complex social and environmental questions. To address global challenges, a holistic understanding of the integrated assessment of natural and social systems is indispensable. Small, developing, and least developed countries, in particular, stand to benefit from optimizing sustainable development by synthesizing knowledge across the sciences, thus contributing to maximizing the impact of both cropland sciences and the education required to generate that knowledge in the countries concerned (Helldén et al., 2023).

INNOVATIVE APPROACHES TO RESEARCH DISSEMINATION

Cambodia has been trying to use innovative approaches to promote research dissemination. One particular area of study that may be explored is the relationship between policy papers and academic papers. Specifically, it is worth investigating the types of policy papers that require certain demands and the types of academic papers that are based on specific theories (Heng & Sol, 2021; Heng et al., 2023). Additionally, it is important to determine which sorts of papers are only suitable for publication and require extensive preparation, while other policy papers may not require the same level of preparation or subsequent steps. Each of the two policy papers and one learning poster paper is integral to the whole process of promoting cooperation, influence, reuse, and participation (Heng, 2024).

Besides critical demonstrations, other innovative communication methods could also be used to make research efforts more useful, especially for implementation. In general, it is summarized that there should be some refocusing on research to a greater emphasis on the engagement of stakeholders and on the dissemination of research outputs to all those who can use them usefully (Hum & Heng, 2024).

The question is creating necessary dissemination tools or media for those aims (Heng et al., 2022). Overall innovative communication methods to apply for both university and center or research institute could include (1) active policy engagements and relevant dissemination processes with both the supply side and demand side, (2) specific or/and annual meetings with both obsessive potential downstream and obsessed upstream stakeholders, (3) research aimed exclusively for beyond academe outputs, and (4) technologies for both research with practice and research and learning assistance such as for locating suitable documents or training support.

In conclusion, it may be beneficial to examine in more detail how that equilibrium might be optimally achieved. Currently, the prevailing approach in research is to focus on practice-based publications that necessitate multidisciplinary collaboration, primarily in a team setting. Peer review may also be conducted for academic researchers, and academic researchers can also be incentivized to carry out research (Sok & Bunry, 2021).

Utilization of Open-Access Journals

All publishers publishing journal articles in Cambodia are open-access, and some universities, such as the Royal University of Phnom Phnom Penh, support also authors to publish their articles open-access by providing funds for article processing charges when the authors publish their works with high-quality journals. Mwambari et al. (2022) noted that the advantages of open-access journals are far-reaching and profound. They democratize access and speed up the spreading of knowledge, helping to avoid the huge cost of subscriptions, thereby making it possible for more scholars in a wider range of institutions to enter the discourse and ultimately increasing the depth and breadth of scholarship. According to Heng and Sol (2021), fostering researchers to find relevant literature resources more rapidly, in a much more cost-effective and convenient manner for their research needs, open access significantly accelerates information application to innovation. Thus, it encourages greater application of knowledge to societal ends and mitigates the harms to society that are due to the misallocation of resources to gain access to proprietary knowledge.

Moreover, most high-quality open-access content is still available in largely subscribed proprietary systems. As a result, subscription fees for such proprietary journals are constantly increasing and unaffordable to many lower-income countries and even multinational institutions (Siler & Frenken, 2020). Therefore, over-teaching or over-using these proprietary contents in HEIs, without making a serious effort to utilize open-access scholarly knowledge, not only causes heavy expenditure on the tuition fees paid by the students and library subscription fees paid by the institution and the tax team but also dilutes the academic quality and rigidity and then jeopardizes the advanced utilization of national intellectual capital and its innovative technology to support the sustainable development of nations (Nelson & Bérubé, 2020).

Although not easily achieved over a few years due to variation of readiness and interest of different faculties, now HEIs, government, and university management should timely take multi-disciplinary collaborative action to lead their scholars toward adapting the knowledge culture more efficiently from over-teaching to research-oriented learning and then to become self-educators in fulfilling their work needs (Ursić et al., 2022). They should convincingly overcome the laziness and/or the mindset of seeking a quick look at the published research from traditional proprietary journals rather than proactively self-discover from multi-databases and freely accessible journal contents, and gradually change inappropriate research and teaching practices into a higher standard of academic culture and excellence (Soulard et al., 2020).

Engagement with the Public and Policymakers

Public universities, especially those with medical schools or health science colleges, are generally more engaged, albeit some with an ivory-tower approach. Others are gradually waking up to such contributions and forcing themselves, as found, for instance, to open up or work with international or other groups to help expedite its advancement (Heng, 2024; Heng & Sol, 2021). Similarly, awareness has been slowly increasing of consulting support, and to some extent on businesses including joint research initiatives and revenue-bearing collaborations, let alone assisting entrepreneurial talent amongst students through business incubation. Nevertheless, there is room for much more advancement (Heng & Sol, 2021).

When contributions and impact are substantial, the public is more prepared to provide support and be more trusting in other areas of research. Even in the so-called developed countries, university governance and its impact on research have been receiving scant attention. While struggles in their cooperative research exist both in the conceptual and implementation phases, international business collaborations have been felt to have shortcomings in some public universities, including a few of the participating faculty in the collaborations (Heng, 2024; Heng et al., 2023; Sok & Bunry, 2021). HEIs with public obligations have become ever more involved in protecting and valorizing scientific knowledge in fields that are publicly funded, through intellectual property rights or patents (Leon-Roa et al., 2024).

CAPACITY BUILDING FOR RESEARCHERS IN CAMBODIA

Cambodia has and continues to receive limited resources for human capacity development. The critique of Cambodia in terms of research development not only highlights the need for more aggressive promotion of research (Heng & Sol, 2022). An even more critical constraint is the shortage of competent human resources that exists at every level necessary for national research to take place, especially for higher-degree research students. Although it is unclear what the exact population or demographic potential for better student learning might be in Cambodia, it is apparent that considerable resource implications exist just to achieve more as-yet-unidentified goals for HE in the emerging international competition for research involvement by individuals in other economies (Heng & Sol, 2021).

Unusual changes will undoubtedly be required to double the number of research students currently receiving training in Cambodian universities. Furthermore, the current educational changes inhibit rather than facilitate the necessary development of new student research, graduates, and other outcomes in HE. For both traditional and innovative research training outcomes in the development of Cambodia's young people, there is no doubt that traditional methods need substantial reappraisal (Eam et al., 2021). There are growing and under-recognized calls for flexibility in programming, innovative curriculum design, multi-institutional research, graduate experience, and double-degree training to meet new capacity needs within different HE systems. These required changes are rarely high priorities in any area of interest in present-day Cambodian higher research education (Sok & Bunry, 2021).

According to the National Congress in 2024, Cambodia has been providing research management frameworks to HEIs to support and encourage these institutions and researchers to have fully equipped support for research and publication with higher quality, especially to start from action research as an encouragement within the HEIs to other types of research to reach national, regional, and international standards. Besides, MoEYS and other related stakeholders have prepared workshops and training to help Cambodian novice researchers walk on the right track. These projects have also received funding from MoEYS and those stakeholders accordingly (MoEYS, 2024a).

Training and Workshops

Apart from what MoEYS has been doing to develop research endeavors for HE in Cambodia as presented in Tables 1 and 2 and other parts in this chapter, the Centre for Khmer Studies (CKS), based in Siem Reap, Cambodia, has organized several training workshops on research methods and technical skills for both students and local scholars. Participant feedback from over 150 project participants is overwhelmingly positive. It proposes that local researchers and students from the targeted groups see these gatherings as complementary to their educational experiences and appreciate the relevance of the skills presented and the facilitators' commitment to effective knowledge transfer (Peycam, 2020).

Not only CKS but other related stakeholders, including the public and private, have also prepared workshops and training to shape Cambodian novice researchers. For example, the Cambodian Education Forum (CEF), based in Phnom Penh, Cambodia, has led internships for six cohorts up to now in writing edited books. The CEF has also prepared around 20 webinars to allow Cambodian novice researchers to present their findings from multidisciplinary fields (see Sol & Heng, 2022, 2023).

In addition, the Education, Research, and Innovation Center (ERIC) led by Dr. Chan Sokhom, based in Kampot, Cambodia, has offered various courses on research methodology and statistics, utilizing small funds to facilitate the process. The ERIC has also led three cohorts of novice researchers to write original research articles to publish in high-quality journals.

Furthermore, the Cambodian Journal of Educational and Social Sciences (CJESS) plans to conduct various webinars and seminars, as well as provide training on research knowledge development, following the release of its first issue. The plan will come into force soon. The journal team has also aimed to provide Cambodian novice researchers with various platforms through which they can submit their manuscripts to national and international publishers.

Such workshops and training can be further tailored to specific types of research such as action research, evaluative research, development or applied research, etc. In participating in research-related workshops, participants should have hands-on practical, and interactive learning experiences.

- Aiming research at policy objectives and audiences
- Improving the policy relevance of research
- Good research management (good planning, good budgeting, effective time and team management)
- Working with policy research funders
- Types of policy questions
- Producing credible research (critical interrogations for validity and reliability)
- Bridging the separate worlds: policy and research communities (inter alia, transparency, team building, relations with stakeholders, getting involved in the policy process, knowledge and capacity sharing)
- Dissemination strategies

CONCLUSION AND FUTURE DIRECTIONS

This chapter serves as a reflection on HE research and publication, particularly the current efforts to foster research activities among Cambodian universities. It also offers future directions for sustainability and broader research advancement. Utilizing various strategies and programs, Cambodian research institutions, RGC, MoEYS, and related stakeholders have made good progress in terms of enhancing the research capacity of university faculty and reviving a research culture. Their vigor in contributing to economic and social advancement makes HEIs the key players in generating the pipeline of graduates and the backbone of economic development. The chapter reveals visible benefits from the current initiatives and highlights several challenges facing HEIs in Cambodia as they seek to further support research and effectively disseminate the knowledge produced.

Key Takeaways from Cambodia's Research Advancement Efforts

This chapter also presents key takeaways from earlier examinations and provides viewpoints on the Cambodian HE sector. Despite continual discussions of reform specifications associated with scholarly advancements, deeply studying the academic requirements for research productivity is critical. This chapter thus examines the faculty members' profiles, essential research capabilities, and public recognition for research activities. Scholarly pursuits, such as conducting groundbreaking research, solving practical problems, addressing sociocultural issues, and publishing research outputs, are critical faculties for Cambodian university academics as well as students. These faculties contribute greatly to Cambodia's future by nurturing a sustainable life and providing catalysts.

Cambodia's policymakers and HE leaders have paid careful attention to finding ways to foster a persuasive research culture within HEIs, which they believe is necessary for setting Cambodia's universities on a course toward various advances. It is important to ensure that faculty members satisfy their research expectations as a means to produce scholars. However, poor research productivity is a common phenomenon in Cambodian government and private HEIs. The goal of advancing HEIs in Cambodia signifies an emphasis on increasing the rate of the number of graduates. Notwithstanding, recognition of Cambodia's HEIs' quality, in terms of fostering and generating scholarly outputs, has emerged as a critical concern, especially when harsh consciousness on the poor research productivity of academics and other stakeholders is raised within the government, diverse development strategies, and the international community.

Recommendations for Sustainable Research Growth

Several points can be counted to improve research productivity in Cambodia. However, three essential points should be considered. The first goal should be to strengthen a research culture among Cambodia's educators, training them in that research philosophy and methodologies that allow them to find their own educational problems and then help to solve them. The end goal of such professional support would be to help Cambodia's HEIs achieve a stringent and robust form of educational practice, one that is data-driven, research-enhanced, evidence-based, and research-informed.

A second objective would be to develop research strength within HEIs themselves, using what already we know from practice in the scholarly communities of the developed nations. For instance, a careful approach to research development and capabilities worldwide would involve better alignment and focus of research priorities, a concentration at higher degree levels on discipline-relevant and useful research skills, the mentoring of new research staff, the establishment of clearer research roles and career paths, the provision of measurable incentives for productive research behavior, a combination of directed larger grant support and smaller project funding, and the fostering of academic linkages and collaboration with local industry.

The final but most important thing is seeking more funds to support research in Cambodia. RGC, MoEYS, and other related stakeholders need to make more partners and seek funds to support research in Cambodia; otherwise, the government's policies cannot be achieved.

REFERENCES

Abbass, K., Qasim, M. Z., Song, H., Murshed, M., Mahmood, H., & Younis, I. (2022). A review of the global climate change impacts, adaptation, and sustainable mitigation measures. *Environmental Science and Pollution Research International*, 29(28), 42539–42559. https://link.springer.com/article/10.1007/s11356-022-19718-6. DOI: 10.1007/s11356-022-19718-6 PMID: 35378646

Alberio, M., & Soubirou, M. (2022). How can a cooperative-based organization of indigenous fisheries foster resilience to global changes? Lessons learned by coastal communities in eastern Québec. *Environmental Policy and Governance*, 32(6), 546–559. DOI: 10.1002/eet.2025

Ayres, D. M. (2000). Tradition, modernity, and the development of education in Cambodia. *Comparative Education Review*, 44(4), 440–463. DOI: 10.1086/447629

Busza, J. (2004). Participatory research in constrained settings: Sharing challenges from Cambodia. *Action Research*, 2(2), 191–208. DOI: 10.1177/1476750304043730

Chamnan, S., & Ford, N. (2018). Higher education in Cambodia: Perspectives of students and faculty. *Journal of Asian Public Policy*, 11(2), 199–213.

Chea, C. (2020). Enhancing research culture in Cambodian universities. *Asia Pacific Education Review*, 21(4), 623–635.

Chen, C. Y., Sok, P., & Sok, K. (2007). Benchmarking potential factors leading to education quality: A study of Cambodian higher education. *Quality Assurance in Education*, 15(2), 128–148. DOI: 10.1108/09684880710748901

Chin, P., Em, S., & Pham, S. (2023). Enhancing research knowledge and productivity through professorship titles in Cambodia: Striking a balance. *Jurnal As-Salam*, 7(2), 133–143. DOI: 10.37249/assalam.v7i2.653

Clayton, T. (1998). Building the new Cambodia: Educational destruction and construction under the Khmer Rouge, 1975–1979. *History of Education Quarterly*, 38(1), 1–16. DOI: 10.2307/369662

Clayton, T. (2005). Re-orientations in moral education in Cambodia since 1975. *Journal of Moral Education*, 34(4), 505–517. DOI: 10.1080/03057240500410236

Dy, S. S. (2004). Strategies and policies for basic education in Cambodia: Historical perspectives. *International Education Journal*, 5(1), 90–97.

Eam, P. (2018). Current state, key challenges and ways forward for Cambodian research capacities: A review of four studies. *Cambodia Education Review*, 2(1), 108–143.

Eam, P., Keo, B., Leng, P., Song, S., & Khieng, S. (2021). Correlates of STEM major choice: A quantitative look at Cambodian university freshmen. *Research in Science & Technological Education*, 39(2), 206–224. DOI: 10.1080/02635143.2019.1682987

Em, S. (2023a, August 6). *The issue of title acquisition through proxy writing in Cambodian academia*. Cambodianess.

Em, S. (2023b, December 27). *Concerns of professor nominations and self-nominated professors in Cambodia*. English Academic Essay (EAE).

Em, S. (2024). Groundbreaking issue of the Cambodian Journal of Educational and Social Sciences (CJESS). [CJESS]. *Cambodian Journal of Educational and Social Sciences*, 1(1), 1–25. DOI: 10.69496/cjess.v1i1.4

Em, S., Khan, S., & Nun, N. (2022). Education system in Cambodia: A brief review from the prehistoric period to the present, and an education strategic plan for the future. *FIRE: Forum for International Research in Education*, 7(3), 141–164. DOI: 10.32865/fire202273320

Evans, C., Kandiko Howson, C., Forsythe, A., & Edwards, C. (2021). What constitutes high-quality higher education pedagogical research? *Assessment & Evaluation in Higher Education*, 46(4), 525–546. DOI: 10.1080/02602938.2020.1790500

Gonzalez, D., Abdel Sattar, R., Budhathoki, R., Carrard, N., Chase, R. P., Crawford, J., Halcrow, G., Kozole, T., MacArthur, J., Nicoletti, C., Toeur, V., Basnet, M. P., Chhetri, A., Gurung, H., Yadav, A., Vourchnea, P., & Willetts, J. (2022). A partnership approach to the design and use of a quantitative measure: Co-producing and piloting the WASH gender equality measure in Cambodia and Nepal. *Development Studies Research*, 9(1), 142–158. DOI: 10.1080/21665095.2022.2073248

Greenhow, C., & Galvin, S. (2020). Teaching with social media: Evidence-based strategies for making remote higher education less remote. *Information and Learning Science*, 121(7/8), 513–524. DOI: 10.1108/ILS-04-2020-0138

Harris, R. (2015). The Impact of research on development policy and practice: This much we know. In Chib, A., May, J., & Barrantes, R. (Eds.), *Impact of Information Society Research in the Global South*. Springer., DOI: 10.1007/978-981-287-381-1_2

Hayes, S., Jopling, M., Hayes, D., Westwood, A., Tuckett, A., & Barnett, R. (2021). Raising regional academic voices (alongside data) in Higher Education (HE) debate. *Postdigital Science and Education, 3*, 242-260. https://link.springer.com/article/10.1007/s42438-020-00131-6

Helldén, D., Sok, S., Chea, T., Nordenstedt, H., Kuruvilla, S., Alvesson, H. M., & Alfvén, T. (2023). Sustainable development goals and multisectoral collaborations for child health in Cambodia: A qualitative interview study with key child health stakeholders. *BMJ Open*, 13(11), 1–9. DOI: 10.1136/bmjopen-2023-073853 PMID: 37989366

Heng, K. (2022, February 21). *Cambodian academics' conceptions of research and the research-teaching nexus: A qualitative analysis*. Cambodian Education Forum.

Heng, K. (2024). Challenges and developments in university research in Cambodia: A case study of two universities. *Higher Education*, 87(6), 1593–1613. https://link.springer.com/article/10.1007/s10734-023-01080-2. DOI: 10.1007/s10734-023-01080-2

Heng, K. (2024). Challenges and developments in university research in Cambodia: A case study of two universities. *Higher Education*, 87(6), 1593–1613. https://link.springer.com/article/10.1007/s10734-023-01080-2. DOI: 10.1007/s10734-023-01080-2

Heng, K., Hamid, M. O., & Khan, A. (2022). Academics' conceptions of research and the research-teaching nexus: Insights from Cambodia. *International Journal of Educational Development*, 90, 102569. DOI: 10.1016/j.ijedudev.2022.102569

Heng, K., Hamid, M. O., & Khan, A. (2023). Research engagement of academics in the Global South: The case of Cambodian academics. *Globalisation, Societies and Education*, 21(3), 322–337. DOI: 10.1080/14767724.2022.2040355

Heng, K., & Heng, K. (2023). The implementation of the professorial ranking system in Cambodia: Opportunities and challenges. *Cambodian Journal of Educational Research*, 3(2), 62–77. DOI: 10.62037/cjer.2023.03.02.04

Heng, K., & Sol, K. (2021). Academic research in Cambodia: Progress, challenges, and ways forward. *Cambodian Journal of Educational Research*, 1(2), 6–23. DOI: 10.62037/cjer.2021.01.02.02

Heng, K., & Sol, K. (2022). Education: Key to making Cambodia great again. *Cambodia Development Center, 4*(3), 1-18.

Heng, K., Sol, K., & Em, S. (2023). COVID-19 and digital transformation of Cambodian higher education: opportunities, challenges, and the way forward. In Ordóñez de Pablos, P., Zhang, X., & Almunawar, M. N. (Eds.), *Handbook of research on education institutions, skills, and jobs in the digital era* (pp. 307–327). IGI Global., DOI: 10.4018/978-1-6684-5914-0.ch018

Heng, K., Sol, K., Kaing, S., & Ros, V. (2020). *Introduction: The need to promote a strong research and publication culture in Cambodia*. Cambodian Education Forum.

Heng, K., Sol, K., Kaing, S., & Ros, V. (2021). *Introduction: Engaging and supporting Cambodian youth to write for publication in English*. Cambodian Education Forum.

Heng, M. (2022). Cambodia's gender policy: An analysis of women and girls' education strategies under the Neary Rattanak IV Strategic Plan (2014-2018). *Cambodian Journal of Educational Research*, 2(1), 37–61. DOI: 10.62037/cjer.2022.02.01.03

Hum, C., & Heng, K. (2024). English in Cambodia. In Moody, A. J. (Ed.), *The Oxford Handbook of Southeast Asian Englishes* (pp. 186–201). Oxford University Press., DOI: 10.1093/oxfordhb/9780192855282.013.8

Indrawati, S. M., & Kuncoro, A. (2021). Improving competitiveness through vocational and higher education: Indonesia's vision for human capital development in 2019–2024. *Bulletin of Indonesian Economic Studies*, 57(1), 29–59. DOI: 10.1080/00074918.2021.1909692

Kiernan, B. (2002). Introduction: Conflict in Cambodia, 1945-2002. *Critical Asian Studies*, 34(4), 483–495. DOI: 10.1080/1467271022000035893

Kim, S., Sok, P., & Eng, S. (2020). Aligning research with national development goals in Cambodia. *Higher Education Policy*, 33(2), 237–256.

Leon-Roa, C., Zuñiga-Collazos, A., Castillo, H. S. V., Guarin, H. P., Franco, C. M. G., Gómez, D. C. R., & Acosta, E. B. G. (2024). Valorization of research results for knowledge and technology transfer in public higher education institutions. *Journal of Open Innovation*, 10(1), 1–10. DOI: 10.1016/j.joitmc.2024.100245

Li, J., & Pilz, M. (2023). International transfer of vocational education and training: A literature review. *Journal of Vocational Education and Training*, 75(2), 185–218. DOI: 10.1080/13636820.2020.1847566

Marginson, S. (2011). Higher education and public good. *Higher Education Quarterly*, 65(4), 411–433. DOI: 10.1111/j.1468-2273.2011.00496.x

Ministry of Education, Youth and Sport [MoEYS]. (2014). *Education strategic plan 2014-2018*.

Ministry of Education., Youth and Sport [MoEYS]. (2024a). *Education Congress 2024: Reviewing what has been done in 2022-2023 and directions for 2023-2024.*

Ministry of Education, Youth and Sport [MoEYS]. (2024b). *Guidelines for professorship choosing, evaluation, and provision in the education sector.*

Mwambari, D., Ali, F. A., & Barak, C. (2022). The impact of open access on knowledge production, consumption, and dissemination in Kenya's higher education system. *Third World Quarterly*, 43(6), 1408–1424. DOI: 10.1080/01436597.2022.2056010

Neau, V. (2003). The teaching of foreign languages in Cambodia: A historical perspective. *Language, Culture and Curriculum*, 16(3), 253–268. DOI: 10.1080/07908310308666673

Nelson, C., & Bérubé, M. (2020). Introduction: A report from the front. In M. Bérubé & C. Nelson (Eds.), *Higher education under fire* (pp. 1-32). Routledge. DOI: 10.4324/9781003061489-1

Nhem, D. (2023). So, is there a teaching-research nexus in Cambodia? Evidence from EFL students at one public university. *Research in Post-Compulsory Education*, 28(1), 48–69. DOI: 10.1080/13596748.2023.2166692

Ntabe, A. C. (2020). *Rethinking the practice of community engagement in health research: The case of the tenofovir trials in Cambodia and Cameroon* (Master's thesis). https://core.ac.uk/download/pdf/304374342.pdf

Petraki, E., & Khat, K. (2022). Challenges and constraints in the design of an ESP course in Cambodia: Implications for higher education institutions. *Asia Pacific Journal of Education*, 42(2), 260–275. DOI: 10.1080/02188791.2020.1798738

Peycam, P. (2020). An integrated capacity-building program. In Peycam, P. (Ed.), *Cultural renewal in Cambodia* (pp. 115–150). Brill., DOI: 10.1163/9789004437357_009

Rany, S., Zain, A. N. M., & Jamil, H. (2012). Cambodia's higher education development in historical perspectives (1863-2012). *International Journal of Learning and Development*, 2(2), 224–241. DOI: 10.5296/ijld.v2i2.1670

Salmi, J. (2009). *The challenge of establishing world-class universities*. World Bank Publications. DOI: 10.1596/978-0-8213-7865-6

Seng, C., Carlon, M. K. J., & Cross, J. S. (2020). Research self-efficacy of Cambodian undergraduate students at province-based universities. *The International Journal of Social Education*, 9(2), 155–190. DOI: 10.17583/rise.2020.4267

Siler, K., & Frenken, K. (2020). The pricing of open access journals: Diverse niches and sources of value in academic publishing. *Quantitative Science Studies*, 1(1), 28–59. DOI: 10.1162/qss_a_00016

Sok, S., & Bunry, R. (2021). Internationalization of higher education in Cambodia: Toward an agenda for higher education development. *International Journal of Comparative Education and Development*, 23(3), 193–211. DOI: 10.1108/IJCED-08-2020-0049

Sol, K., & Heng, K. (2022). The Cambodian Education Forum at two years old: Keeping the momentum going. *Cambodian Journal of Educational Research*, 2(1), 1–7. DOI: 10.62037/cjer.2022.02.01.01

Sol, K., & Heng, K. (2023). The Cambodian Journal of Educational Research's 3rd anniversary: Reflecting on its development. *Cambodian Journal of Educational Research*, 2(2), 1–6. DOI: 10.62037/cjer.2023.03.02.01

Soulard, E. C., Walker, J., & Petrakis, R. E. (2020). Implementation of a surface water extent model in Cambodia using cloud-based remote sensing. *Remote Sensing (Basel)*, 12(6), 1–19. DOI: 10.3390/rs12060984

Sovansophal, K., & Shimizu, K. (2020). A review on STEM enrollment in higher education of Cambodia: Current status, issues, and implications of initiatives. *Journal of International Development and Cooperation*, 26(1), 123–134. DOI: 10.15027/48743

Sun, P., Doh, J. P., Rajwani, T., & Siegel, D. (2021). Navigating cross-border institutional complexity: A review and assessment of multinational nonmarket strategy research. *Journal of International Business Studies*, 52(9), 1818–1853. https://link.springer.com/article/10.1057/s41267-021-00438-x. DOI: 10.1057/s41267-021-00438-x PMID: 34176993

Ullah, M., Shahid, A., Din, I. U., Roman, M., Assam, M., Fayaz, M., Ghadi, Y., & Aljuaid, H. (2022). Analyzing Interdisciplinary Research Using Co-Authorship Networks. *Complexity*, 2022(1), 1–13. DOI: 10.1155/2022/2524491

UNESCO. (2019). *Higher education in Cambodia: Challenges and perspectives.*

Ursić, L., Baldacchino, G., Bašić, Ž., Sainz, A. B., Buljan, I., Hampel, M., Kružić, I., Majić, M., Marušić, A., Thetiot, F., Tokalić, R., & Markić, L. V. (2022). Factors influencing interdisciplinary research and industry-academia collaborations at six European universities: A qualitative study. *Sustainability (Basel)*, 14(15), 9306. DOI: 10.3390/su14159306

World Bank. (2018). *Cambodia's quest for knowledge: Investing in research and innovation.*

Żuk, P., & Szulecki, K. (2020). Unpacking the right-populist threat to climate action: Poland's pro-governmental media on energy transition and climate change. *Energy Research & Social Science*, 66, 101485. DOI: 10.1016/j.erss.2020.101485

KEYWORDS AND DEFINITIONS

Academic Excellence: Academic excellence refers to the proven capacity to successfully perform, accomplish, and surpass expectations in educational endeavors.

Higher Education: Higher education refers to tertiary education that follows secondary schooling. The learning aspects occur in universities and further education colleges and often encompass both undergraduate and postgraduate studies.

Professorship: Professorship is the rank or position of a university professor.

Research Advancement: Research advancement encompasses the major progress or breakthroughs made in a certain area of research, resulting in the acquisition of new information, enhanced methodologies, or novel applications that greatly contribute to the comprehension or resolution of intricate issues.

Scholarly Pursuits: Scholarly pursuits encompass activities or studies undertaken for educational objectives, including attending courses, conducting research, and producing hypothetical papers.

Chapter 3
Empowering Academicians:
Strategies to Boost Research Output in Self-Financing Colleges in Tamil Nadu

V. Mohana Sundari
https://orcid.org/0000-0003-3144-4600
SRM Institute of Science and Technology, Kattankulathur, India

R. Vijayraja
SRM Institute of Science and Technology, Kattankulathur, India

G. Kumar
https://orcid.org/0000-0002-2946-3434
SRM Institute of Science and Technology, Kattankulathur, India

ABSTRACT

This research explores strategies to boost research output among academicians in Tamil Nadu's self-financing colleges. Despite higher education's research emphasis, these professionals face challenges in conducting and disseminating research. Identifying effective empowerment strategies is crucial for enhancing the research culture. The study aims to link academician empowerment to improved research performance by examining personal, professional, technological, institutional, and community factors. It highlights the balancing act between teaching, administrative duties, and research responsibilities, compounded by limited resources like time and infrastructure. Using a quantitative approach with 197 participants, the research employs analysis, exploring relationships Findings underscore the positive impact of empowerment on research performance, with implications for policymakers and

DOI: 10.4018/979-8-3373-0806-7.ch003

administrators to enhance environments and academic quality in these institutions, despite limitations in sample representativeness and the study's focus on self-financing colleges within Tamil Nadu.

INTRODUCTION:

Higher education institutions are increasingly emphasizing research as a critical component of academic excellence and global competitiveness. However, academicians in self-financing colleges face unique challenges that can hinder their research productivity. These challenges include heavy teaching loads, administrative responsibilities, limited funding, inadequate research infrastructure, and a lack of institutional support. This research aims to investigate strategies that can empower academicians in self-financing colleges in Tamil Nadu to overcome these challenges and enhance their research output. By understanding the factors that contribute to academician empowerment and their impact on research performance, this study seeks to provide actionable recommendations for policymakers, administrators, and academicians.

The concept of empowerment in the context of academia encompasses various dimensions, including personal motivation, professional development, access to technological resources, institutional support, and the influence of research communities. These dimensions collectively shape the ability of academicians to engage in research activities effectively. Personal motivation and self-efficacy are critical for initiating and sustaining research efforts, while professional development opportunities and networking can provide the necessary skills and connections. Access to technology and institutional support, such as funding and research infrastructure, are essential for conducting research. Additionally, collaboration and support within research communities can enhance research quality and output.

Literature Review

Here's a literature review summary based on previous work related to empowering academicians and enhancing research output in higher education, especially in the context of self-financing colleges:

Previous studies have identified several challenges that academicians in self-financing colleges encounter, including heavy teaching loads, administrative responsibilities, limited research funding, inadequate infrastructure, and a lack of institutional support (Agarwal & Garg, 2018). These challenges can significantly impact their research productivity and overall academic performance.

Research indicates that various factors contribute to academician empowerment and research productivity. These include personal motivation, professional development opportunities, access to technological resources, institutional support, and active participation in research communities (Suri, 2011; Kothari, 2004). These dimensions collectively shape the ability of academicians to engage effectively in research activities.

Strategies aimed at empowering academicians have been shown to positively influence academic performance and research output. Effective empowerment strategies encompass fostering intrinsic motivation, providing continuous professional development opportunities, ensuring access to adequate research infrastructure, and creating a supportive institutional environment (Hair et al., 2019; Anderson & Shattuck, 2012).

The use of quantitative research methods, such as surveys and statistical analysis, is prevalent in studying factors affecting academician empowerment and research productivity. These methods allow for rigorous examination of relationships between variables like personal motivation, professional development, institutional support, and academic performance (Creswell & Creswell, 2017; Fraenkel et al., 2012).

SEM has been employed to understand complex relationships among various dimensions affecting academician empowerment and research performance. Studies using SEM have shown how factors like personal motivation, professional development, and institutional support interact to influence overall academic outcomes in higher education settings (Hair et al., 2019).

Empirical studies have demonstrated significant correlations between empowerment factors and academic performance in different educational contexts. These studies highlight the importance of comprehensive strategies that address multiple dimensions of empowerment to enhance research productivity and overall academic success (Anderson & Shattuck, 2012).

Purpose of the Study:

The purpose of this study is to explore and identify strategies that empower academicians in self-financing colleges to enhance their research output. The study examines various dimensions including personal, professional, technological, institutional, and community-based factors. By analyzing the relationship between these factors and academician empowerment, the study aims to provide insights into how these dimensions can be leveraged to improve research performance. The ultimate goal is to develop a framework that can be used by self-financing colleges to create a supportive research environment that fosters academic excellence.

Specifically, this study seeks to address the following research questions:

What are the key challenges faced by academicians in self-financing colleges that impact their research productivity?

How do personal, professional, technological, institutional, and community-based factors influence academician empowerment?

What is the relationship between academician empowerment and research performance?

What strategies can be implemented to enhance academician empowerment and research output in self-financing colleges?

Discussion about Self-Financing College Academicians:

Academicians in self-financing colleges often juggle multiple responsibilities, including teaching, administrative duties, and research. This multitasking can significantly impact their research productivity. In addition to these responsibilities, academicians in self-financing colleges face several other challenges. Limited access to funding and research grants, inadequate research infrastructure, and a lack of institutional support are common issues that hinder their research endeavors. Furthermore, the pressure to publish in high-impact journals and the lack of collaborative opportunities can also affect their research productivity. Understanding these challenges is essential for developing targeted strategies that can empower academicians and enhance their research output.

Self-financing colleges typically operate with limited financial resources, which can result in fewer research grants and funding opportunities for academicians. This financial constraint can hinder their ability to conduct high-quality research. Additionally, the lack of dedicated research facilities and access to advanced research tools and technologies further exacerbates the problem. Academicians in these institutions often have to rely on their personal resources or seek external collaborations to carry out their research.

The administrative burden placed on academicians in self-financing colleges is another significant challenge. The need to fulfill various administrative responsibilities, such as student counseling, curriculum development, and departmental management, can consume a substantial portion of their time and energy. This leaves them with limited time to focus on research activities. Moreover, the high teaching load, often with large class sizes, can further reduce the time available for research.

Despite these challenges, academicians in self-financing colleges possess a strong motivation to engage in research. Personal factors such as intrinsic motivation, passion for the subject, and a desire for professional growth drive them to pursue research activities. Professional factors, including opportunities for professional development, collaboration with peers, and participation in conferences and workshops, also play a crucial role in enhancing their research productivity.

Research Design:

The research adopts a quantitative approach targeting academicians in self-financing colleges across Tamil Nadu. A stratified sampling technique is employed to ensure representation from various colleges. The study involves a sample size of 197 participants, which is considered sufficient to provide meaningful insights into the research question. Data is collected through structured questionnaires designed to capture information on personal, professional, technological, institutional, and community-based factors that influence academician empowerment and research performance.

The questionnaire is divided into several sections, each focusing on different dimensions of academician empowerment. The personal dimension includes questions related to motivation, self-efficacy, and personal goals. The professional dimension covers aspects such as professional development opportunities, networking, and collaboration. The technology dimension assesses access to research tools, IT support, and the use of technology in research. The institutional dimension examines the level of institutional support, funding, and research infrastructure. The community-based dimension explores collaboration, peer support, and involvement in research forums.

The data collected through the questionnaires is subjected to rigorous statistical analysis using SPSS and AMOS software. Descriptive statistics are used to summarize the data, while inferential statistics such as t-tests and ANOVA are used to compare means across different groups. Correlation analysis is employed to examine the relationships between variables, and regression analysis is used to identify predictors of academician empowerment and research performance. Structural equation modelling (SEM) is used to assess the overall model fit and to understand the complex relationships between the independent and dependent variables.

Statistical Tools Used:

Statistical analysis is conducted using SPSS and AMOS software. Descriptive statistics are used to summarize the data, while inferential statistics such as t-tests and ANOVA are used to compare means across different groups. Correlation analysis is employed to examine the relationships between variables, and regression analysis is used to identify predictors of academician empowerment and research performance. Structural equation modelling (SEM) is used to assess the overall model fit and to understand the complex relationships between the independent and dependent variables.

Descriptive statistics provide an overview of the demographic characteristics of the participants, including their age, gender, academic rank, and years of experience. T-tests and ANOVA are used to compare the mean scores of different groups based on these demographic characteristics. For example, t-tests can be used to compare the mean scores of male and female academicians, while ANOVA can be used to compare the mean scores of academicians at different academic ranks.

Correlation analysis is used to examine the relationships between the various dimensions of academician empowerment and research performance. For example, the correlation between personal motivation and research productivity can be assessed to determine the strength and direction of the relationship. Regression analysis is used to identify the key predictors of academician empowerment and research performance. For example, regression analysis can be used to determine the extent to which professional development opportunities predict research productivity.

Structural equation modelling (SEM) is used to assess the overall model fit and to understand the complex relationships between the independent and dependent variables. SEM allows for the simultaneous examination of multiple relationships between variables, providing a comprehensive understanding of the factors that influence academician empowerment and research performance. SEM also allows for the assessment of model fit, ensuring that the proposed model accurately represents the data.

Constructs for the Study:

The study identifies several key constructs for analysis. The independent variables include personal dimension (e.g., motivation, self-efficacy), professional dimension (e.g., professional development, networking), technology dimension (e.g., access to research tools, IT support), institutional dimension (e.g., institutional support, funding), and research forum community (e.g., collaboration, peer support). The dependent variable is academician performance (e.g., number of publications, quality of research output).

The personal dimension includes factors such as motivation, self-efficacy, and personal goals. Motivation refers to the intrinsic and extrinsic factors that drive academicians to engage in research activities. Self-efficacy refers to the belief in one's ability to successfully conduct research. Personal goals refer to the specific research objectives and aspirations of academicians.

The professional dimension includes factors such as professional development opportunities, networking, and collaboration. Professional development opportunities refer to the availability of training, workshops, and conferences that enhance research skills. Networking refers to the ability to connect with other researchers and professionals in the field. Collaboration refers to the extent to which academicians work with peers and colleagues on research projects.

The technology dimension includes factors such as access to research tools, IT support, and the use of technology in research. Access to research tools refers to the availability of software, databases, and other resources necessary for conducting research. IT support refers to the availability of technical assistance and support for using research

Figure 1. Conceptual framework

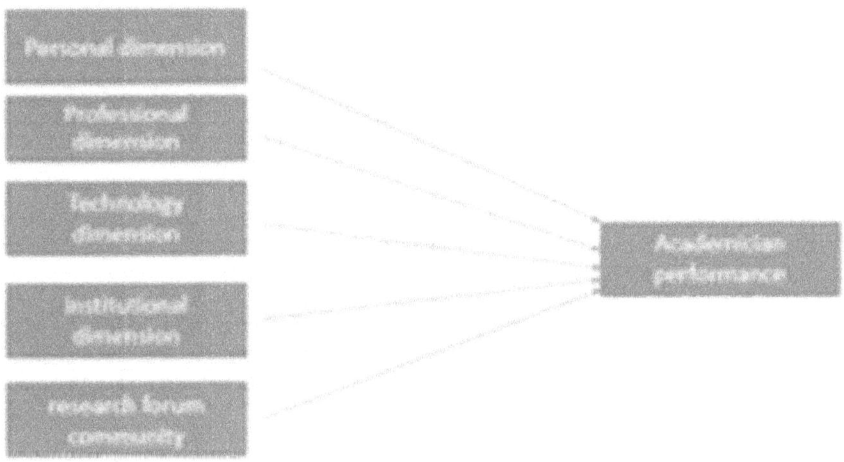

Data Analysis

Table 1. Reliability test

Items	Cronbach's Alpha
Personal Dimension	.725
Professional Dimension	.792
Technology Dimension	.721
Institutional Dimension	.708
Research Forum Community	.720

The Cronbach's Alpha value is used to check the reliability values of tested questionnaire dimensions. The tested study dimensions are > 0.700, which reflects good internal consistency. This value is more than 0.700. This indicates a very strong internal consistency among the variables

Table 2. Correlation Analysis

Correlations								
		PD	**PRD**	**TD**	**ID**	**RFC**	**AE**	**AP**
PD	Pearson Correlation	1	.736**	.706**	.671**	.763**	.734**	.743**
	Sig. (2-tailed)		.000	.000	.000	.000	.000	.000
	N	197	197	197	197	197	197	197
PRD	Pearson Correlation	.736**	1	.743**	.742**	.732**	.776**	.766**
	Sig. (2-tailed)	.000		.000	.000	.000	.000	.000
	N	197	197	197	197	197	197	197
TD	Pearson Correlation	.706**	.743**	1	.704**	.770**	.772**	.746**
	Sig. (2-tailed)	.000	.000		.000	.000	.000	.000
	N	197	197	197	197	197	197	197
ID	Pearson Correlation	.671**	.742**	.704**	1	.681**	.706**	.726**
	Sig. (2-tailed)	.000	.000	.000		.000	.000	.000
	N	197	197	197	197	197	197	197
RFC	Pearson Correlation	.763**	.732**	.770**	.681**	1	.732**	.728**
	Sig. (2-tailed)	.000	.000	.000	.000		.000	.000
	N	197	197	197	197	197	197	197
AE	Pearson Correlation	.734**	.776**	.772**	.706**	.732**	1	.802**
	Sig. (2-tailed)	.000	.000	.000	.000	.000		.000
	N	197	197	197	197	197	197	197
AP	Pearson Correlation	.743**	.766**	.746**	.726**	.728**	.802**	1
	Sig. (2-tailed)	.000	.000	.000	.000	.000	.000	
	N	197	197	197	197	197	197	197

**. Correlation is significant at the 0.01 level (2-tailed).

All dimensions (personal, professional, technological, institutional, and research forum community) are significantly positively correlated with academician performance and empowerment. The highest correlation is between academician empowerment (AE) and academician performance (AP), suggesting that empowerment is a critical factor in improving research performance. This indicates that improvements in any of these dimensions are likely to enhance both empowerment and performance of academicians.

Table 3. Descriptive Statistics

Descriptive Statistics								
	N	Mean	Std. Deviation	Variance	Skewness		Kurtosis	
	Statistic	Statistic	Statistic	Statistic	Statistic	Std. Error	Statistic	Std. Error
PD	197	19.1523	3.64316	13.273	-1.587	.173	2.465	.345
PRD	197	18.9239	3.63364	13.203	-1.722	.173	2.775	.345
TD	197	18.6599	3.52847	12.450	-1.711	.173	2.640	.345
ID	197	18.8934	3.35544	11.259	-1.521	.173	2.575	.345
RFC	197	19.0152	3.46628	12.015	-1.650	.173	2.415	.345
AE	197	19.1827	3.29921	10.885	-1.851	.173	3.135	.345
AP	197	19.0051	3.52179	12.403	-1.704	.173	2.913	.345
Valid N (listwise)	197							

Interpretation: The descriptive statistics reveal high mean scores across all dimensions, suggesting that participants generally have a positive perception of these factors. The negative skewness values indicate that the responses are skewed towards higher scores, and the positive kurtosis values suggest a relatively peaked distribution, implying consistency in responses.

Regression Analysis

Table 4. Model Summary

					Change Statistics				
Model	R	R Square	Adjusted R Square	Std. Error of the Estimate	R Square Change	F Change	df1	df2	Sig. F Change
1	.859[a]	.738	.730	1.83074	.738	89.220	6	190	.000

a. Predictors: (Constant), AE, ID, PD, RFC, TD, PRD

The regression model explains 73.8% of the variance in academician performance, which is substantial. The high R-squared value indicates that the model fits the data well, and the significant F-statistic suggests that the overall model is significant.

Table 5. ANOVAa

Model		Sum of Squares	df	Mean Square	F	Sig.
1	Regression	1794.188	6	299.031	89.220	.000[b]
	Residual	636.807	190	3.352		
	Total	2430.995	196			

a. Dependent Variable: AP
b. Predictors: (Constant), AE, ID, PD, RFC, TD, PRD

In the table, ANOVA result 'p' value is significant (Which is less than 0.05) reflect that dependent variable Academic performance is more reliable.

Table 6. Coefficientsa

Model		Unstandardized Coefficients		Standardized Coefficients			Collinearity Statistics	
		B	Std. Error	Beta	t	Sig.	Tolerance	VIF
1	(Constant)	.215	.837		.257	.797		
	PD	.157	.063	.162	2.487	.014	.325	3.075
	PRD	.145	.068	.150	2.136	.034	.281	3.565
	TD	.109	.069	.109	1.586	.114	.290	3.453
	ID	.160	.064	.152	2.510	.013	.374	2.676
	RFC	.072	.070	.071	1.030	.304	.293	3.414
	AE	.345	.075	.323	4.625	.000	.282	3.542

a. Dependent Variable: AP

- **Personal Dimension (PD):** Significant positive predictor of academician performance. Academicians' personal factors such as motivation and self-efficacy are crucial for enhancing their research output.
- **Professional Dimension (PRD):** Significant positive predictor. Opportunities for professional development, networking, and collaboration are important for improving academician performance.
- **Technological Dimension (TD):** Although correlated with performance, it is not a significant predictor when other variables are considered. This suggests technology alone may not be sufficient without supportive personal and professional environments.
- **Institutional Dimension (ID):** Significant positive predictor. Institutional support, resources, and infrastructure are vital for improving research performance.

- **Research Forum Community (RFC):** Not a significant predictor in the regression model, indicating that community support alone may not directly impact performance without other supportive factors.
- **Academician Empowerment (AE):** The strongest predictor of academician performance. Empowerment initiatives that encompass personal, professional, technological, and institutional support are crucial for enhancing research output.

CFA

Figure 2. CFA

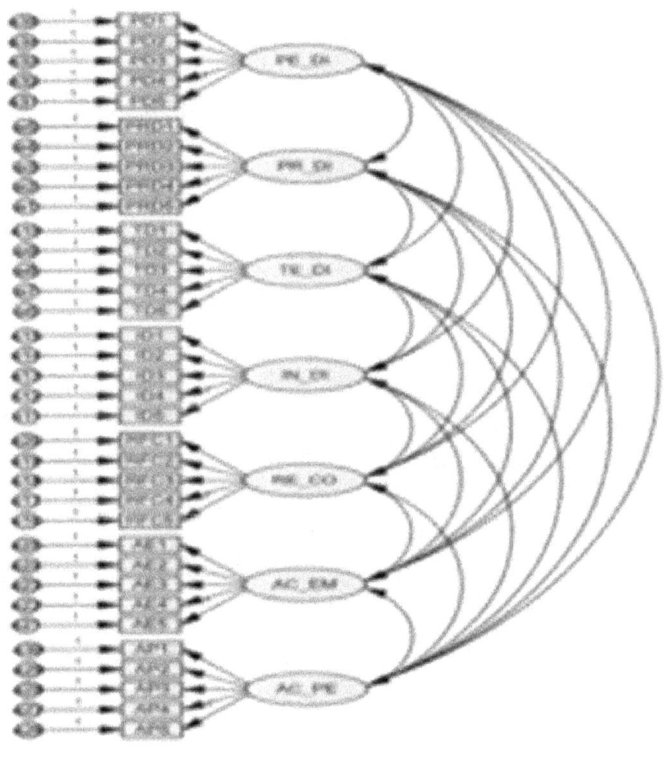

Table 7. Fit indices of the CFA

Fit indices	Model value
Chi-square value/DF	2.942
Goodness of fit index(GFI)	0.905
Adjusted Goodness of fit index(AGFI)	0.972
Normed fit index(NFI)	0.914
RMSEA	0.066
Parsimony normed fit index (PNFI)	0.892

Confirmatory Factor Analysis (CFA) supports the validity and reliability of the constructs, with fit indices indicating a good model fit (χ^2/df = 2.942, GFI = 0.905, AGFI = 0.972, NFI = 0.914, RMSEA = 0.066, PNFI = 0.892). These results suggest that a multifaceted approach to empowering academicians, encompassing personal motivation, professional development, institutional support, and an overall supportive environment, is vital for boosting research output. SEM

Figure 3. SEM

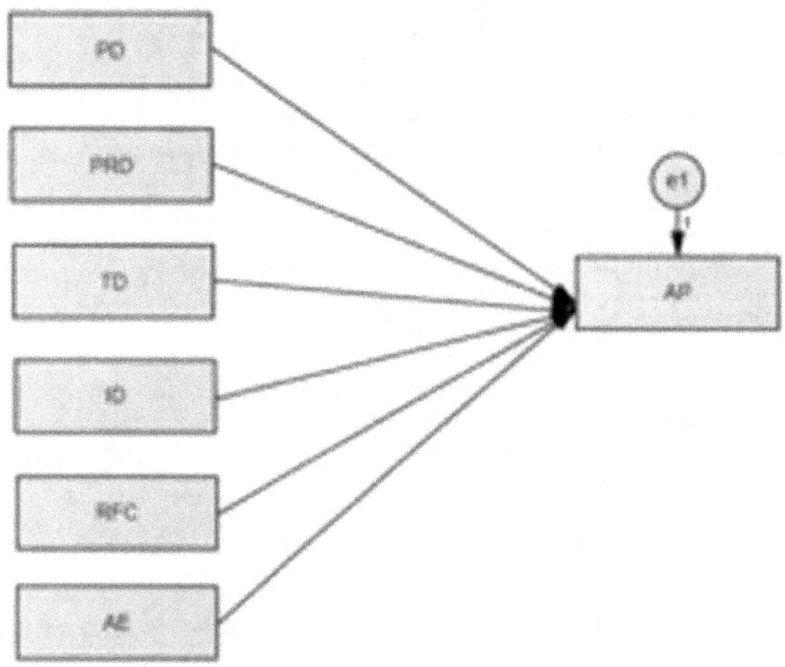

SEM Results:

The SEM analysis confirms the relationships among the variables and helps understand the direct and indirect effects of the independent variables on academician performance. The model fit indices (such as CFI, TLI, RMSEA) should be checked to ensure the model's adequacy, though these details are not provided here.

Interpretation:

The SEM results support the findings from the regression analysis, indicating that personal, professional, institutional, and empowerment factors are significant predictors of academician performance. The direct and indirect effects shown in the SEM provide a comprehensive understanding of how these dimensions interact to influence research output.

Discussion

Personal Dimension (PD):

The personal dimension, including factors such as self-efficacy, motivation, and time management, significantly impacts academician performance. The positive correlation and regression coefficient indicate that academicians who feel confident and motivated are more likely to produce high-quality research.

Professional Dimension (PRD):

Professional development opportunities, mentorship, and networking significantly enhance academician performance. The strong correlation and significant regression coefficient suggest that continuous professional growth and collaboration are essential for research productivity.

Technological Dimension (TD):

While technology access is correlated with performance, it is not a significant predictor in the regression model. This implies that technology should be integrated with other dimensions, such as professional development and institutional support, to be effective in enhancing research output.

Institutional Dimension (ID):

Institutional support, including funding, resources, and a conducive research environment, significantly impacts academician performance. The positive regression coefficient indicates that institutions need to provide adequate support and infrastructure to facilitate research activities.

Research Forum Community (RFC):

The community aspect, including participation in research forums and collaborative networks, shows a strong correlation with performance but is not a significant predictor in the regression model. This suggests that community support needs to be part of a broader strategy that includes personal, professional, and institutional support.

Academician Empowerment (AE):

Empowerment is the strongest predictor of academician performance, highlighting the importance of initiatives that boost confidence, provide opportunities for growth, and create a supportive environment. Empowerment encompasses various dimensions and is crucial for enhancing research output.

CONCLUSION

The study "Empowering Academicians: Strategies to Boost Research Output in Self-Financing Colleges in Tamil Nadu" provides a comprehensive analysis of the factors influencing academic research performance. The findings underscore the importance of a multidimensional approach to empowerment, incorporating personal, professional, technological, institutional, and community-based support.

The high mean scores for these dimensions indicate that academicians have a positive perception of these factors, recognizing their importance in enhancing research capabilities. Significant correlations between these dimensions and both academician empowerment and performance highlight the interconnectedness of these factors in driving research productivity.

Regression analysis reveals that personal motivation, professional development, and institutional support are significant predictors of academician performance, emphasizing the need for targeted strategies to enhance these areas. Interestingly, while technological support and community involvement are crucial, their impact is most effective when integrated with other empowerment strategies.

In conclusion, the study highlights the critical role of a supportive and resource-rich environment in enhancing the research output of academicians in self-financing colleges. Policymakers and administrators should implement holistic strategies that address personal motivation, professional development, and institutional support. Additionally, academicians should actively engage in professional growth opportunities and leverage institutional resources to maximize their research potential. Future research should explore these dimensions in different contexts and over time to gain deeper insights into the dynamics of academic empowerment and performance. By focusing on these key areas, self-financing colleges in Tamil Nadu can foster a conducive environment for research excellence, ultimately contributing to the advancement of knowledge and academic success.

For Future Research:

Future studies should explore the interplay of these dimensions in different contexts and over time to develop a comprehensive understanding of academician empowerment. Longitudinal studies and research in diverse settings can provide deeper insights into the factors that influence research performance.

This detailed interpretation provides a comprehensive understanding of the statistical analysis and highlights the key factors that influence academician performance in self-financing colleges in Tamil Nadu.

REFERENCES

Agarwal, R., & Garg, A. (2018). Research productivity in higher education institutions: Challenges and strategies for academicians. *The Journal of Educational Research*, 10(2), 45–60.

Anderson, T., & Shattuck, J. (2012). Design-based research: A decade of progress in education research? *Educational Researcher*, 41(1), 16–25. DOI: 10.3102/0013189X11428813

Bandura, A. (1997). *Self-efficacy: The exercise of control.* W. H. Freeman.

Bowen, G. A. (2009). Document analysis as a qualitative research method. *Qualitative Research Journal*, 9(2), 27–40. DOI: 10.3316/QRJ0902027

Chen, D. H., & Dahlman, C. J. (2005). *The knowledge economy, the KAM methodology and World Bank operations.* World Bank Institute.

Creswell, J. W., & Creswell, J. D. (2017). *Research design: Qualitative, quantitative, and mixed methods approaches* (5th ed.). SAGE Publications.

Deci, E. L., & Ryan, R. M. (2008). Facilitating optimal motivation and psychological well-being across life's domains. *Canadian Psychology*, 49(1), 14–23. DOI: 10.1037/0708-5591.49.1.14

Eccles, J. S., & Wigfield, A. (2002). Motivational beliefs, values, and goals. *Annual Review of Psychology*, 53(1), 109–132. DOI: 10.1146/annurev.psych.53.100901.135153 PMID: 11752481

Field, A. (2018). *Discovering statistics using IBM SPSS Statistics* (5th ed.). SAGE Publications.

Fraenkel, J. R., Wallen, N. E., & Hyun, H. H. (2012). *How to design and evaluate research in education* (8th ed.). McGraw-Hill.

Hackman, J. R., & Oldham, G. R. (1976). Motivation through the design of work: Test of a theory. *Organizational Behavior and Human Performance*, 16(2), 250–279. DOI: 10.1016/0030-5073(76)90016-7

Hair, J. F., Black, W. C., Babin, B. J., & Anderson, R. E. (2019). *Multivariate data analysis* (8th ed.). Cengage Learning.

Herzberg, F., Mausner, B., & Snyderman, B. B. (1959). *The motivation to work* (2nd ed.). John Wiley & Sons.

Kothari, C. R. (2004). *Research methodology: Methods and techniques* (2nd ed.). New Age International.

Locke, E. A., & Latham, G. P. (2002). Building a practically useful theory of goal setting and task motivation. *The American Psychologist*, 57(9), 705–717. DOI: 10.1037/0003-066X.57.9.705 PMID: 12237980

Maslow, A. H. (1943). A theory of human motivation. *Psychological Review*, 50(4), 370–396. DOI: 10.1037/h0054346

Saldaña, J. (2015). *The coding manual for qualitative researchers* (3rd ed.). SAGE Publications.

Suri, H. (2011). Purposeful sampling in qualitative research synthesis. *Qualitative Research Journal*, 11(2), 63–75. DOI: 10.3316/QRJ1102063

Tinto, V. (1975). Dropout from higher education: A theoretical synthesis of recent research. *Review of Educational Research*, 45(1), 89–125. DOI: 10.3102/00346543045001089

Yin, R. K. (2018). *Case study research and applications: Design and methods* (6th ed.). SAGE Publications.

Chapter 4
The Significance of and the Justification for Developing an Own Research Profile

Andile Alred Mdikana
University of South Africa, South Africa

ABSTRACT

Any researcher needs to develop a sound research profile to boost his or her standing in the scholarly and research community. Developing a research profile starts with an idea that is brainstormed to develop a research focus. It also includes defining a research niche and stating the rationale and the aims and objectives of the research niche (University of Birmingham). It is to be hoped that emerging academics would, through this input, be in a better position to develop appealing, well thought out, well-organized and professional research profiles.

BACKGROUND

Working as a researcher can be a daunting career choice, particularly when one is just starting out. To establish a research identity is of paramount significance. One of the tools to establish it is an own research profile (Almomani, 2022). It not only takes time and considerable effort, but also requires the expertise of a mentor to develop a sound research profile (Griffiths, Thompson & Hrynie, 2010). A mentor must be a seasoned scholar and academic as well as a kind, supportive, and nurturing person. Some emerging scholars have been proactive in developing their research profiles by attending workshops and conferences to showcase their work.

DOI: 10.4018/979-8-3373-0806-7.ch004

In addition to that, they have been active on social media to ensure that their work is disseminated across the globe (Fauth, 2008).

The following issues are critical in developing an own research profile. First, a research profile should indicate a clear niche area and a research focus. Second, it should include a list of workshops and conferences that have been attended. Third, it should contain a list of publications. Finally, if a research profile is that of an established scholar, it should mention the number of people mentored. One way of becoming a mentor is to do an extensive literature survey to learn about best practices for mentoring emerging academics.

Countless scholars have been headhunted based on their sound research profiles (Bazeley, 2006) because they have increased their visibility through their scholarly work and have ORCID numbers and have published in OPEN ACCESS journals and scholarly books. A research profile is meant to be read by a wider audience (Almomani, 2022). Developing an own research profile should not be rushed since it requires time, effort, planning and management. Developing an own research profile is a must for every academic (Fauth, 2008). It is high time that institutions of Higher Learning make the development of an own research profile compulsory. Scholars' global visibility benefits not only themselves, but also the universities or institutions they work for. Their universities establish an international reputation as first-class institutions and attract post-graduate students and post-doctoral fellows. Such scholars are often invited as guest speakers to meetings and conferences (Aksnes, Langfeldt & Wouters, 2019).

So far, the positive aspects of developing an own research profile have been described. Some challenges of establishing an own research profile will now be highlighted. To develop a solid research profile can be a daunting task for the emerging researcher. On the one hand, budding researchers should demonstrate a willingness to grow and, on the other, there should be a goodness of fit between mentor and mentee. Some emerging academics are reluctant to seek help in developing their own-research profiles (Leta, Glanzel & Thijs, 2006). Their reluctance can be attributed to a fear of the unknown. They should realize that asking for help is, in fact, a sign of strength. But part of it is understandable since not all mentors have the requisite interpersonal skills and personality.

CRITICAL ASPECTS OF DEVELOPING AN OWN RESEARCH PROFILE

This section will discuss the critical aspects of developing an own research profile such as career building; getting funds; finding collaborators; and increasing your impact and engagement:

Career building, as well as establishing your professional research identity, is a long-term process (Miller, Taylor & Bedeian, 2011). You can start by writing articles or attending conferences culminating in publications in reputable journals and scholarly books. Career building may also include networking with members of the scholarly community and getting involved in external examinations for master's and doctoral students (Miller, Taylor & Bedeian, 2011). When called upon to externally examine an examination, an up-and-coming researcher should take the opportunity with enthusiasm. This may culminate in a rich and appealing research profile. Again, the role of the mentor in it cannot be underestimated. Universities across the globe call for greater recognition of the role of mentors in career building (Shinton Consulting, 2013). They have even put formal structures and systems in place so that senior academics help junior academics establish their research identities and develop their own research profiles.

A sound research profile is a useful tool in accessing funding or grants. One of the conditions for funding is a research profile discussing an academic's research identity by highlighting his or her research achievement, engaged scholarship and academic citizenship (Marginson, 2010). Because scholarly funders want value for money, researchers are expected to deliver results and effectively communicate their findings. A well-thought-out and impactful research profile will go a long way in securing funds. A poorly and ill-considered research profile, on the other hand, will, most of the time, lead to the rejection of a research funding application.

A sound research profile not only makes it easy to find collaborators across the globe (Thomas, 2009) but also increases the visibility of one's research. Collaboration with scholars across the globe often involves research partnerships and cooperation with international research organizations. It entails key-knowledge sharing and the advancement of research enterprise (Thomas, 2009).

More than enough research evidence shows that collaboration increases the global research visibility and marketability of researchers' work (Marginson, 2010). A research profile makes it easier for the members of the international scholarly community to locate one another. Collaboration is key to research skills growth, development, and personal growth.

A well-thought-out research profile increases impact and engagement (Scott, 1998). Owing to one's visibility in the print and electronic media, one might be approached by not only non-government organizations, but also government and policy makers to act as a consultant or to render research and development services. Therefore, every academic should develop an own research profile. Certain institutions of higher learning require every academic to have a research profile as a matter of principle and policy (Zumwalt & Craig, 2005). This has yielded some results seeing that academics have started to take research seriously and develop research profiles to maximize their research identities.

DEVELOPING A SOUND RESEARCH PROFILE

Any researcher needs to develop a sound research profile to boost his or her standing in the scholarly and research community. Developing a research profile starts with an idea that is brainstormed to develop a research focus. It also includes defining a research niche and stating the rationale and the aims and objectives of the research niche (University of Birmingham). This section will shed light on ways to develop a research profile. It is to be hoped that emerging academics would, through this input, be in a better position to develop appealing, well thought out, well-organized and professional research profiles.

Because the 4th Industrial Revolution is upon us, all academics are encouraged to utilize various forms of information communication technologies (Van Hoof, Estrella, Eljuri & Torres, 2013). This means that they should develop an online presence. Their research profiles need to be clear and to the point and must highlight their academic and scholarly achievements. Social media fanatics do not have time to read long passages. An online presence should target academics, scholars, policy makers, research institutions, civil society, and government. Every academic must make sure that he or she has a professional social-media profile on Twitter, LinkedIn, ResearchGate, and Facebook. These platforms can be an effective way to share your work and to keep up with current research in your discipline.

The power of social media should not be underestimated (Visser-Wijnveen, Van Driel, Van der Rijst, Verloop & Visser, 2010). Scores of researchers use social media to follow the latest trends in research, share their research work, and recruit scholars of good standing on behalf of their institutions. Therefore, a research profile should be well crafted and stimulate interest. This will make it easier for social media to capture the essence of what the academic tries to communicate in terms using own research profile. The development of a research profile, especially for social-media purposes, should not be rushed. It is advisable to ask a mentor for review and opinion. Some academics have received employment offers through social media because it was believed that they could add value to institutions of higher learning.

A market with specific research needs must be decided on in the development of a research profile. This will enable academics to share their research outputs globally. A well-planned research profile is a marketing tool that can enable academics to move on to greener pastures (Waal, Verbruggen & Wright, 2010). An academic, therefore, needs to be patient in developing a research profile. Research evidence suggests that the development of a sound and impactful research profile leads to increased job and funding opportunities (Zumwalt & Craig, 2005). There is a growing trend among academics to develop their own research profiles. Research mentoring seems to play a role in it. In fact, research mentoring should be compulsory as it yields superior results in terms of research and development and culminates in a sound research profile. Many institutions have a Development and Communications office that communicates to the broader community the research their academics conduct. It is, therefore, important to work closely with this office.

Research has shown that well-developed research profiles need time, effort, and patience. Webster (2024) observes that it took him some time to develop a sound research profile. He further contends that discipline and motivation play a vital role in developing a sound research profile. Webster also thanks his mentor and peers for their inputs. Owing to his research profile, he has attracted a considerable number of research collaborators. He is a sought-after researcher, scholar, and academic. Zumwalt and Craig (2005) indicate that their research profiles have enabled them to attract research funding and grants. They are invited all over the world to give talks, lectures, and presentations and to facilitate workshops on research-related matters. This has also presented an opportunity for him to be a guest speaker at many international conferences. He owes it to a well-conceptualized research profile and thanks the peers and the mentors that have supported him all the way. A research profile is necessary if an academic wishes to create a research visibility of own work.

INSTITUTIONAL RESEARCH PROFILE

Institutions across the world are starting to realize the value and the significance of institutional profiles (Marginson, 2010). In most cases, institutional profiles reflect individual researcher profiles, increase the prestige of institutions and attract new talent. An institution can develop a sound institutional research profile if it has structures and systems in place to support it. Once again, the role of mentors cannot be over-emphasized. The mentorship system should form an integral part of developing an institutional research profile reflective of its reputable researchers. If the institutional research profile is taken seriously, every academic would be encouraged to develop a research profile. According to one school of thought, an own research profile induces a researcher to conduct more research to improve his or her research, productivity and research visibility (Almomani, 2022). Institutions of higher learning still have a long way to go in terms of developing their institutional research profiles. A great deal of advocacy work and lobbying needs to be done. The idea of institutional research profiles still needs to be explored in higher-education institutions in South Africa and other parts of the world. It is to be hoped that the input given here will serve as a motivation for institutional research support.

CHARACTERISTICS OF AN OWN RESEARCH PROFILE

A research profile indicates who a particular academic is (Griffiths, Thompson & Hrynie, 2010). It reflects the academic's research interests, achievements and contributions in the scholarly community. Characterized by diligence over time, it is a marketing tool for funding and moving to greener pastures. A research profile is normally organized in a simple way. It is easy to read and has been developed gradually over time. Budding academics who draft a research profile are advised to seek the advice and guidance of senior researchers and mentors (Fauth, 2008). A sound research profile will ensure access to funding and employment opportunities because they will be sought after. Building up an own-research profile requires great determination to succeed in academia. Every academic is therefore challenged to develop an own research profile.

CRITICAL ISSUES TO CONSIDER WHEN DEVELOPING AN OWN RESEARCH PROFILE

The first issue an academic developing a research profile has to consider is what he or she wants to achieve (Bazeley, 2006). It can be to share his or her own research with others in the field or to promote him- or herself. This involves appealing to the wider research market of scholars. Once his or her objective has been identified and described, an academic would be in a better situation to communicate with fellow scholars, funders and people who might be interested in his or her work. A clear definition of what the academic wants to achieve makes it easier for the audience to follow his or her research work. Clearly defined objectives usually yield positive results. What an academic strives to achieve defines his or research identity and demonstrates scholarly and academic growth. A good mentor will encourage you in terms of what you want to achieve. There should be a goodness of fit between the mentor and the mentee to successfully realize this objective.

The second issue to consider is a definition of the audience (Bazeley, 2006). An academic need to consider platforms where he or she can exchange views with fellow scholars. In the context of this chapter, emerging scholars and post-graduate students would be your target market. According to Almomani (2022), identifying the relevant audience is key to your research visibility through a sound research profile. Bazeley (2006) argues that a mismatched audience is an indication of a poorly constructed research profile. A well-conceptualized research profile guarantees scholarly success. It has been proven over the years that identifying your audience tends to yield superior results (Bazeley, 2006).

Third, a scholar or academic must never get complacent because research profiles must improve and evolve. For this reason, a regular update of one's research profile is desirable. A research profile should be in line with the current trends and developments. According to Bazeley (2006), continuous improvement increases the visibility and the quality of a research profile. Aksnes, Langfeldt and Wouters (2019) contend that a stale research profile serves no purpose. Research shows that research profiles that keep on improving lead to increased collaboration and funding opportunities (Miller, Taylor & Bedeian, 2011). One should not be pessimistic; there is always room for improvement. Scholars must move with the times in this ever-changing world. Growth and development should be embraced with the seriousness it deserves.

Fourth, academics struggle to define their research niche and focus research in areas they haven't yet explored sufficiently. A research profile should unequivocally state the academic's research interest (Bazeley, 2006) to attract like-minded academics. A research profile should clearly state your research focus and niche. Global scholars look for academics who have similar research interests and with whom they can collaborate. According to Leta, Glanzel and Thijs (2006), scholars seek partnerships with like-minded scholars. Research is a contested terrain, and an academic is recognized by the type of research he or she has published. In fact, partnerships are becoming the norm in the global scholarly community (Marginson, 2010) and are leading to the massification of research articles and scholarly books.

Fifth, time management is critical in developing a sound research profile. A balance between teaching and research in general should be maintained and time should be set aside to develop one's own research profile and have a family life. It is advisable that researchers set aside time to improve their research profiles. Time management includes discipline and diary management. An academic should set aside time to work on his or her research and research profile. Research productive scholars tend to manage their time effectively (Miller, Taylor & Bedeian, 2011).

Because time is critical, emerging scholars should learn from their elders about time management.

Finally, these critical issues can be addressed through collaborative consultancy. An academic should not work in isolation but should ask their peers and mentors' support. In this way, growth and development are considered not as a one-way process but as a collective process. Gone are the days when academics suffered in isolation; these days they are involved in co-operative research learning. Collaboration normally yields good results as information and ideas are exchanged. Research shows that collaborative consultancy leads to research productivity and visibility (Marginson, 2010). Today institutions of higher learning worldwide encourage their academics to engage in collaborative consultation and offer incentives to those who are involved in it. Some institutions have even made it compulsory.

DEVELOPMENT OF AN OWN RESEARCH PROFILE AS A PRAGMATIC EXERCISE

When they develop their research profiles, researchers need to be pragmatic. Pragmatism is an approach that acknowledges the fact that the world can be interpreted in many ways (Marginson, 2010). It addresses a range of issues, including valuing experience; writing for publication; managing time; learning from others; considering a sideways move; finding an academic mentor; becoming a reviewer; attending conferences; and seeking research funding. Developing a research profile is a lengthy process, but that there is growing interest in a more clearly defined career pathway to support aspiring researchers. Pragmatism is a concerted effort that defines an academic. It is a rational process that leads to effective research profile development. A pragmatically developed research profile is more likely to be an efficient marketing tool.

DISTINGUISHING ONE'S OWN RESEARCH PROFILE

A research profile can turn a researcher into a household name in the research community (Almomani, 2022). A researcher's reputation relies on the quality of his or her research work. Esteem is won by publishing in international scholarly books and reputable journals, which are research impactful. Furthermore, networking is key to research profile development. In their research profiles, academics must distinguish themselves from others by highlighting their research publications and interest in the field they work in. A research profile should also indicate your research experience. As a researcher, one needs to not only understand the field one works in, but also be knowledgeable about what is currently happening in it and what still needs to be achieved in it now and in the future (Almomani, 2022). It is worthwhile to look at the research profiles of researchers in one's field and the general organization of their profiles.

One way to distinguish oneself as an academic is to have a clear definition of the research market (Fauth, 2008). Through a well-developed research profile, you can stand out and have scores of academics wanting to collaborate with you. To establish your presence, it is important to be a regular contributor to scholarly books and reputable journals. Global scholars who want expertise must know where and how to find a certain academic. The research terrain is highly competitive and contested. A well-thought-out research profile makes it easier to attract research funding and collaborators. Emerging academics struggling to come up with a carefully planned and appealing research profile must seek the assistance of a mentor. Research mentors are there to support and guide junior academics. It is worth noting that some institutions of higher learning have systems and structures in place to assist emerging academics in developing sound research profiles.

Mentoring is key to developing sound research profiles setting academics apart (Miller, Taylor & Bedeian, 2011). A mentor should support the emerging academic in developing a sound research profile concerning the research interests and needs of the emerging researcher. The support given by the mentor should be such that aspiring researchers develop their own research identity. This will assist them to define their research interest and niche. Mentors should adopt a nurturing approach and have a warm personality. A quality research profile highlights a researcher's achievements, experience and skills. It could attract collaborators, get research funding and build confidence during conferences and seminars.

REACHING OUT TO THE RESEARCH COMMUNITY BY MEANS OF YOUR RESEARCH PROFILE

To maximize the impact of his or her research profile, the academic needs to publish strategically (Fauth, 2008). High-impact journals and indexed scholarly books should have preference because they guarantee a research identity and reputation. This means that time must be invested in locating relevant and authoritative publications. Publication in reputable journals should be reflected in your research profile. It should attract the attention of global scholars who might want to collaborate. Furthermore, it can maximize the impact and visibility of an academic's work. In addition, scholars need to monitor and understand research indicators, which will encourage them to build up a formidable research profile and follow current trends in research outputs. Research indicators will also help them keep up with the latest developments in the research world. Academics need to be active on social media to boost their presence and marketability.

Instead of operating in isolation, scholars should reach out to communities (Aksnes, Langfeldt & Wouters, 2019). As academic citizens, they should engage with their community. They should work closely with government, policy makers and non-government organizations by making their research profiles accessible to them. Academics should attend community meetings where they can showcase their profiles and help meet their community's needs. They can be involved in human and community development by adopting a people-centered approach. In this way, their relevance will be acknowledged. Collaboration between government, society and community-based organizations is central to the developmental state and collective societal growth.

PLANNING THE DEVELOPMENT OF YOUR RESEARCH PROFILE

Academics must know what they want to achieve in the long term. This means clear objectives. They should also have the right audience and target market in mind. The development of a research profile should be preceded by being visible in high-impact scholarly books and journals. Fledgling academics must earn a research and scholarly reputation, which requires a clear research identity, research focus and research niche. Finally, it is important to have teaching and learning reflected in one's research profile. Budding scholars are encouraged to publish from teaching and conduct research on students. It means that there should be a link between teaching and research. Some academics conduct research on teaching and learning to improve not only their teaching strategies but also module and program development.

First, a topic needs to be identified to have a clear focus (Marginson, 2010). Second, one needs to do a preliminary information search by conducting a literature review. A literature review can locate relevant information that could inform the development of the theoretical framework and the findings of previous research. In addition, literature will indicate if the chosen topic is over-researched and if one should instead refocus on other under-researched areas of academic importance. Third, it is important to evaluate the sources. There is sometimes a space to include sources when developing a research profile. Sources should be the work of experts in the field. Drafting a research profile is like writing a research project. Make notes of and highlight the important aspects of findings. Finally, write your research profile. If the steps above are followed, your research profile will be outstanding. Research shows that scholars who have taken the time to develop their research profiles are more likely to get good research results and to be marketable (Fauth, 2008). They may also attract funding and partnerships.

BENEFITS OF DEVELOPING A SOUND RESEARCH PROFILE

Opportunities for networking is one of the most significant benefits of research profile development (Griffiths, Thompson & Hrynie, 2010). Networking opportunities not only allows for the sharing of research, but also opens doors for partnerships and research funding. Academics become more aware of the need to have a research profile and many institutions of higher learning across the globe encourage it. Some have even included the development of research profiles in their performance evaluations. According to Fauth (2008), the development of a research profile leads to new research and development opportunities. A lot of work still needs to be done to conscientize academics about the development of a sound research profile. For some, a research profile is something new (Almomani, 2022). A great deal of groundwork and lobbying is needed. Well-established scholars teach us the value of developing a sound proposal. This is when the expertise of a mentor is needed.

Another benefit of a research profile is greater awareness of the academic's research. This can be achieved by not only attending seminars, conferences, and workshops, but also making oral and poster presentations and giving talks. Attendance of these events will attract the attention of potential collaborators and funders. You might attract academics with similar research interests. Research shows that a well-developed research profile can help to establish oneself as an academic of note (Almomani, 2022) because it reflects one's research experience, research interests, research values and research skills. Academics are encouraged to develop sound research profiles to establish themselves academically.

Making a name in your field means more collaboration (Miller, Taylor & Bedeian, 2011). Joining relevant research community networks helps to network with like-minded researchers and is an opportunity for growth and development. One can learn a great deal from interacting with academics with similar research interests and become known by developing a unique and appealing research profile. An academic establishes himself or herself as an expert in a particular research field by getting published in reputable journals and scholarly books. As it has already been said, research profiles are a marketing tool. More than enough research evidence suggests that a well-crafted research profile increases visibility and international recognition (Bazeley, 2006). Being an academic of note starts with building up a sound research profile. In this manner, you create opportunities for members of scholarly and academic communities to approach you.

Lastly, an emerging academic must believe in him- or herself. Seminars, lectures, and workshops are excellent opportunities for networking and marketing oneself. Fellow scholars are more likely to help budding academics develop an appealing and thoughtful research profile. A developing academic should fearlessly develop his or her research profile. Presenting one's own research is an opportunity for growth and development. Conferences and seminars can cause anxiety in budding academics but they should be willing to learn from experts. Constructive criticism helps them to learn and grow.

CONSEQUENCES OF NOT DEVELOPING A SOUND RESEARCH PROFILE

Failure to develop a sound research profile may result in limited career development (Griffiths, Thompson & Hrynie, 2010). Collaborators, fellow scholars, and research funders look at one's research profile. It is the only way to get to know a developing academic. You may be headhunted and regarded as a sought-after academic on condition that your research profile is well-developed. A research profile constantly evolves over time owing to changes in your circumstances and career opportunities. Failure to develop a sound research profile may harm one's career prospects and limit one's opportunities for future employment. Institutions of higher learning increasingly encourage their academics to develop research profiles. At some of them, a research profile is not only compulsory but also promotion depends on it.

The lack of a research profile may be the cause of disregard and poor visibility in the research community (Miller, Taylor & Bedeian, 2011). Without a sound research profile, one misses the opportunity to showcase one's work, contribute to the advancement of knowledge in an area of expertise, and demonstrate one's capabilities. To deal with this challenge, one should attend conferences, give talks and get to be invited as guest speaker at key conferences. One's visibility can also be improved by being active in social media. Once again, increased visibility can lead to international scholarly recognition.

A poorly conceptualized research profile can cause limited opportunities for collaboration and networking. A good research profile promotes collaboration with other researchers in your field. It can help you to connect with others in your research field and build up a professional network. This is the reason that a research profile should be well-developed and well-conceptualized. Therefore, academics must be helped to develop effective research profiles to maximize collaboration and networking. Academia certainly recognizes the importance of partnerships and networking. The research profile is a valuable tool in this regard.

A poorly developed research profile leads to a poor research impact. Research should lead to knowledge production (Marginson, 2010). Your published research work should be reflected in your research profile. It will also include your achievements, conference presentations, research interests and the leadership roles you have played. Your research profile should demonstrate your ability to publish in leading journals and scholarly books. All successful scholars invest time in conceptualizing and developing their research profiles. Again, the role of research mentors cannot be underestimated. They are there to guide and support fledgling scholars in building up sound research profiles. Institutional support can deter academics from publishing in predator publications, which will damage the integrity and scholarly standing of an academic.

Finally, a poorly developed research profile may result in a lack of funding (Fauth, 2008). Funding agencies and grant programs look at a researcher's publication record and research profile. This means that an academic should publish consistently and continuously in high-impact publications. Because funders look at both the quality and the quantity of your publications, both should be reflected in your research profile. They want value for their money. The ability to attract funding enables the academic to conduct research without worrying about funding. Funds can be used to employ research assistants who can help to collect data, to develop materials, to pay publication fees, to attend national and international conferences, or to conduct workshops. All these activities should be reflected in the research profile.

CHALLENGES OF DEVELOPING A RESEARCH PROFILE

The following challenges in developing a sound research profile have been identified:

1) Time consuming

Since putting together a decent research profile can be time consuming, time management is essential. Researchers need to set aside time for building up their research profiles. This involves maintaining a balance between their teaching and learning responsibilities, their research, and their family life. Time management also necessitates a diary in which research profile writing can be prioritized. A well-managed timetable yields superior results and research profile productivity. A considerable amount of time should be invested in research profile development to attract funding and improve networking and collaboration. Good time management is crucial in developing a quality research profile and minimizes mistakes and a lack of focus.

2) Structuring and organizing a research profile

A research profile needs to have a sound structure and should be presented in an organized way. Structure and organization give meaning and a sense of direction. If the details are structured and organized, the message is clearly communicated. It also avoids the use of jargon. An unstructured and unorganized research profile is difficult to read and the cause of a lack of interest. The use of diagrams can clearly communicate a message. An organized and well-structured research profile makes it easy to understand its message.

3) Lack of discipline

The development of a research profile needs a great deal of discipline. An academic must know when to do what. Discipline involves setting goals and following the steps to achieve them. It requires a vision and mission, and means sticking to your plans and making every effort to realize them.

4) Lack of mentoring support

Most institutions of higher learning do not have mentoring systems for emerging academics. This can affect their ability to develop a sound research profile. The relationship between a mentor and a mentee must be good because the former should guide, support and nurture the latter.

5) Lack of peer support

It is well documented that some scholars are not supportive of their peers because they want to climb the career ladder alone. They are in competition with their peers and display no spirit of togetherness and cooperative learning. This can be overcome by fostering a team spirit and teamwork.

6) Lack of institutional support

Regrettably, some institutions of higher learning still ignore the significance of good research profiles. This is evidenced by their inability to put structures in place to support up-and-coming academics in developing their own research profiles. Here and there, some institutions of higher learning are beginning to see the light. Institutional support is vital in that resources are provided to the researcher.

THINKING-LONG-TERM

It is important to view the development of own research profile as a long-term plan. This includes setting realistic goals, staying focused and being strategic (Miller, Taylor & Bedeian,2011). One may aspire to go for promotion. This takes careful planning and is something normally realised with a considerable lead time of at least three to five years unless one has already has a strong research profile and good supporting evidence about teaching and professional service ((Marginson, 2010). Academicians need to be able to demonstrate a coherent trajectory of research achievement over a few years and projected quality research output ((Fauth, 2008). It is important to become known as a leading expert in an area – not only nationally but eventually, internationally. In fact, it is an expectation for promotion to full professor in most universities. An academician might have limited time, so an academician must work out on what the research profile is going to look like and be strategic about who one researches with, which journals one publishes in and which conferences one attend (Griffiths, Thompson & Hrynie, 2010). Seek out successful colleague researchers and ask their advice about how best to apply for external research grants, or how to get entry to publish in certain journals, or which conferences one should attend.

Network with colleagues in your field nationally and internationally and invite them to co-publish. Think cleverly about how you can work towards multiple outputs from your research through your collaborations (Bazeley, 2006). It is important to view the development of own research profile as a long-term plan. This includes setting realistic goals, staying focused and being strategic. For example, work out where you aim to be in, say, five years' time. An academician may aspire to go for promotion (note (Almomani, 2022). This takes careful planning and is something normally realised with a considerable lead time of at least three to five years unless one already has a strong research profile and good supporting evidence about research, teaching and professional service (Aksnes, Langfeldt & Wouters, 2019). An academician needs to be able to demonstrate a coherent trajectory of research achievement over a few years and projected quality-research.

Although it may be exciting to research a diversity of topics, to go to certain conferences in exotic locations and to be approached by a textbook publisher to write a textbook, the significant question you have to ask yourself is – do these count towards building a credible research (Bazeley, 2006). It is important to become known as a leading expert in an area – not only nationally but eventually, internationally. In fact, it is an expectation for promotion to full professor in most universities. Time is limited to develop a formidable and sound research profile, so work out what your research profile is going to look like and be strategic about who you research with, which journals you publish in and which conferences you attend (Almomani,

2022). This may entail saying NO to. Seek out successful colleague researchers and ask their advice about how best to apply for external research grants, or how to get entry to publish in certain journals, or which conferences you should attend because they have highest standing. Ask them to look at your profile and output and provide honest feedback about what you should do (Waal, Verbruggen & Wright, 2010)).

Connecting your teaching and research is also an advantageous approach. Time is best used if teaching and research are NOT poles apart (Zumwalt & Craig, 2005). An academician is employed because of expertise in a discipline area. That area is likely your passion also (xxxx). So, connect as much as possible your teaching and research profile. As an academician, demonstrate that you bring contemporary research in the area to the course(s) you teach, which can only benefit students (Fauth, 2008). Likewise, supervise postgraduate students in 'your' research area. There is research to support the fact that active researchers who infuse teaching with their own research create a more vibrant and pertinent teaching-learning experience for themselves and their students (Van Hoof, Estrella, Eljuri & Torres, 2013).

Share with your students the research you are engaged in and demonstrate its value and contribution to knowledge. It can be a struggle to successfully juggle the teaching, service, management, and research activities expected of you as an academic (Shinton Consulting, 2013). An essential key to successfully managing the myriad of responsibilities and ensuring research happens is to have a plan and stick to it.

FUTURE RESEARCH DIRECTIONS

Almost nothing has been written on the development of research profiles from a scholarly perspective. The little outdated information on the subject signifies a knowledge gap that necessitates more research on the significance and justification of a research profile. Future research should focus on the following aspects: i) the critical aspects of developing an own research profile; ii) developing an own research profile; iii) critical issues when developing a research profile; iv) development of an own research profile as a pragmatic exercise; v) distinguishing one's own research profile; vi) reaching out to the community by means of one's research profile; vii) planning the development of a research profile; viii) benefits of developing an own research profile; and ix) the consequences of not developing a research profile. Future research should be empirical because of the scarcity of research about this subject. The outcome of such empirical research can result in more research articles from a scholarly point of view.

CONCLUSION

This work is probably the first of its kind. It has shed some light on what constitutes a research profile and how it is developed. The development of a research profile, as an academic and scholarly exercise, is of particular significance. The benefits of research profile development have been brought to the attention of academics. Critical issues in developing a research profile have been indicated. Research profiles are a gateway to the future as they promote the visibility of academics in the scholarly community. A well-developed research profile creates opportunities for research growth and the ability to attract more funding. Funders look at your research profile in which your achievements, skills and research interests are highlighted, when they consider your application for funding. Institutional support for developing sound research profiles is on the increase. Research profiles are even used for promotion purposes. It is to be hoped that this chapter will provide some direction towards the development of a sound research profile. This author also hopes that it will raise more research questions that will culminate in new scholarly insights. Finally, it is to be hoped that the insights expressed in this chapter will ignite scholarly debates and interventions regarding the development of research profiles in the scholarly community.

REFERENCES

Aksnes, D., Langfeldt, L., & Wouters, P. (2019). Citations, citation indicators, and research quality: An overview of basic concepts and theories. *SAGE Journals.*https://journals.sagepub.com/doi/10.1177/2158244019829575

Almomani, S. (2022). *How to Build a Standout Research Profile?* Aug 8, 2022. *eContent Pro*

Bazeley, P. (2006) From Vocational College to New University: The Research Profile and Development Needs of Academic Staff in a Period of Transition. Pages 121-132 | Published online: 01 Nov 2006.

Fauth, G. C. (2008). Women in Educational Administration: A Research Profile. *The Educational Forum*, •••, 49.

Griffiths, V., Thompson, S. & Hrynie, L. (2010). Developing a research profile: mentoring and support for teacher educators. *Professional Development in Education.* (30), Issue 1-2.

Leta, J., Glanzel, W., & Thijs, B. (2006). Science in Brazil. Part 2: Sectoral and institutional research profiles. *Scientometrics*, 67(1), 1–2. DOI: 10.1007/s11192-006-0051-y

Marginson, S. (2010). Higher education in the global knowledge economy. *Procedia: Social and Behavioral Sciences*, 2(5), 6962–6980. DOI: 10.1016/j.sbspro.2010.05.049

Miller, A. N., Taylor, S. G., & Bedeian, A. G. (2011). Publish or perish: Academic life as management faculty lives it. *Career Development International*, 16(5), 422–445. DOI: 10.1108/13620431111167751

Scott, P. (1998). Massification, internationalization, and globalization. In Scott, P. (Ed.), *The globalization of higher education* (pp. 108–129). Open University Press.

Shinton Consulting. (2013). *Guide to building a research profile.* Sussex Researcher Careers. https://sussexresearchercareers.org/wp-content/uploads/2017/08/guide-to-building-a-research-profile.pdf

Thomas, H. (2009). Business Schools and Management Research: A UK Perspective. *Journal of Management Development*, 28(8), 660–667. DOI: 10.1108/02621710910985432

University of Birmingham. (n.d.) *Five ways to raise your research profile.* University of Birmingham. https://intranet.birmingham.ac.uk/as/libraryservices/library/research/5-ways-improve-research-impact.aspx

Van Hoof, H., Estrella, M., Eljuri, M., & Torres, L. (2013). Ecuador's system of higher education in times of change. *Journal of Hispanic Higher Education*, 12(4), 345–355. DOI: 10.1177/1538192713495060

Visser-Wijnveen, G.J., Van Driel, J.H., Van der Rijst, R.M., Verloop, N. & Visser, A. (2010). The ideal research-teaching nexus in the eyes of academics: building profiles. Pages 195-210 | Received 24 Mar 2009, Accepted 29 Nov 2009, Published online: 02 Mar 2010

Waal, T., Verbruggen, A., & Wright, T. (2010). University research for sustainable development: Definitions and characteristics explored. *Journal of Cleaner Production*, 18(7), 629–636. DOI: 10.1016/j.jclepro.2009.09.017

Webster, J. (2024). Building A Research Profile: Why And How? Published Jan 24, 2024.

Chapter 5
Research Supervision in Higher Education

Doris Chasokela
https://orcid.org/0009-0001-5983-8508
National University of Science and Technology, Zimbabwe

Agnes Maradze
Lupane State University, Zimbabwe

ABSTRACT

This chapter explores the concept of research supervision in higher education, focusing on the roles and responsibilities of supervisors, the supervisory relationship, and the key factors that impact the effectiveness of research supervision. The chapter starts by defining research supervision and its significance for supervisors and supervisees. It then delves into the complex dynamics of the supervisory relationship, examining the different roles and responsibilities of supervisors, the importance of communication and feedback, and the challenges that can arise in this relationship. Finally, the chapter provides practical guidance and strategies for enhancing the effectiveness of research supervision, drawing on current research and best practices in the field.

INTRODUCTION

Effective research supervision is a critical component of higher education, playing a vital role in shaping the next generation of researchers and academics (Johnson, 2015). The supervisor-student relationship is complex and multifaceted, influenced by factors such as institutional context, discipline-specific norms, and personal relationships. In this chapter, the dynamics of research supervision in higher education,

DOI: 10.4018/979-8-3373-0806-7.ch005

highlight the key challenges and opportunities that arise in this critical aspect of academic life. Research supervision is a dynamic and iterative process that involves multiple stakeholders, including supervisors, supervisees, and other team members (Gazzola, et al., 2014; Mainstone & Wonnacott, 2021). The supervisor's role is to provide guidance, support, and feedback to their supervisees while navigating the complex web of relationships within the institution. Effective supervision requires a deep understanding of the research process, discipline-specific methodologies, and the institution's academic culture. Moreover, supervisors must balance their research agendas with the needs and goals of their supervisees, often juggling competing demands from different stakeholders.

The supervisor-supervisee relationship is built on trust, communication, and mutual respect (Beinart, 2014; Lyon & Potkar, 2011). However, this relationship can be influenced by various factors that can impact its effectiveness. Power imbalances (Arczynski & Morrow, 2017), language barriers, cultural differences (Killian, 2013), and conflicting expectations can all contribute to tension and conflict (Tsui, et al., 2014) in the supervisory relationship. Effective supervisors must navigate these complexities while maintaining open communication channels and providing constructive feedback. According to Gray (2010) and Parker-Jenkins (2018), the supervisor's primary responsibility is to provide expert guidance and mentorship to their supervisees, helping them to develop their research skills and knowledge. This involves setting clear expectations, establishing goals and timelines, and providing regular feedback and support. Supervisors are also responsible for ensuring that their supervisees adhere to ethical standards, meet institutional requirements, and comply with funding agency regulations. Effective communication is critical in any supervisory relationship. Supervisors must be able to communicate clearly and concisely, providing feedback that is timely, constructive, and specific. This involves not only conveying information but also listening actively to the concerns and needs of supervisees. Effective communication can help build trust and confidence in the supervisory relationship, facilitating the development of supervisees' research skills and independence.

Conflict is an inevitable part of any supervisory relationship therefore supervisors must be able to manage conflict effectively by addressing issues promptly, empathetically, and constructively. This involves active listening, open communication, and a willingness to negotiate and find mutually acceptable solutions. Effective supervisors recognize that conflict can be an opportunity for growth and learning, rather than a source of tension or division. Research supervision is a dynamic process that requires flexibility, adaptability, and creativity. Effective supervisors employ various strategies to enhance their supervisory practice, including peer mentoring programs, research training workshops, and collaboration with colleagues. They also recognize the importance of self-care and prioritize their professional development to ensure they are equipped to meet the evolving needs of their supervisees.

Building effective supervisory relationships requires a deep understanding of the needs and goals of supervisees. Effective supervisors take the time to get to know their supervisees as individuals, understanding their strengths, weaknesses, motivations, and aspirations. They also recognize the importance of building trust and rapport, fostering an environment where supervisees feel supported, valued, and empowered to take risks. Effective supervision involves striking a balance between guidance and independence. Supervisors must encourage supervisees to take ownership of their research projects while providing guidance and support when needed. This involves setting clear boundaries, establishing realistic expectations, and recognizing milestones achieved along the way. In conclusion, research supervision is a complex process that requires effective communication, conflict management skills, strategic planning strategies for enhancing supervision practice building effective supervisory relationships promoting student independence and autonomy.

Understanding Research Supervision in Higher Education

Research supervision is a critical component of higher education, playing a vital role in shaping the next generation of researchers and academics. As the foundation of research output and knowledge creation, effective supervision is essential for ensuring the quality and rigor of research conducted by graduate students and early-career researchers. In higher education institutions, research supervision is not only about providing guidance and mentorship but also about fostering independent thinking, critical inquiry, and innovative problem-solving. Despite its importance, research supervision remains a complex and multifaceted process, influenced by factors such as institutional context, discipline-specific norms, personal relationships, and cultural differences (Motshoane & McKenna, 2021). Research supervision takes place within an institution with its own set of policies, procedures, and cultural norms. Supervisors must be aware of these factors and adapt their supervisory practice accordingly. Research supervision varies across disciplines, with different expectations around methodology, terminology, and communication styles. Supervisors must be familiar with the specific norms and conventions of their discipline.

The supervisor-supervisee relationship is built on trust, communication, and mutual respect. Effective supervisors must be able to establish strong working relationships with their supervisees while maintaining objectivity and professionalism. Research supervision often involves power imbalances between supervisors and supervisees. Supervisors must be aware of these power dynamics and take steps to mitigate them. Research supervision takes place in a globalized academic environment where cultural differences can impact communication styles, values, and expectations. Supervisors must be sensitive to these differences and adapt their supervisory practices accordingly. By recognizing these factors, supervisors can develop a deeper understanding of the complexities involved in research supervision and adapt their practice to meet the evolving needs of their supervisees. (Hamid, et al., 2024; Stynes, et al., 2024; Veles & Danaher, 2024; Wanyama & Eyamu, 2021).

The Complex Dynamics of Supervision

Research supervision is a dynamic and iterative process that involves multiple stakeholders, including supervisors, supervisees, and other team members. The supervisor's role is to provide guidance, support, and feedback to their supervisees while navigating the complex web of relationships within the institution. Effective supervision requires a deep understanding of the research process, discipline-specific methodologies, and the institution's academic culture. Moreover, supervisors must balance their research agendas with the needs and goals of their supervisees, often juggling competing demands from different stakeholders. The dynamics of supervision can be further complicated by issues such as power imbalances, language barriers, and cultural differences. (Odularu & Akande, 2024; Molaba, et al., 2024)

Power imbalances can occur in research supervision when there is a significant difference in status, authority, or experience between the supervisor and supervisee. This can create a dynamic where the supervisee feels intimidated or disempowered, leading to difficulties in expressing their ideas, asking questions, or seeking feedback. Power imbalances can arise due to various factors such as a senior researcher may have more experience, expertise, and institutional authority, giving them a significant advantage in the supervisory relationship. Gender-based power imbalances can affect the dynamics of supervision, with women often being marginalized or silenced in male-dominated fields. Supervisors from dominant cultures may hold more power and influence than those from minority cultures. Researchers from prestigious institutions may have more influence and authority than those from less prestigious institutions. To mitigate power imbalances, supervisors should establish open and transparent communication channels, encourage active listening and respect for diverse perspectives, foster a culture of mutual respect and inclusivity, be aware of their own biases and privileges, and create opportunities for supervisees to take ownership of their research and decision-making.

Language barriers can significantly impact the effectiveness of research supervision. Supervisors and supervisees may come from different linguistic backgrounds, leading to difficulties in communication, understanding, and collaboration. Language barriers can occur due to supervisors and supervisees not sharing a common first language, leading to misunderstandings and miscommunications. Research requires specialized terminology and jargon, which can be challenging for non-native speakers to understand. Cultural differences in non-verbal cues, such as facial expressions and body language, can lead to misunderstandings. To overcome language barriers, supervisors should use clear and simple language, provide written materials and resources in multiple languages, encourage supervisees to ask questions and clarify any misunderstandings, use visual aids and diagrams to facilitate understanding, and be patient and empathetic when communicating with non-native speakers.

Cultural differences can also affect the dynamics of research supervision. Cultural norms, values, and practices can influence how supervisors and supervisees approach research, communication, and collaboration. Cultural differences can occur due to researchers from different countries may have different cultural norms, values, and expectations. Research disciplines often have their own unique cultures and subcultures that can affect the supervisory relationship. The culture of the institution where research is conducted can also influence the supervisory relationship. To navigate cultural differences, supervisors should be aware of their own cultural biases and assumptions. learn about the cultural background and values of their supervisees, foster an inclusive and respectful environment, encourage open communication and feedback, and be flexible and adaptable when encountering cultural differences. By recognizing and addressing these complexities, supervisors can create a more inclusive and effective research supervision environment that supports the growth and development of their supervisees.

The Supervisor's Role and Responsibilities

The supervisor's primary responsibility is to provide expert guidance and mentorship to their supervisees, helping them to develop their research skills and knowledge. This involves setting clear expectations, establishing goals and timelines, and providing regular feedback and support. Supervisors are also responsible for ensuring that their supervisees adhere to ethical standards, meet institutional requirements, and comply with funding agency regulations. In addition to these technical aspects, supervisors play a crucial role in shaping their supervisees' research identities, providing opportunities for professional development, and fostering a sense of community within the institution. Ultimately, the supervisor's role extends beyond mere guidance to one of partnership and collaboration with their supervisees. This shift in perspective recognizes that research is a co-creative process that requires mutual respect, trust, and open communication between the supervisor and supervisee. A collaborative approach acknowledges that supervisees are not just recipients of guidance, but active contributors to the research process who bring their expertise, perspectives, and ideas to the table. By embracing this partnership, supervisors can create a dynamic where they and their supervisees work together to identify research questions, design methods, analyze data, and draw conclusions. This collaborative mindset fosters a sense of ownership and agency among supervisees, enabling them to take greater responsibility for their research and develop the skills and confidence they need to succeed as independent researchers. As a result, the supervisor-supervisee relationship evolves into a true partnership, characterized by mutual respect, trust, and a shared commitment to producing high-quality research that advances knowledge in their field. (Rusdiana, 2024; Wu, et al., 2024; Makina, 2024)

Effective Communication and Feedback

Communication is important for success in dissertation writing, not just any communication but effective communication. Effective communication has a crucial role in postgraduate supervision, which is an essential component of higher education. This is because it entails the sharing of significant data that is essential to the supervision procedure. Mncina et al. (2024) provide evidence to bolster the idea that communication is important in postgraduate supervision because it eliminates any possible misunderstandings or ambiguities that could develop during information sharing. Within this particular context, effective communication allows supervisors to engage in meaningful dialogues with their supervisees, thereby facilitating the provision of clear instructions, constructive criticism, and opportunities for critical thinking (Cekiso, et al., 2019). Furthermore, Govender (2023) highlighted the importance of emotional support through communication, positing that such support plays a pivotal role in assisting students in navigating and mitigating anxiety and burnout that may manifest throughout their studies (Chugh, et al., 2022). This demonstrates how effective communication stabilizes the supervision process, by allowing both sides to express their opinions regarding the ongoing study.

Effective communication is informed by different types of communication, such as verbal, non-verbal, written, and visual, with each playing a unique role in enhancing message delivery to the recipient. Adedokun & Oyetunde-Joshua (2024) and Dar & Dar (2019) argue that despite messages being brief at times, communication can be effective if the shared messages use common, simple, and precise language for clarity. Although effective communication is a critical component in dissertation or thesis supervision, its relevance in postgraduate supervision has received a lot of attention due to eminent challenges that have surfaced as a result of race, gender, personal beliefs, and religion (Chugh, et al., 2022; Ali, et al., 2019). Therefore, for communication to be effective in postgraduate supervision, both the supervisor and supervisee ought to adopt an appropriate model of communication that encourages a high level of interaction. Communication models facilitate the communication process and provide a basis for effective and efficient communication during human interactions (Mncina, et al., 2024)

Drew (2023) posits that there are 3 categories of communication models and these are Transactional, Interactive, and Linear models. According to Jan Mendoza (2020) cited in Mncina, et al., 2024, the linear model of communication involves one-way communication from the sender to the recipient, without any form of feedback. Conversely, the interactive and transactional models underscore the significance of reciprocal communication and feedback as a vital component of efficacy (Ghosh, 2022) . In contrast, the transactional model permits each participant to start the conversation by having them switch places at some point throughout the exchange (Jan-Mendoza, 2020). This implies that to improve involvement and facilitate successful supervision, both the supervisee and the supervisor have the chance to start a conversation (Mncina, et al., 2024). Jan Mendoza (2020) states that the transactional category includes several models, including Barnlund's Transactional Model of Communication and Dance's Helical Model. Because it encourages quick feedback between communicators, Barnlund's transactional model of communication is the best model for supervision, this is also supported by (Chanty, 2023). This strategy is especially appropriate for the supervisee and supervisor since it involves communicators switching places during the communication process and immediate feedback will facilitate the research project's smooth advancement.

It is important to keep in mind that when the supervisee and supervisor get along well, constructive criticism will be accepted favorably and the study will go forward more quickly. The importance of feedback is acknowledged at all educational levels, albeit the type and tone of such feedback may differ. Feedback that is prompt, supportive, and of excellent quality is necessary for facilitating and assisting PhD students (Saeed, et al., 2021). Feedback comes in a variety of forms, including written, verbal, formal, informal, evaluative, descriptive, and self-assessed. It can be provided electronically or through a physical medium and it serves as a tool for students to assess their knowledge, abilities, and comprehension to choose their outcome (Carless, 2020). Cekiso et al. (2019) think that effective feedback practices facilitate the growth of self-reflection and promote communication. It is also noted by the authors that it improves student performance and imparts knowledge to them as well. Learning is improved if feedback is a two-way process between the assessor and the student. A study by Saeed, AI Qunayeer, & AL-Jaberi (2021) concluded that a majority of the research students valued high-quality feedback and supervisor feedback played a significant role in influencing student satisfaction.

It has been noted by various studies (Carless, 2020; Mahlangu, 2021; Saeed, et al., 2021; Chugh, et al., 2022; Mncina, et al., 2024) that learning and achievement can be hampered by ineffective feedback, which can also strain the connection between the supervisor and the student. Giving feedback to research students can be challenging as noted by (Ali, et al., 2019) due to several factors, such as the candidate, stage of the program, type of program (full-time, part-time), course enrolment, and supervisory position. However effective communication can counter these challenges as it gives room for effective feedback.

Managing Conflict and Negotiating Supervision

Conflict in both human and organizational life is practically unavoidable. One of the biggest obstacles to people's interactions is effectively handling disagreement since it arises from so many various sources. Ahmadi et al. (2020) defined conflict as, "an interpersonal dispute or dissension between two or more people that results from differences in viewpoints, rivalry, unfavorable views, vague role expectations, or a lack of communication. Interpersonal conflict is caused by a variety of individual variances, such as differences in personality, culture, attitude, values, perceptions, and other areas. There are several reasons why conflicts occur such as divergences in objectives, anticipations, principles, and approaches to effectively manage the circumstances are unavoidable. Research has indicated that disputes can arise between students and supervisors in higher education.

In many instances, conflict arises from unmet or unmanaged expectations by one or both parties (Japheth, et al., 2023). It is good practice in the early stages of a supervisory relationship, in the interest of avoiding conflict at a later stage, to have a conversation that spells out the roles of both student and supervisor, general expectations around working hours, style of supervision that is envisaged and how this might develop as the student's experience and autonomy grows and the frequency of meetings, where and in what format (in-person, online, etc.) these will take place (UC, 2020). It is a good idea for both student and supervisor to be clear about what their expectations might be, and what milestones or other deliverables might be required during the writing period which varies depending on the level of study. Conflicts are easily managed if both parties practice open dialogue and discuss concerns as they arise, treating each other with respect (Trinity College, 2024). In most institutions, there is a written policy on how supervision must be conducted. Some institutions come up with a supervisor–research–student agreement which provides a good basis for understanding the roles and responsibilities of both parties and provides prompts for discussing matters relating to research and learning. The agreement template often includes open fields for students and supervisors to mutually agree on the type and frequency of communications, and how they propose to raise and resolve matters as they arise within the relationship. These agreements are used as a measure to minimize and manage conflict between the parties and therefore supervisors are mandated to use these right at the beginning of the supervision.

When a Conflict Arises

'Although the word "conflict" can have a negative connotation, many "conflicts" are just miscommunications that can be settled amicably and early on amongst the involved parties. Supervisors ought to voice their concerns in an impartial, non-emotional, and judgment-free manner (Ahmadi, et al., 2020). Supervisors should prioritize the issue over the student. To help the student comprehend how to enhance his or her work, the supervisor must actively listen to the student and offer pertinent questions (Ahmadi, et al., 2020). According to Cardilini, et al., 2021), putting such details and expectations in the agreement before any disputes develop can facilitate honest discussions when they are necessary. Referring back to the procedure that both parties agreed upon for bringing up and resolving disputes is crucial. From Mbogo & Wambua (2020), perspective, the best course of action is usually to discuss any concerns immediately and as soon as possible with the other side and this can be done during a meeting, which enables the other party to more accurately assess the situation based on body language and tone of voice since tone can be tricky to accurately decipher over text or email.

When a problem comes up, it has to be put on the agenda for the following supervisory meeting. If that isn't suitable, another meeting should be set up to discuss the issue. It is a good idea to decide after the meeting who will send out an email to all attendees summarizing the actions or future steps that were agreed upon. The email should only address what has to be done moving forward and not a recap of the conversation, as this could unintentionally bring up the subject again. When voicing a problem, it's critical to be precise, tactful, inquisitive, and focused on taking action. Various studies (Atibuni et al., 2017; Cekiso, et al., 2019; Mbogo & Wambua, 2020; Mahlangu, 2021) have suggested that it can be feasible for both sides to decide to meet in front of a third party if the disagreement gets heated or acrimonious. This third party could take on the role of facilitator, chairperson, or observer. Both the supervisor and the student may find comfort in the presence of a third party who may serve as a "witness" to the meeting. On the other hand, any expected undesirable behavior, like aggressiveness, might be discouraged by having a facilitated talk with another person present (Trinity College, 2024). At the beginning of the meeting, it should be made clear what the third party's function will be in this specific situation.

Strategies for Enhancing Research Supervision

To improve research supervision, a few key tactics are required. Providing constructive feedback that is problem-focused and looks for methods to enhance performance is a useful tactic to apply (Japheth, et al., 2023). It is known that unfavorable comments are person-focused and can be depressing rather than motivating. In this context, feedback should therefore have a clear goal because imprecise comments might leave students perplexed and unsure of what to do. This is especially true when evaluating written work (Carless, 2020). According to Chugh, et al., (2022), supervisors should address any shortcomings in the series that need to be fixed in addition to the work that is being done well. In doing so, the supervisor must be explicit about the mistakes made, the necessary steps to fix them, and the anticipated outcomes. It is always a good tactic when the supervisor pays close attention to the student, giving the student a chance to express his or her thoughts on constructive criticism, allowing the student to clarify any points the supervisor raises with questions, and then recommending appropriate action.

Japheth et al. (2023) contend that the student needs to be ready to accept the supervisor's helpful feedback constructively. Students should examine their mindsets and understand that the supervisor's criticism is not meant to be derogatory but ought to be seen as a component of the educational process. Another strategy identified by Mncina, et al., (2024) that enhances research supervision is that students must pay close attention to what is being said, especially if it seems negative or critical, and then reflect to the supervisor what they have learned. Research supervision becomes easier for both if the student can take praise with gratitude since constructive criticism is a sign of a good rapport between the student and the supervisor.

Supervisors may effectively engage in multiple supervision models over time and between students to improve research supervision (Japheth, et al., 2023). This idea is further supported by the research of Mncina, et al., (2024) who, using the interpersonal teacher behavior model developed by Wubbels, et al. (1985) discovered that students gain various advantages from managing and affiliating with their supervisor. The ability of a supervisor to flexibly adapt their actions and their role in the student-supervisor relationship based on the student's current knowledge, skills, attitudes, expectations, circumstances affecting their learning, and changing needs over time is critical for effective dissertation or thesis supervision, according to (Saeed, et al., 2021) This requires supervisors to make use of their key service areas in addition to their roles and motives to establish a genuine connection with their students in a safe and supportive environment.

Building Effective Supervisory Relationships

Completing a thesis or dissertation successfully depends on having a strong supervising connection. Relationship problems or breakdowns that occur while pursuing a degree might have a significant impact on finishing it. Numerous research studies have addressed the importance of the supervisor-student relationship quality in connection to the research process, student happiness, and timely completion of graduate and postgraduate studies (Vereijken., et al. 2018; Cekiso, et al., 2019; Yende, 2021). Lee (2018) states that research students who have positive working relationships with their PhD or master's supervisors are more likely to make good progress and feel satisfied with their work. Katikireddi & Reilly (2017) on the other hand think that the practice of research supervisory work is a facilitative process that includes instructional duties and activities. These studies have demonstrated the need for an increased understanding of the duties, expectations, and interests of research students in supervision to sustain an effective and healthy relationship between the two parties.

Yende (2021) posits that to address some of the issues related to the supervisor-student relationship, a thorough examination of the graduate experience and supervisory pedagogy is required to define and convey the postgraduate supervisor's role and satisfy postgraduate students' expectations. Carton (2023) proposed a list of supervisory roles and attitudes that include director, facilitator, advisor, guider (who suggests a schedule for writing up and providing feedback on progress), and manager (who routinely checks progress, monitors study, provides systematic feedback, and plans work) which will enhance the relationship with the student.

Promoting Student Independence and Autonomy

Becoming autonomous learners is especially important for research students who finalize their coursework and start writing their dissertation or thesis. The characteristics of students being autonomous learners are essential to survive in a disruptive period, especially in education (Prihandoko, et al., 2021). Research advisors and higher education institutions have different ways to encourage students to finish their dissertation, thesis, or doctorate and most of the support comes in the form of advice and encouragement (Janeiro & Gonzalez, 2015). Howard et al. (2018) believed that coming up with deadlines, setting goals, and how feedback must be handled would assist students in dissertation or thesis completion. Moon (2005) cited in (Janeiro & Gonzalez, 2015) argues that students' critical thinking is developed through the ability to engage in sustained and systematic inquiry. In this manner, students can be seen as producers of knowledge, not the norm where they are always the consumers of such. This shift according to Janeiro & Gonzalez, 2015) is described as 'contextual knowing' – where knowledge is seen as constructed and where teachers are 'seen as facilitators and partners in the process of the development of knowledge. In this context, student autonomy and independence are advocated for when there is a shift from teacher-directed to self-directed learning and students 'take charge' of their learning. Dissertation or thesis writing is often promoted as an effective medium to advance autonomous (or independent) learning among undergraduates and postgraduate students (Prihandoko et al. 2021).

Janeiro & Gonzalez (2015) posit that a complicated web of interrelated elements determines the degree to which students show a willingness to exert control or take charge of their learning, during the dissertation process. For comments on their work, students typically turn to their dissertation or thesis supervisors and other unofficial support systems (such as friends and family). Potential conflicts that may occur from the supervisor's perceived need to both facilitate and promote independent learning while also scaffolding and supporting dissertation students are personified in the interaction between the student and the supervisor. Rewriting might be discussed again in a dialogical approach thanks to the supervisory procedure. "Dialogues of participation," as defined by Janeiro & Gonzalez (2015), link this kind of practice to a facilitative procedure that can foster the growth of the student voice.

Cardilini et al. (2021) contend that power dynamics and conflicts characterize the multifaceted nature of the supervisory relationship. These connections can eliminate chances for students to grow in their capacity for autonomy. However, students' autonomy and independence in writing their theses and dissertations must be encouraged. Students' self-control and autonomy may increase with motivation and empowerment. According to Janeiro & Gonzalez (2015) study, supervisors who facilitate student-teacher dialogue about writing foster a more dialogic, as opposed to transmissive, dialogue in which the supervisors pose questions and offer reflection points for the students to consider, rather than providing "answers" or making explicit recommendations for textual changes. Conversely, the lack of communication throughout the feedback-giving process can impede students' growth in self-reliance, ability to successfully revise texts, and ability to create new information (Saeed, et al., 2021).

Effective Communication and Feedback

Communication is important for success in dissertation writing, not just any communication but effective communication. Effective communication has a crucial role in postgraduate supervision, which is an essential component of higher education. This is because it entails the sharing of significant data that is essential to the supervision procedure. Mncina et al. (2024) provide evidence to bolster the idea that communication is important in postgraduate supervision because it eliminates any possible misunderstandings or ambiguities that could develop during information sharing. Within this particular context, effective communication allows supervisors to engage in meaningful dialogues with their supervisees, thereby facilitating the provision of clear instructions, constructive criticism, and opportunities for critical thinking (Cekiso, et al., 2019). Furthermore, Govender (2023) highlighted the importance of emotional support through communication, positing that such support plays a pivotal role in assisting students in navigating and mitigating anxiety and burnout that may manifest throughout their studies (Chugh, et al., 2022). This demonstrates how effective communication stabilizes the supervision process, by allowing both sides to express their opinions regarding the ongoing study.

Effective communication is informed by different types of communication, such as verbal, non-verbal, written, and visual, with each playing a unique role in enhancing message delivery to the recipient. Adedokun & Oyetunde-Joshua (2024) and Dar & Dar (2019) argue that despite messages being brief at times, communication can be effective if the shared messages use common, simple, and precise language for clarity. Although effective communication is a critical component in dissertation or thesis supervision, its relevance in postgraduate supervision has received a lot of attention due to eminent challenges that have surfaced as a result of race, gender, personal beliefs, and religion (Chugh, et al., 2022; Ali, et al., 2019). Therefore, for communication to be effective in postgraduate supervision, both the supervisor and supervisee ought to adopt an appropriate model of communication that encourages a high level of interaction. Communication models facilitate the communication process and provide a basis for effective and efficient communication during human interactions (Mncina, et al., 2024)

Drew (2023) posits that there are 3 categories of communication models and these are Transactional, Interactive, and Linear models. According to Mncina, et al. 2024, the linear model of communication involves one-way communication from the sender to the recipient, without any form of feedback. Conversely, the interactive and transactional models underscore the significance of reciprocal communication and feedback as a vital component of efficacy (Ghosh, 2022). In contrast, the transactional model permits each participant to start the conversation by having them switch places at some point throughout the exchange (Jan-Mendoza, 2020). This implies that to improve involvement and facilitate successful supervision, both the supervisee and the supervisor have the chance to start a conversation (Mncina, et al., 2024). Jan Mendoza, (2020) states that the transactional category includes several models, including Barnlund's Transactional Model of Communication and Dance's Helical Model. Because it encourages quick feedback between communicators, Barnlund's transactional model of communication is the best model for supervision, this is also supported by (Chanty, 2023). This strategy is especially appropriate for the supervisee and supervisor since it involves communicators switching places during the communication process and immediate feedback will facilitate the research project's smooth advancement.

It is important to keep in mind that when the supervisee and supervisor get along well, constructive criticism will be accepted favorably and the study will go forward more quickly. The importance of feedback is acknowledged at all educational levels, albeit the type and tone of such feedback may differ. Feedback that is prompt, supportive, and of excellent quality is necessary for facilitating and assisting PhD students (Saeed, et al., 2021). Feedback comes in a variety of forms, including written, verbal, formal, informal, evaluative, descriptive, and self-assessed. It can be provided electronically or through a physical medium and it serves as a tool for students to assess their knowledge, abilities, and comprehension to choose their outcome (Carless, 2020). Cekiso et al. (2019) think that effective feedback practices facilitate the growth of self-reflection and promote communication. It is also noted by the authors that it improves student performance and imparts knowledge to them as well. Learning is improved if feedback is a two-way process between the assessor and the student. A study by Saeed et al. (2021) concluded that a majority of the research students valued high-quality feedback and supervisor feedback played a significant role in influencing student satisfaction.

It has been noted by various studies (Carless, 2020; Mahlangu, 2021; Saeed, et al., 2021; Chugh, et al., 2022; Mncina, et al., 2024) that learning and achievement can be hampered by ineffective feedback, which can also strain the connection between the supervisor and the student. Giving feedback to research students can be challenging as noted by (Ali, et al., 2019) due to several factors, such as the candidate, stage of the program, type of program (full-time, part-time), course enrolment, and supervisory position. However effective communication can counter these challenges as it gives room for effective feedback.

Managing Conflict and Negotiating Supervision

Conflict in both human and organizational life is practically unavoidable. One of the biggest obstacles to people's interactions is effectively handling disagreement since it arises from so many various sources. Ahmadi et al. (2020) defined conflict as, "an interpersonal dispute or dissension between two or more people that results from differences in viewpoints, rivalry, unfavorable views, vague role expectations, or a lack of communication. Interpersonal conflict is caused by a variety of individual variances, such as differences in personality, culture, attitude, values, perceptions, and other areas. There are several reasons why conflicts occur such as divergences in objectives, anticipations, principles, and approaches to effectively manage the circumstances are unavoidable. Research has indicated that disputes can arise between students and supervisors in higher education.

In many instances, conflict arises from unmet or unmanaged expectations by one or both parties (Japheth, et al., 2023). It is good practice in the early stages of a supervisory relationship, in the interest of avoiding conflict at a later stage, to have a conversation that spells out the roles of both student and supervisor, general expectations around working hours, style of supervision that is envisaged and how this might develop as the student's experience and autonomy grows and the frequency of meetings, where and in what format (in-person, online, etc.) these will take place (UC, 2020). It is a good idea for both student and supervisor to be clear about what their expectations might be, and what milestones or other deliverables might be required during the writing period which varies depending on the level of study. Conflicts are easily managed if both parties practice open dialogue and discuss concerns as they arise, treating each other with respect (Trinity College, 2024). In most institutions, there is a written policy on how supervision must be conducted. Some institutions come up with a supervisor–research–student agreement which provides a good basis for understanding the roles and responsibilities of both parties and provides prompts for discussing matters relating to research and learning. The agreement template often includes open fields for students and supervisors to mutually agree on the type and frequency of communications, and how they propose to raise and resolve matters as they arise within the relationship. These agreements are used as a measure to minimize and manage conflict between the parties and therefore supervisors are mandated to use these right at the beginning of the supervision.

When a Conflict Arises

Although the word "conflict" can have a negative connotation, many "conflicts" are just miscommunications that can be settled amicably and early on amongst the involved parties. Supervisors ought to voice their concerns in an impartial, non-emotional, and judgment-free manner (Ahmadi, et al., 2020). Supervisors should prioritize the issue over the student. To help the student comprehend how to enhance his or her work, the supervisor must actively listen to the student and offer pertinent questions (Ahmadi, et al., 2020). According to Cardilini et al. (2021), putting such details and expectations in the agreement before any disputes develop can facilitate honest discussions when they are necessary. Referring back to the procedure that both parties agreed upon for bringing up and resolving disputes is crucial. From Mbogo & Wambua, (2020) perspective, the best course of action is usually to discuss any concerns immediately and as soon as possible with the other side and this can be done during a meeting, which enables the other party to more accurately assess the situation based on body language and tone of voice since tone can be tricky to accurately decipher over text or email.

When a problem comes up, it has to be put on the agenda for the following supervisory meeting. If that isn't suitable, another meeting should be set up to discuss the issue. It is a good idea to decide after the meeting who will send out an email to all attendees summarizing the actions or future steps that were agreed upon. The email should only address what has to be done moving forward and not a recap of the conversation, as this could unintentionally bring up the subject again. When voicing a problem, it's critical to be precise, tactful, inquisitive, and focused on taking action.

Various studies (Atibuni et al. 2017; Cekiso et al. 2019; Mbogo & Wambua, 2020; Mahlangu, 2021) have suggested that it can be feasible for both sides to decide to meet in front of a third party if the disagreement gets heated or acrimonious. This third party could take on the role of facilitator, chairperson, or observer. Both the supervisor and the student may find comfort in the presence of a third party who may serve as a "witness" to the meeting. On the other hand, any expected undesirable behavior, like aggressiveness, might be discouraged by having a facilitated talk with another person present (Trinity College, 2024). At the beginning of the meeting, it should be made clear what the third party's function will be in this specific situation.

Strategies for Enhancing Research Supervision

To improve research supervision, a few key tactics are required. Providing constructive feedback that is problem-focused and looks for methods to enhance performance is a useful tactic to apply (Japheth, et al., 2023). It is known that unfavorable comments are person-focused and can be depressing rather than motivating. In this context, feedback should therefore have a clear goal because imprecise comments might leave students perplexed and unsure of what to do. This is especially true when evaluating written work (Carless, 2020). According to Chugh et al. (2022), supervisors should address any shortcomings in the series that need to be fixed in addition to the work that is being done well. In doing so, the supervisor must be explicit about the mistakes made, the necessary steps to fix them, and the anticipated outcomes. It is always a good tactic when the supervisor pays close attention to the student, giving the student a chance to express his or her thoughts on constructive criticism, allowing the student to clarify any points the supervisor raises with questions, and then recommending appropriate action.

Japheth et al. (2023) contend that the student needs to be ready to accept the supervisor's helpful feedback constructively. Students should examine their mindsets and understand that the supervisor's criticism is not meant to be derogatory but ought to be seen as a component of the educational process. Another strategy identified by Mncina et al. (2024) that enhances research supervision is that students must pay close attention to what is being said, especially if it seems negative or critical, and then reflect to the supervisor what they have learned. Research supervision becomes easier for both if the student can take praise with gratitude since constructive criticism is a sign of a good rapport between the student and the supervisor.

Supervisors may effectively engage in multiple supervision models over time and between students to improve research supervision (Japheth, et al., 2023). This idea is further supported by the research of Mncina et al. (2024) who, using the interpersonal teacher behavior model developed by Wubbels et al. (1985) discovered that students gain various advantages from managing and affiliating with their supervisor. The ability of a supervisor to flexibly adapt their actions and their role in the student-supervisor relationship based on the student's current knowledge, skills, attitudes, expectations, circumstances affecting their learning, and changing needs over time is critical for effective dissertation or thesis supervision, according to (Saeed, et al., 2021) This requires supervisors to make use of their key service areas in addition to their roles and motives to establish a genuine connection with their students in a safe and supportive environment.

Building Effective Supervisory Relationships

Completing a thesis or dissertation successfully depends on having a strong supervising connection. Relationship problems or breakdowns that occur while pursuing a degree might have a significant impact on finishing it. Numerous research studies have addressed the importance of the supervisor-student relationship quality in connection to the research process, student happiness, and timely completion of graduate and postgraduate studies (Vereijken, et al., 2018; Cekiso, et al., 2019; Yende, 2021). Lee (2018) states that research students who have positive working relationships with their PhD or master's supervisors are more likely to make good progress and feel satisfied with their work. Katikireddi & Reilly (2017) on the other hand think that the practice of research supervisory work is a facilitative process that includes instructional duties and activities. These studies have demonstrated the need for an increased understanding of the duties, expectations, and interests of research students in supervision to sustain an effective and healthy relationship between the two parties.

Yende (2021) posits that to address some of the issues related to the supervisor-student relationship, a thorough examination of the graduate experience and supervisory pedagogy is required to define and convey the postgraduate supervisor's role and satisfy postgraduate students' expectations. Carton (2023) proposed a list of supervisory roles and attitudes that include director, facilitator, advisor, guider (who suggests a schedule for writing up and providing feedback on progress), and manager (who routinely checks progress, monitors study, provides systematic feedback, and plans work) which will enhance the relationship with the student.

Promoting Student Independence and Autonomy

Becoming autonomous learners is especially important for research students who finalize their coursework and start writing their dissertation or thesis. The characteristics of students being autonomous learners are essential to survive in a disruptive period, especially in education (Prihandoko, et al., 2021). Research advisors and higher education institutions have different ways to encourage students to finish their dissertation, thesis, or doctorate and most of the support comes in the form of advice and encouragement (Janeiro & Gonzalez, 2015). Howard et al. (2018) thought that coming up with deadlines, setting goals, and how feedback must be handled would assist students in dissertation or thesis completion. Janeiro & Gonzalez (2015) argues that students' critical thinking is developed through the ability to engage in sustained and systematic inquiry. In this manner, students can be seen as producers of knowledge, not the norm where they are always the consumers of such. This shift according to Janeiro & Gonzalez (2015) is described as 'contextual knowing' – where knowledge is seen as constructed and where teachers are 'seen as facilitators and partners in the process of the development of knowledge. In this context, student autonomy and independence are advocated for when there is a shift from teacher-directed to self-directed learning and students 'take charge' of their learning. Dissertation or thesis writing is often promoted as an effective medium to advance autonomous (or independent) learning among undergraduates and postgraduate students (Prihandoko, et al., 2021).

Janeiro & Gonzalez (2015) posit that a complicated web of interrelated elements determines the degree to which students show a willingness to exert control or take charge of their learning, during the dissertation process. For comments on their work, students typically turn to their dissertation or thesis supervisors and other unofficial support systems (such as friends and family). Potential conflicts that may occur from the supervisor's perceived need to both facilitate and promote independent learning while also scaffolding and supporting dissertation students are personified in the interaction between the student and the supervisor. Rewriting might be discussed again in a dialogical approach thanks to the supervisory procedure. "Dialogues of participation," as defined by Janeiro & Gonzalez (2015), link this kind of practice to a facilitative procedure that can foster the growth of the student voice.

Cardilini et al. (2021) contend that power dynamics and conflicts characterize the multifaceted nature of the supervisory relationship. These connections can eliminate chances for students to grow in their capacity for autonomy. However, students' autonomy and independence in writing their theses and dissertations must be encouraged. Students' self-control and autonomy may increase with motivation and empowerment. According to Janeiro & Gonzalez (2015) study, supervisors who facilitate student-teacher dialogue about writing foster a more dialogic, as opposed to transmissive, dialogue in which the supervisors pose questions and offer reflection points for the students to consider, rather than providing "answers" or making explicit recommendations for textual changes. Conversely, the lack of communication throughout the feedback-giving process can impede students' growth in self-reliance, ability to successfully revise texts, and ability to create new information (Saeed, et al., 2021).

CONCLUSION

Research supervision in higher education is a complex and multifaceted process that requires a deep understanding of the intricate dynamics between supervisors and supervisees. Effective research supervision is not just about guiding students or junior researchers, but about fostering a collaborative partnership that nurtures their growth, creativity, and autonomy. By recognizing and addressing the complexities of power imbalances, language barriers, and cultural differences, supervisors can create an inclusive and supportive environment that promotes the development of their supervisees. Ultimately, the supervisor's role extends beyond mere guidance to one of partnership and collaboration, where both parties work together to produce high-quality research that advances knowledge in their field. As we move forward in an increasingly global and interconnected world, we must prioritize research supervision as a critical component of higher education, one that empowers students and junior researchers to become independent thinkers, innovative problem-solvers, and future leaders in their fields. By doing so, we can ensure that the next generation of researchers is equipped to tackle the complex challenges facing our world today and tomorrow.

REFERENCES

Adedokun, T., & Oyetunde-Joshua, F. (2024). Navigating the academic odyssey: Exploring the role of supervisors in supporting postgraduate students. *Journal of Culture and Values in Education*, 7(1), 1–18. DOI: 10.46303/jcve.2024.1

Ahmadi, F., Shamsi, A., & Mohammadi, N. (2020). Using intelligent interaction to manage student–supervisor conflict: A qualitative study. *Journal of Education and Health Promotion*, 9(18). PMID: 32154313

Akparep, J. Y., Jengre, E., & Amoah, D. A. (2017). Demystifying the Blame Game in the Delays of Graduation of Research Students in Universities in Ghana: The Case of University for Development Studies. *European Journal of Business and Innovation Research*, 5(1), 34–50.

Ali, J., Ullah, H., & Sanauddin, N. (2019). Postgraduate Research Supervision: Exploring the Lived Experience of Pakistani Postgraduate Students. *FWU Journal of Social Sciences*, 3(1), 14–25.

Arczynski, A. V., & Morrow, S. L. (2017). The complexities of power in feminist multicultural psychotherapy supervision. *Journal of Counseling Psychology*, 64(2), 192–205. DOI: 10.1037/cou0000179 PMID: 27918171

Atibuni, D. Z., Kibanja, G. M., Olema, D. K., & Ssenyonga, J. (2017). Challenges and strategies of research engagement among Master of Education students in Uganda. *International Journal of Educational Policy Research and Review*, 4(3), 19–25.

Beinart, H. (2014). Building and sustaining the supervisory relationship. *The Wiley international handbook of clinical supervision*, 255-281.

Cardilini, A. P., Risely, A., & Richardson, M. F. (2021). Supervising the PhD: Identifying common mismatches in expectations between candidate and supervisor to improve research training outcomes. *Higher Education Research & Development*, •••, 1–15.

Carless, D. (2020). Longitudinal perspectives on students' experiences feedback: A need for teacher-student partnership. *Higher Education Research & Development*, 39(3), 425–498. DOI: 10.1080/07294360.2019.1684455

Carton, D. J. (2023, May 4). *Thing 8: Building Professional Relationships in Research Supervision*. Retrieved from May 4, 2023: https://www.23thingsinternational.com/post/thing-9-building-professional-relationships-in-research-supervision#:~:text=the%20fundamental%20touchstone%20for%20all,course%20of%20their%20research%20programm

Cekiso, M., Tshotsho, B., Masha, R., & Saziwa, T. (2019). Supervision Experiences of Postgraduate Research Students at One South African Higher Education Institution. *South African Journal of Higher Education, 33(3), 8-25., 33*(3), 8-25. DOI: 10.20853/33-3-2913

Chanty. (2023, June 8). *Implementing the Transactional Model of Communication for Better Collaboration.* Retrieved from Chanty: https://www.chanty.com/blog/transactional-model-of-communication/

Chugh, R., Macht, S., & Harreveld, B. (2022). Supervisory feedback to postgraduate research students: A literature review, Assessment & Evaluation in. *Assessment & Evaluation in Higher Education*, 47(5), 683–697. DOI: 10.1080/02602938.2021.1955241

Dar, M. A., & Dar, H. M. (2019). contribution of effective communication in educational settings: A critical review. *International Journal of Research in all Subjects in Multi Languages, 7(8) 13-20., 7*(20), 13-20.

Drew, C. (2023, July 22). *Home: All 8 Models of Communication, Explained!* Retrieved May 06, 2024, from Helpful Professor.com: https://helpfulprofessor.com/communication-models/

Gazzola, N., De Stefano, J., Theriault, A., & Audet, C. (2014). Positive experiences of doctoral-level supervisors-in-training conducting group format supervision: A qualitative investigation. *British Journal of Guidance & Counselling*, 42(1), 26–42. DOI: 10.1080/03069885.2013.799263

Ghosh, U. (2022). Role of communication in education. *Journal for Research Trends and Innovation*, 7(6), 326–329.

Govender, N. (2023). Govender, N. (2023). Attributes and Relationships in the Supervisor-Supervisee Partnership: An autoethnographic Study. In Samuel, M. A., & Mariaye, H. (Eds.), *Transforming Postgraduate in Africa. In M. Samuel, & H. Mariaye*. Transforming Postgraduate in Africa.

Gray, D. E. (2010). Towards the lifelong skills and business development of coaches: An integrated model of supervision and mentoring. *Coaching (Abingdon, UK)*, 3(1), 60–72. DOI: 10.1080/17521880903559705

Hamid, M. O., Hanna, B., Gannaway, D., & Nguyen, T. T. T. (2024). The ethics and epistemology of researching higher degree by research supervision: An encounter with institutional ethics review. *International Journal of Research & Method in Education*, •••, 1–14. DOI: 10.1080/1743727X.2023.2299973

Howard, P. J., Gorzycki, M., Desa, G., & Allen, D. D. (2018). Academic reading: Comparing students' and faculty perceptions of its value, practice, and pedagogy. *Journal of College Reading and Learning*, 48(3), 189–209. DOI: 10.1080/10790195.2018.1472942

Ireland assignments. (2020, April 23). *Home: blog/dissertation writing guide/ Methods Of Communicating Effectively With The Ph.D. Dissertation Supervisor.* Retrieved from Ireland Assignment Help.com: https://www.irelandassignmenthelp .com/blogs/methods-communicating-effectively-phd-supervisor/

Jan Mendoza, C. (2020). *Home: Academia: Communication and Communication Models.* Retrieved from Academia.edu website: https://www.academia.edu/ 44018864/Communication_and_Communication_sssss

Janeiro, M. G., & González, M. G. (2015). Developing Students' Autonomy And Self-Regulation Through A Co-Teaching Research Methods Experience. *Contemporary Issues in Education Research*, 8(1), 25–32. DOI: 10.19030/cier.v8i1.9059

Japheth, N., Ssentamu, P., & Wambua, B. (2023). Strategies Used for Effective Research Supervision in the Completion of Postgraduate Studies in Selected Universities of Uganda. *Journal of the National Council for Higher Education Vol. 10, Issue 2, 2023 1, 10*(2).

Johnson, W. B. (2015). *On being a mentor: A guide for higher education faculty.* Routledge. DOI: 10.4324/9781315669120

Katikireddi, S. V., & Reilly, J. (2017). Characteristics of good supervision: A multi-perspective qualitative exploration of the Masters in Public Health dissertation. *Journal of Public Health (United Kingdom), 39(3), pp.625 632. doi:, 39*(3), 625-632.DOI: 10.1093/pubmed/fdw

Killian, K. D. (2013). Differences making a difference: Cross-cultural interactions in supervisory relationships. In *Integrating gender and culture in family therapy training* (pp. 61–103). Routledge.

Lee, A. (2018). How can we develop supervisors for the modern doctorate?', Studies in Higher Education. *Taylor & Francis, 33*(5), 878-890.

Lyon, R. E. C., & Potkar, K. A. (2011). The supervisory relationship. In *Counselor supervision* (pp. 41–78). Routledge.

Mahlangu, V. (2021). Exploring Challenges of Supervising Postgraduate Students in Open Distance Learning in Higher Education Settings. In Books, B. C. (Ed.), *New Challenges to Education: Lessons from Around the World.* Bulgarian Comparative Education Society.

Mainstone, F., & Wonnacott, J. (2021). The integrated supervision model: The 4× 4× 4. In *The Routledge International Handbook of Social Work Supervision* (pp. 321–333). Routledge.

Makina, A. (2024). A Critical Reflection of the Feedback Practices in the Postgraduate Supervision in an ODL Institution. *Reflections on Postgraduate Supervision and Academic Development in an African Distance Education Environment*, 49.;Mbogo, R. W., & Wambua, J. M. (2020). Supervision Challenges and Delays in Completion of PhD Programmes in Public and Private Universities: Experiences of Supervisors and Graduate Students in Kenya. *European Journal of Education Studies*, 6(7), 261–279. DOI: 10.5281/zenodo.3653866

Mncina, T. C., Letsie, H., Nkhi, S. E., & Mofana, M. (2024). Effective Communication in Postgraduate Supervision: Shaping Experiences and Overcoming Challenges. *Interdisciplinary Journal of Education Research*, 6(6), 1–21. DOI: 10.38140/ijer-2024.vol6.13

Molaba, M., Gololo, S. S., & Seeletse, S. M. (2024). Optimizing postgraduate supervision using multilevel modeling techniques in a constrained higher education environment. *International Journal of Research in Business and Social Science (2147-4478), 13*(2), 375-381.

Motshoane, P., & McKenna, S. (2021). Crossing the border from candidate to supervisor: The need for appropriate development. *Teaching in Higher Education*, 26(3), 387–403. DOI: 10.1080/13562517.2021.1900814

Odularu, O. I., & Akande, J. O. (2024). Reflection of experiences with academic supervisors, supervisees, and issues of power. *South African Journal of Higher Education*, 38(2), 254–273. DOI: 10.20853/38-2-5706

Parker-Jenkins, M. (2018). Mind the gap: Developing the roles, expectations, and boundaries in the doctoral supervisor–supervisee relationship. *Studies in Higher Education*, 43(1), 57–71. DOI: 10.1080/03075079.2016.1153622

Prihandoko, L. A., Anggawirya, A., & Rahman, F. (2021). Students' Perceptions Towards Autonomous Learners Concept in Academic Writing Classes: Sequential Mixed-Method. *Advances in Social Science, Education and Humanities Research*, 603, 487–491.

Rusdiana, A. (2024). Educational Supervision and Evaluation: Semester II Academic Year 2023/2024.

Saeed, M. A., & Qunayeer, A. I. S. H., & AL-Jaberi, M. A. (2021, April). Exploring Supervisory Feedback Formulation on Academic Writing of Research Proposals and Postgraduates' Responses to Feedback: A Case Study. *Sage publications Open access*, 1-15. doi:, P., Muntean, C. H., Sheridan, F., Thornbury, E., & Pathak, P. (2024, March). Research Supervision Framework: A Student's Experience. In *2024 IEEE World Engineering Education Conference (EDUNINE)* (pp. 1-5). IEEE.DOI: 10.1177/ Stynes

TrinityCollege. (2024, March 8). *Current Students>Postgraduate Research Students>Conflict IN Supervision*. Retrieved from Trinity College Dublin: https://www.tcd.ie/graduatestudies/students/research/conflict-resolution/

Tsui, M. S., O'Donoghue, K., & Ng, A. K. (2014). Culturally competent and diversity-sensitive clinical supervision. *The Wiley international handbook of clinical supervision*, 238-254.

UC. (2020). *cumming.ucalgary.ca.proskills.resources*. Retrieved from University of Calgary: https://cumming.ucalgary.ca/sites/default/files/teams/13/proskills/resources/resolving_conflicts_in_the_graduate_student-supervisor_relationship.pdf

Veles, N., & Danaher, P. A. (2024). Transformative research collaboration as third space and creative understanding: Learnings from higher education research and doctoral supervision. *Research Papers in Education*, 39(1), 50–66. DOI: 10.1080/02671522.2022.2089212

Vereijken, M., van der Rijst, R., van Driel, J., & Dekker, F. (2018). Factors of effective postgraduate student-supervisor relationships at selected universities in South Africa. [JAE]. *Journal of African Education*, 2(2), 135–155.

Wanyama, S. B., & Eyamu, S. (2021). Perceived organizational support, graduate research supervision, and research completion rate. *Employee Relations*, 43(6), 1414–1430. DOI: 10.1108/ER-05-2020-0205

Wu, S., Oubibi, M., & Bao, K. (2024). How supervisors affect students' academic gains and research ability: An investigation through a qualitative study. *Heliyon*, 10(10), e31079. DOI: 10.1016/j.heliyon.2024.e31079 PMID: 38779017

Yende, S. J. (2021). Factors of effective postgraduate student-supervisor relationships at selected universities in South Africa. [JAE]. *Journal of African Education*, 2(2), 135–155. DOI: 10.31920/2633-2930/2021/v2n2a6

KEY TERMS AND DEFINITIONS

Academic Advising: The process of providing guidance and support to students in making informed decisions about their academic choices, including course selection, program requirements, and career options.

Collaboration: The process of working together with others towards a common goal or objective, often involving shared decision-making, coordination, and mutual support.

Communication: The process of exchanging information or ideas between individuals or groups through various means, such as verbal or written communication.

Feedback: Information or reactions that are given to someone about their performance or behavior, intended to help them learn from their experiences and improve their skills.

Higher Education: A level of education that involves study beyond secondary education, typically consisting of undergraduate and graduate programs at universities and colleges.

Mentorship: A relationship between an experienced person (mentor) and a less experienced person (mentee) where the mentor provides guidance, support, and encouragement to help the mentee develop their skills and achieve their goals.

Research Methodology: The systematic approach used to conduct research, including the selection of research questions, study design, data collection and analysis techniques, and data interpretation.

Research Supervision: The process of guiding and supporting a student or junior researcher in conducting research, typically resulting in the production of a thesis, dissertation, or other scholarly work.

Student Supervision: The process of guiding and supporting students in their academic work, typically involving individualized attention and guidance on research projects, thesis writing, and academic goals.

Supervision: The process of overseeing and guiding the work of someone, typically a student or junior researcher, to ensure that they complete their tasks and achieve their goals.

Thesis Supervisor: A senior academic or researcher who oversees and guides a student or junior researcher in writing their thesis, providing feedback and guidance on research methodology, data analysis, and writing style.

Thesis Writing: The process of writing a thesis, which is a lengthy and comprehensive document that presents original research and arguments in a specific field or discipline.

Chapter 6
Research Supervision in African Higher Education

Manjari Sharma
https://orcid.org/0000-0003-1459-165X
Christ University, India

Sharad Gupta
https://orcid.org/0000-0002-4994-3523
Christ University, India

ABSTRACT

This chapter delves into the challenges and transformative potential of research supervision within the African higher education landscape. It provides practical insights into effective supervision strategies, mentoring frameworks, and innovative methodologies aimed at fostering high-quality research proposals and dissertations among postgraduate students. By exploring culturally relevant supervision frameworks through the lens of the Africanisation concept, the chapter highlights the importance of integrating indigenous knowledge systems into the academic discourse. It addresses key issues such as the gap between traditional and modern supervision models, limited mentorship programs, and cross-cultural challenges. The chapter offers solutions including structured mentorship programs, supervisor training, and enhanced research infrastructure, aiming to empower supervisors and postgraduate students alike, ultimately contributing to the development of a more inclusive and supportive research environment.

DOI: 10.4018/979-8-3373-0806-7.ch006

1. INTRODUCTION

1.1 Importance of Research Supervision

Research supervision is a fundamental aspect of academic development, especially at the postgraduate level. It significantly influences the quality and impact of research outputs. Effective supervision is essential for guiding students through the complexities of research methodologies and academic writing, while also fostering critical thinking, independence, and academic integrity (Manathunga, 2012). The role of a supervisor extends beyond mere oversight, as it involves mentoring, coaching, and providing intellectual and emotional support (Lee, 2008). This holistic approach is crucial for nurturing the next generation of scholars, who will contribute to the advancement of knowledge and address societal challenges through innovative research (Pearson & Brew, 2002).

The importance of research supervision is further emphasized by its impact on the timely completion of postgraduate degrees and the quality of research publications. Studies have shown that effective supervision correlates with higher rates of thesis completion and publication outputs (Halse & Malfroy, 2010). Moreover, the supervisor-student relationship is critical in shaping students' academic identities and career trajectories. A positive supervision experience can inspire students to pursue academic careers and contribute to the scholarly community (Grant, 2003).

Research supervision also plays a crucial role in maintaining the standards of academic research. Supervisors act as gatekeepers of academic quality, ensuring that research proposals and dissertations adhere to ethical guidelines, methodological rigor, and scholarly relevance (Delamont, Atkinson, & Parry, 2000). This oversight is particularly important in an era where academic integrity is constantly challenged by issues such as plagiarism and research misconduct (Morris & Laipple, 2015). Effective supervision is essential for upholding the integrity of research and maintaining the trust and credibility of scholarly work.

1.2 Scope and Objectives

This chapter aspires to delve into the intricacies and transformative potential of research supervision within the African higher education milieu. It seeks to provide all-encompassing insights into effective supervision strategies, mentoring frameworks, and innovative methodologies that can bolster the quality of research proposals and dissertations among postgraduate students. The purview of this chapter encompasses:

1. **Analyzing the Obstacles**: Identifying the pivotal obstacles faced by higher education institutions in Africa concerning research supervision, including the dearth of mentorship models, publication tribulations, and inadequate resources.
2. **Exploring Best Practices**: Scrutinizing best practices in research supervision from both global and local vantage points to discern strategies that can be adapted to the African context.
3. **Developing Culturally Relevant Frameworks**: Proposing supervision models that fuse African cultural contexts, capitalizing on the concept of Africanisation to cultivate inclusive and efficacious supervision relationships.
4. **Evaluating Mentorship Models**: Assessing a variety of mentorship models and their efficacy in improving research outcomes and academic development.
5. **Providing Policy Recommendations:** Offering policy recommendations to facilitate the implementation of effective research supervision practices in African universities.

The primary objectives of this chapter are to:

1. Enhance the understanding of the importance and impact of research supervision in academic development.
2. Highlight the unique challenges and opportunities within the African higher education context.
3. Propose innovative and culturally relevant frameworks for research supervision.
4. Provide actionable recommendations for policymakers, educators, and institutions to improve research supervision practices.

1.3 Conceptual Framework

The goal of Africanisation is to develop culturally relevant supervision frameworks that incorporate African perspectives, values, and contexts into academic practices and curricula (Le Grange, 2016). This approach challenges the dominance of Western paradigms in academia and seeks to create a more inclusive and representative educational environment.

In the context of research supervision, Africanisation involves recognizing and integrating the cultural, social, and historical contexts of African students into the supervision process (Higgs, 2016). This can enhance the relevance and impact of research by ensuring that it addresses local issues and contributes to the development of African societies. Culturally relevant supervision frameworks can also improve the supervisor-student relationship by fostering mutual understanding and respect (Kaya & Seleti, 2013).

One of the key aspects of Africanisation in research supervision is the inclusion of indigenous knowledge systems and methodologies. Indigenous knowledge systems offer alternative ways of understanding and investigating phenomena, which can enrich the research process and outcomes (Odora Hoppers, 2002). By valuing and incorporating these systems, supervisors can help students develop research projects that are not only academically rigorous but also culturally meaningful and socially impactful.

Moreover, Africanisation can address some of the structural challenges faced by African higher education institutions. For instance, it can promote the development of mentorship models that reflect African communal values and practices, thereby enhancing the support network available to postgraduate students (Mgqwashu, 2016). Additionally, it can foster a sense of identity and belonging among students, which is crucial for their academic motivation and success (Oyewole, 2010).

Implementing Africanisation in research supervision requires a paradigm shift in how supervision is conceptualized and practiced. It calls for a departure from the traditional, hierarchical model of supervision towards a more collaborative and egalitarian approach (Keane, 2016). This shift can empower students to take ownership of their research and contribute actively to the academic community.

2. BACKGROUND AND CONTEXT

Higher education institutions in Africa face unique challenges in research supervision, driven by historical, cultural, and structural factors. This section examines the current state of research supervision in African universities, the obstacles faced by supervisors and students, and the potential benefits of adopting culturally relevant supervision frameworks.

2.1 Higher Education in Africa

Higher education in Africa has undergone significant transformations over the past few decades, marked by rapid expansion, increased enrollment, and a growing emphasis on research and development (Teferra & Altbach, 2004). Despite these advancements, African universities continue to grapple with numerous challenges that impede effective research supervision. These include inadequate funding, limited access to resources, and insufficient training for supervisors (Bunting et al., 2014; Cloete et al., 2015).

The expansion of higher education has led to a surge in the number of postgraduate students, creating a strain on existing supervision capacities (World Bank, 2010). This strain is exacerbated by the brain drain phenomenon, where qualified academics migrate to developed countries in search of better opportunities, leaving behind a shortage of experienced supervisors (Mouton, 2010; Teferra, 2000). As a result, many postgraduate students in Africa face challenges in finding supervisors who can provide the necessary guidance and support for their research endeavors (Adeniyi & Taiwo, 2011).

In addition to these structural issues, the cultural context of African higher education presents unique challenges and opportunities for research supervision. Traditional African educational practices emphasize communal learning and the transmission of knowledge through oral traditions (Adeyemi & Adeyinka, 2003). Integrating these practices into the formal academic setting can enhance the relevance and impact of research, but it also requires a shift in the traditional supervision paradigms (Crossman, 2004).

2.2 Challenges in Research Supervision

The challenges in research supervision in African higher education institutions are multifaceted, encompassing both systemic and interpersonal dimensions. Systemic challenges include the lack of institutional support for research activities, limited access to research funding, and inadequate infrastructure (Sawyerr, 2004; Oanda & Jowi, 2013). These issues are compounded by the fact that many African universities operate under resource-constrained environments, where research is often not prioritized over teaching and administrative duties (Kotecha, Wilson-Strydom, & Fongwa, 2012).

At the interpersonal level, the supervisor-student relationship is critical to the success of the research supervision process. Effective supervision requires a balance of guidance, support, and autonomy, which can be challenging to achieve in contexts where supervisors are overburdened and under-resourced (Lee, 2008; Lessing & Schulze, 2002). Additionally, the hierarchical nature of the academic relationship can sometimes hinder open communication and the development of a collaborative research environment (Kam, 1997; Manathunga, 2007).

Mentorship is another critical aspect of research supervision that is often lacking in African higher education institutions. Effective mentorship involves not only providing academic guidance but also supporting the personal and professional development of students (Crisp & Cruz, 2009; Ehrich, Hansford, & Tennent, 2004). However, many supervisors lack the training and resources to fulfill this role adequately, resulting in gaps in the mentorship provided to postgraduate students (McCallin & Nayar, 2012; Wisker, 2012).

The lack of structured mentorship programs in African universities further exacerbates these challenges. While informal mentorship relationships do exist, they are often ad-hoc and dependent on the individual supervisor's commitment and capacity (Shannon, 1995; Roberts, 2000). This lack of formalized mentorship structures can lead to inconsistencies in the quality of supervision and support that students receive (Heath, 2002; Ives & Rowley, 2005).

2.3 Significance of Culturally Relevant Frameworks

Given the unique cultural and structural challenges faced by African higher education institutions, there is a growing recognition of the need for culturally relevant supervision frameworks. The concept of Africanisation offers a promising approach to addressing these challenges by integrating African perspectives, values, and knowledge systems into the research supervision process (Le Grange, 2016; Higgs, 2016).

Africanisation involves recognizing the importance of indigenous knowledge systems and methodologies, which have historically been marginalized in the academic discourse (Odora Hoppers, 2002; Dei, 2000). By incorporating these systems into the research supervision process, supervisors can help students develop research projects that are not only academically rigorous but also culturally meaningful and socially impactful (Semali & Kincheloe, 1999; Kaya & Seleti, 2013).

Furthermore, culturally relevant supervision frameworks can enhance the supervisor-student relationship by fostering mutual understanding and respect. In traditional African societies, education is viewed as a communal responsibility, and learning is often facilitated through mentorship and apprenticeship (Adeyemi & Adeyinka, 2003; Crossman, 2004). By adopting these principles, supervisors can create a more inclusive and supportive research environment that aligns with the cultural contexts of their students (Mgqwashu, 2016; Oyewole, 2010).

Implementing Africanisation in research supervision also requires a shift in the traditional, hierarchical model of supervision towards a more collaborative and egalitarian approach (Keane, 2016; Grant, 2003). This shift can empower students to take ownership of their research and contribute actively to the academic community, fostering a sense of identity and belonging that is crucial for their academic motivation and success (Manathunga, 2012; Pearson & Brew, 2002).

Moreover, culturally relevant frameworks can address some of the structural challenges faced by African higher education institutions. For instance, they can promote the development of mentorship models that reflect African communal values and practices, thereby enhancing the support network available to postgraduate students (Kotecha, Wilson-Strydom, & Fongwa, 2012; McCallin & Nayar, 2012). Additionally, they can foster greater engagement with local communities and stakeholders, ensuring that research is grounded in the realities and needs of African societies (Sawyerr, 2004; Teferra, 2004).

3. LITERATURE REVIEW

3.1 Theoretical Perspectives on Research Supervision

Research supervision is underpinned by various theoretical frameworks that inform the practices and dynamics between supervisors and students. One prevalent theory is the Apprenticeship Model, which views the supervisor as a master and the student as an apprentice. This model emphasizes skill acquisition through close, hands-on mentorship (Delamont, Parry, & Atkinson, 2000). However, this hierarchical approach can sometimes stifle student autonomy and creativity (Grant, 2003; Lee, 2008).

An alternative is the Cognitive Apprenticeship Model, which incorporates the traditional apprenticeship's practical elements but also emphasizes the development of critical thinking and problem-solving skills (Collins, Brown, & Newman, 1989). This model supports students in becoming independent researchers, capable of navigating complex research landscapes (Pearson & Brew, 2002).

Another influential framework is the Social Constructivist Model, which posits that knowledge is co-constructed through social interactions (Vygotsky, 1978). This perspective aligns well with collaborative supervision approaches, where the supervisor and student engage in dialogue and jointly explore research problems (Lave & Wenger, 1991; Lee, 2012). This model is particularly relevant in culturally diverse settings, as it acknowledges the importance of context and the multiplicity of perspectives in the learning process (McCallin & Nayar, 2012).

The Situated Learning Theory, proposed by Lave and Wenger (1991), further emphasizes learning as a social process that occurs within a community of practice. This theory highlights the importance of engaging students in real-world research communities, fostering their professional identity and integration into the academic community (Wenger, 1998).

3.2 Mentorship and Research Productivity

Mentorship plays a crucial role in enhancing research productivity. Effective mentors provide not only academic guidance but also emotional and career support, which are critical for the holistic development of postgraduate students (Crisp & Cruz, 2009). Studies have shown that mentored students tend to have higher research outputs, greater job satisfaction, and better career progression compared to their non-mentored peers (Paglis, Green, & Bauer, 2006; Johnson, Rose, & Schlosser, 2007).

Mentorship effectiveness is often linked to the quality of the mentor-mentee relationship. High-quality relationships characterized by mutual respect, trust, and open communication can significantly enhance the mentee's academic and professional outcomes (Allen & Eby, 2003; Ragins & Kram, 2007). However, disparities in mentorship experiences can arise due to factors such as gender, race, and cultural background, necessitating the need for inclusive and culturally sensitive mentorship practices (Thomas, 2001; Santos & Reigadas, 2004).

Structured mentorship programs have been shown to be particularly effective in improving research productivity. Such programs provide a formal framework for mentorship, ensuring consistency and support across different departments and faculties (Ehrich, Hansford, & Tennent, 2004). These programs often include mentor training, mentee orientation, and regular assessment of the mentorship process, which help in maintaining high standards and addressing any issues that may arise (McCallin & Nayar, 2012).

3.3 Best Practices in Supervision

Best practices in research supervision encompass a range of strategies designed to support the academic and professional development of postgraduate students. One key practice is the establishment of clear expectations and goals at the outset of the supervision relationship (Gurr, 2001). This includes agreeing on timelines, milestones, and the roles and responsibilities of both the supervisor and the student (Wisker, 2012).

Regular, structured meetings are another critical component of effective supervision. These meetings provide opportunities for the supervisor to monitor progress, provide feedback, and address any challenges the student may be facing (Lessing & Schulze, 2002; De Kleijn et al., 2012). Feedback should be constructive, specific, and aimed at fostering the student's intellectual growth and research skills (Halse & Malfroy, 2010).

Encouraging student independence is also a hallmark of good supervision. This involves providing students with the autonomy to make decisions about their research while offering support and guidance as needed (Pearson & Brew, 2002). Supervisors should aim to develop students' critical thinking, problem-solving, and self-management skills, which are essential for their future careers (Lee, 2008).

Cultural competence is increasingly recognized as an essential skill for supervisors, particularly in diverse and international academic environments (Henze, 2010). Supervisors should be aware of cultural differences and strive to create an inclusive and supportive environment for all students (Montgomery, 2010). This includes being sensitive to different communication styles, learning preferences, and cultural norms (Smith, 2010).

3.4 Cross-Cultural Supervision

Cross-cultural supervision presents unique challenges and opportunities. Supervisors must navigate cultural differences in communication styles, expectations, and academic practices (Huang, 2012). Understanding and respecting these differences is crucial for building a positive and productive supervision relationship (Ryan & Viete, 2009).

Effective cross-cultural supervision involves adopting a flexible and adaptive approach. This includes being open to different ways of thinking and working, and being willing to adjust supervision styles to meet the needs of students from diverse cultural backgrounds (Manathunga, 2007). Building cultural awareness and sensitivity can help supervisors better support their students and create a more inclusive academic environment (Hernandez, 2010).

Language barriers can also pose significant challenges in cross-cultural supervision. Supervisors need to be patient and supportive, providing additional resources and support to help students overcome these barriers (Evans & Green, 2007). Encouraging students to engage with their peers and participate in academic communities can also help them build confidence and improve their language skills (Montgomery, 2010).

Cross-cultural supervision can be enriched by integrating indigenous knowledge systems and methodologies. This approach recognizes the value of diverse perspectives and can enhance the relevance and impact of research (Dei, 2000). Supervisors who are open to incorporating these perspectives into their students' research can help them produce work that is both academically rigorous and culturally meaningful (Odora Hoppers, 2002).

3.5 Strategies and Frameworks for Inclusive Supervision

Inclusive supervision strategies are essential for supporting the diverse needs of postgraduate students. These strategies involve creating a supportive and respectful environment where all students feel valued and included (Johnson-Bailey & Cervero, 2004). Supervisors should be proactive in addressing any issues of bias or discrimination and promoting equity and inclusion within their research groups (Smith, 2010).

One effective framework for inclusive supervision is the Community of Practice model, which emphasizes collaborative learning and mutual support (Wenger, 1998). This model encourages supervisors and students to work together as a community, sharing knowledge and resources to achieve common goals (Lave & Wenger, 1991). By fostering a sense of belonging and collective responsibility, this approach can enhance student engagement and success (Tinto, 1993).

Another strategy is to implement structured mentorship programs that provide additional support for underrepresented groups. These programs can include peer mentoring, professional development workshops, and networking opportunities, which help students build the skills and connections they need to succeed in academia (Nettles & Millett, 2006; Campbell & Campbell, 2007).

Supervisors should also be trained in cultural competence and inclusive teaching practices. This training can help them better understand the needs of their diverse student body and develop strategies to support their academic and personal growth (Henze, 2010). Regular reflection and feedback can help supervisors continuously improve their practice and ensure they are meeting the needs of all students (Montgomery, 2010).

3.6 Comparative Studies of Supervision Approaches

Comparative studies of traditional and modern approaches to research supervision provide valuable insights into their effectiveness. Traditional supervision models, characterized by a hierarchical and directive approach, have been criticized for being rigid and exclusionary (Grant, 2003; Lee, 2008). In contrast, modern supervision models emphasize collaboration, flexibility, and student autonomy, which can lead to better student outcomes and higher levels of satisfaction (Pearson & Brew, 2002; Manathunga, 2007).

Studies have shown that modern supervision approaches, such as the Co-Supervision Model, where students have multiple supervisors, can enhance the quality of supervision by providing a broader range of expertise and support (Mainhard et al., 2009; Pilbeam & Denyer, 2009). This model also reduces the burden on individual supervisors and fosters a more collaborative research environment (Firth & Martens, 2008).

Another modern approach is the Team Supervision Model, which involves a group of supervisors working together to support a cohort of students (Manathunga, 2012). This model promotes interdisciplinary collaboration and provides students with access to a diverse range of perspectives and skills (Agu & Odimegwu, 2014). It also helps to mitigate the risks associated with relying on a single supervisor, such as mismatched expectations or supervisory conflicts (McCallin & Nayar, 2012).

Comparative studies also highlight the importance of adapting supervision models to the specific context and needs of the students. For example, the Personalized Supervision Model tailors the supervision process to the individual student's background, preferences, and research goals (Halse & Malfroy, 2010). This approach can enhance student satisfaction and academic success by providing a more tailored and responsive supervision experience (Ives & Rowley, 2005).

4. MAIN FOCUS OF THE CHAPTER

4.1 Comprehensive Strategies for Research Supervision

The main focus of this chapter is to present comprehensive strategies for effective research supervision within the African higher education context. Effective supervision is crucial for ensuring high-quality research outputs and fostering the academic and professional development of postgraduate students. The strategies discussed here are informed by best practices in supervision, tailored to address the unique challenges faced by African higher education institutions.

4.1.1 Establishing Clear Expectations and Goals

One of the foundational strategies for effective research supervision is the establishment of clear expectations and goals at the outset of the supervision relationship (Gurr, 2001). This involves creating a detailed supervision agreement that outlines the roles and responsibilities of both the supervisor and the student, as well as specific milestones and timelines for the research project (Wisker, 2012). Clear expectations help prevent misunderstandings and ensure that both parties are aligned in their objectives, which is critical for maintaining a productive and harmonious working relationship (Hockey, 1996; Murphy, Bain, & Conrad, 2007).

4.1.2 Regular and Structured Meetings

Regular, structured meetings are essential for effective research supervision. These meetings provide opportunities for supervisors to monitor progress, provide feedback, and address any challenges that the student may be facing (Lessing & Schulze, 2002; De Kleijn et al., 2012). Structured meetings should have a clear agenda, with specific topics to be discussed and action items to be followed up on (Grant, 2003). This approach helps ensure that meetings are productive and focused, and that both the supervisor and the student are clear about the next steps in the research process (Lee, 2008).

4.1.3 Constructive and Specific Feedback

Providing constructive and specific feedback is another critical component of effective research supervision. Feedback should be aimed at fostering the student's intellectual growth and research skills, rather than simply pointing out flaws or deficiencies (Halse & Malfroy, 2010). Constructive feedback helps students understand what they need to improve and how to achieve their goals, which is essential for their development as independent researchers (Pearson & Brew, 2002; Manathunga, 2007).

4.1.4 Encouraging Student Independence

Encouraging student independence is crucial for developing their critical thinking, problem-solving, and self-management skills (Pearson & Brew, 2002). Supervisors should provide students with the autonomy to make decisions about their research while offering support and guidance as needed (Lee, 2008). This approach helps students build confidence in their abilities and prepares them for their future careers as independent researchers (Murphy, Bain, & Conrad, 2007).

4.1.5 Cultural Competence in Supervision

Cultural competence is increasingly recognized as an essential skill for supervisors, particularly in diverse and international academic environments (Henze, 2010). Supervisors should be aware of cultural differences and strive to create an inclusive and supportive environment for all students (Montgomery, 2010). This includes being sensitive to different communication styles, learning preferences, and cultural norms (Smith, 2010). Cultural competence can help supervisors build strong, respectful relationships with their students, which is crucial for effective supervision (Ryan & Viete, 2009).

4.2 Innovative Frameworks for Research Supervision

In addition to these strategies, this chapter proposes innovative frameworks for research supervision that align with culturally relevant practices and leverage the concept of Africanisation. These frameworks aim to create more inclusive and effective supervision relationships, which are critical for addressing the unique challenges faced by African higher education institutions.

4.2.1 Culturally Relevant Supervision Frameworks

Culturally relevant supervision frameworks incorporate African perspectives, values, and knowledge systems into the research supervision process (Le Grange, 2016; Higgs, 2016). These frameworks recognize the importance of indigenous knowledge systems and methodologies, which have historically been marginalized in the academic discourse (Odora Hoppers, 2002; Dei, 2000). By valuing and integrating these systems, supervisors can help students develop research projects that are not only academically rigorous but also culturally meaningful and socially impactful (Semali & Kincheloe, 1999; Kaya & Seleti, 2013).

4.2.2 Integrating Indigenous Knowledge Systems

Integrating indigenous knowledge systems into the research supervision process can enhance the relevance and impact of research (Dei, 2000). Indigenous knowledge systems offer alternative ways of understanding and investigating phenomena, which can enrich the research process and outcomes (Odora Hoppers, 2002). Supervisors who are open to incorporating these perspectives into their students' research can help them produce work that is both academically rigorous and culturally meaningful (Semali & Kincheloe, 1999; Kaya & Seleti, 2013).

4.2.3 Collaborative and Egalitarian Supervision Models

Collaborative and egalitarian supervision models promote a more inclusive and supportive research environment (Manathunga, 2007; Lee, 2008). These models emphasize collaboration and mutual respect between the supervisor and the student, rather than a hierarchical relationship (Lave & Wenger, 1991). By fostering a sense of partnership and shared responsibility, these models can enhance student engagement and motivation, leading to better research outcomes (Pearson & Brew, 2002; Manathunga, 2012).

4.2.4 Mentorship Programs Reflecting African Communal Values

Mentorship programs that reflect African communal values and practices can provide additional support for postgraduate students (Mgqwashu, 2016). These programs emphasize the importance of community and collective responsibility in the learning process, which aligns with traditional African educational practices (Adeyemi & Adeyinka, 2003; Crossman, 2004). By fostering a sense of belonging and mutual support, these programs can enhance student retention and success (Tinto, 1993; Montgomery, 2010).

4.2.5 Professional Development for Supervisors

Providing professional development opportunities for supervisors is essential for enhancing their skills and knowledge in research supervision (Halse & Malfroy, 2010). This training should include best practices in supervision, cultural competence, and emerging trends in research methodologies (Wisker, 2012; Henze, 2010). Regular reflection and feedback can help supervisors continuously improve their practice and ensure they are meeting the needs of their students (Montgomery, 2010).

4.3 Implementing Structured Mentorship Programs

Structured mentorship programs provide a formal framework for mentorship, ensuring consistency and support across different departments and faculties (Ehrich, Hansford, & Tennent, 2004). These programs often include mentor training, mentee orientation, and regular assessment of the mentorship process, which help in maintaining high standards and addressing any issues that may arise (McCallin & Nayar, 2012).

4.3.1 Mentor Training

Effective mentorship programs begin with comprehensive training for mentors. This training should cover topics such as mentoring best practices, communication skills, and cultural competence (Crisp & Cruz, 2009; Johnson, Rose, & Schlosser, 2007). Training helps mentors understand their roles and responsibilities and equips them with the skills needed to support their mentees effectively (McCallin & Nayar, 2012).

4.3.2 Mentee Orientation

Mentee orientation is another critical component of structured mentorship programs. Orientation sessions provide mentees with an overview of the mentorship program, including its goals, expectations, and available resources (Nettles & Millett, 2006). These sessions help mentees understand what they can expect from their mentors and how to make the most of the mentorship relationship (Campbell & Campbell, 2007).

4.3.3 Regular Assessment and Feedback

Regular assessment and feedback are essential for maintaining the effectiveness of mentorship programs. These assessments should evaluate the progress of the mentorship relationship, identify any issues or challenges, and provide opportunities for both mentors and mentees to give feedback (Ehrich, Hansford, & Tennent, 2004). Continuous assessment helps ensure that the program is meeting its objectives and provides insights for future improvements (McCallin & Nayar, 2012).

5. ISSUES, CONTROVERSIES, AND PROBLEMS

5.1 The Gap Between Traditional and Modern Supervision Models

Research supervision in African higher education is often caught between traditional and modern supervision models. Traditional models, characterized by a hierarchical and authoritative relationship between supervisor and student, have been criticized for being rigid and inflexible (Grant, 2003; Kam, 1997). These models can stifle creativity and independence, limiting students' ability to develop critical thinking and problem-solving skills (Manathunga, 2007). In contrast, modern supervision models emphasize collaboration, flexibility, and student autonomy, which align better with contemporary educational philosophies (Lee, 2008; Pearson & Brew, 2002).

However, transitioning from traditional to modern supervision models is challenging. Many supervisors are accustomed to the traditional approach and may resist adopting new methods that require a more collaborative and egalitarian relationship with students (Halse & Malfroy, 2010). Additionally, institutional structures and policies often support traditional models, making it difficult to implement changes at the systemic level (Boud & Lee, 2005; Deuchar, 2008). This gap between traditional and modern supervision models creates tensions and inconsistencies in the supervision process, which can negatively impact student outcomes (Grant, 2003).

5.2 Limited Availability of Structured Mentorship Programs

The limited availability of structured mentorship programs is another significant issue in research supervision in African higher education institutions. While mentorship is recognized as a critical component of effective supervision, many institutions lack formalized programs to support mentorship relationships (Shannon, 1995; Roberts, 2000). This lack of structure can lead to inconsistencies in the quality of mentorship provided, with some students receiving more support than others depending on their supervisor's commitment and capacity (Heath, 2002; Ives & Rowley, 2005).

Structured mentorship programs provide a framework for consistent and high-quality mentorship by establishing clear guidelines, training programs for mentors, and regular assessments to ensure the mentorship process is effective (Crisp & Cruz, 2009; Ehrich, Hansford, & Tennent, 2004). However, implementing such programs requires significant resources and institutional commitment, which are often lacking in African higher education institutions (Sawyerr, 2004; Oanda & Jowi, 2013).

5.3 Supervisor-Student Dynamics and Challenges in Managing Expectations

The dynamics between supervisors and students play a critical role in the success of the supervision process. Effective supervision requires a balance of guidance, support, and autonomy, which can be challenging to achieve (Hockey, 1996; Murphy, Bain, & Conrad, 2007). Supervisors need to provide enough direction to ensure students stay on track with their research while also allowing them the independence to explore their ideas and develop their skills (Pearson & Brew, 2002).

One of the main challenges in managing supervisor-student dynamics is aligning expectations. Supervisors and students often have different expectations regarding the supervision process, which can lead to misunderstandings and conflicts (Grant, 2005; Lee, 2008). For example, students may expect more hands-on guidance and frequent feedback, while supervisors may expect students to work independently and seek feedback only when necessary (Ives & Rowley, 2005). These mismatched expectations can create frustration and hinder the progress of the research project (Manathunga, 2007).

5.4 Resource Constraints

Resource constraints are a pervasive issue affecting research supervision in African higher education institutions. Many universities operate under severe budgetary limitations, which impact their ability to provide adequate support for research activities (Teferra & Altbach, 2004; Cloete et al., 2015). These constraints affect various aspects of the supervision process, including access to research materials, funding for fieldwork, and availability of adequate research facilities (Sawyerr, 2004).

Resource limitations also affect the availability and quality of training programs for supervisors. Without proper training, supervisors may lack the necessary skills and knowledge to effectively guide their students (McCallin & Nayar, 2012; Wisker, 2012). Additionally, high student-to-supervisor ratios, driven by increasing enrollment and limited faculty, further strain the supervision process, leading to overburdened supervisors and insufficient attention to individual students (Bunting et al., 2014).

5.5 Institutional Policies and Support

Institutional policies and support systems play a crucial role in shaping the research supervision landscape. Policies that are not conducive to effective supervision can create barriers to successful research outcomes. For instance, rigid administrative requirements, bureaucratic processes, and a lack of flexibility in academic policies can hinder the supervision process (Halse & Malfroy, 2010; Lee, 2008).

Moreover, institutional support is essential for fostering a positive supervision environment. This includes providing resources for professional development, creating forums for sharing best practices, and offering administrative support to reduce the burden on supervisors (Pearson & Brew, 2002; Manathunga, 2007). Institutions that fail to provide adequate support for research supervision may struggle to maintain high standards of research and academic integrity (Hockey, 1996; Murphy, Bain, & Conrad, 2007).

5.6 Cross-Cultural Challenges

Cross-cultural challenges are significant in the context of research supervision, particularly in African higher education institutions, which often host students from diverse cultural backgrounds. Cultural differences can impact communication styles, expectations, and approaches to learning and research (Ryan & Viete, 2009; Huang, 2012). Supervisors need to be culturally competent and sensitive to these differences to effectively support their students (Hernandez, 2010; Henze, 2010).

Cultural misunderstandings can lead to conflicts and miscommunications that hinder the supervision process (Evans & Green, 2007; Manathunga, 2007). For instance, students from collectivist cultures may expect a more directive and involved approach from their supervisors, while supervisors from individualist cultures may emphasize independence and self-direction (Montgomery, 2010; Smith, 2010). Addressing these cross-cultural challenges requires supervisors to be adaptable and open to different perspectives and approaches (Dei, 2000; Odora Hoppers, 2002).

6. SOLUTIONS AND RECOMMENDATIONS

6.1 Developing Culturally Relevant Supervision Frameworks

One of the primary solutions to the challenges faced in research supervision in African higher education is the development of culturally relevant supervision frameworks. These frameworks should integrate African perspectives, values, and knowledge systems to create a more inclusive and supportive research environment (Le Grange, 2016; Higgs, 2016). By recognizing and incorporating indigenous knowledge systems, supervisors can guide students in producing research that is both academically rigorous and culturally meaningful (Odora Hoppers, 2002; Dei, 2000).

6.1.1 Integrating Indigenous Knowledge Systems

Incorporating indigenous knowledge systems into the research supervision process can enrich the research and ensure its relevance to local contexts. This approach involves acknowledging the value of traditional knowledge and methodologies and integrating them with modern academic practices (Semali & Kincheloe, 1999; Kaya & Seleti, 2013). Supervisors should encourage students to explore indigenous knowledge systems and consider them in their research design and methodology, thus bridging the gap between traditional and contemporary knowledge (Odora Hoppers, 2002; Dei, 2000).

6.1.2 Promoting Collaborative and Egalitarian Supervision Models

Promoting collaborative and egalitarian supervision models can help address the hierarchical nature of traditional supervision. These models emphasize mutual respect, shared responsibility, and open communication between supervisors and students (Manathunga, 2007; Lee, 2008). By fostering a partnership approach, supervisors can create a more supportive and empowering research environment, which can enhance student engagement and motivation (Pearson & Brew, 2002; Manathunga, 2012).

6.2 Implementing Structured Mentorship Programs

Structured mentorship programs are essential for providing consistent and high-quality mentorship across different departments and faculties (Ehrich, Hansford, & Tennent, 2004). These programs should include comprehensive training for mentors, orientation sessions for mentees, and regular assessments to ensure the effectiveness of the mentorship process (Crisp & Cruz, 2009; McCallin & Nayar, 2012).

6.2.1 Comprehensive Mentor Training

Effective mentorship programs begin with comprehensive training for mentors. This training should cover mentoring best practices, communication skills, and cultural competence (Johnson, Rose, & Schlosser, 2007). By equipping mentors with the necessary skills and knowledge, institutions can ensure that mentors are well-prepared to support their mentees effectively (McCallin & Nayar, 2012).

6.2.2 Mentee Orientation

Mentee orientation is another critical component of structured mentorship programs. Orientation sessions provide mentees with an overview of the mentorship program, including its goals, expectations, and available resources (Nettles & Millett, 2006). These sessions help mentees understand what they can expect from their mentors and how to make the most of the mentorship relationship (Campbell & Campbell, 2007).

6.2.3 Regular Assessment and Feedback

Regular assessment and feedback are essential for maintaining the effectiveness of mentorship programs. These assessments should evaluate the progress of the mentorship relationship, identify any issues or challenges, and provide opportunities for both mentors and mentees to give feedback (Ehrich, Hansford, & Tennent, 2004). Continuous assessment helps ensure that the program is meeting its objectives and provides insights for future improvements (McCallin & Nayar, 2012).

6.3 Providing Training for Supervisors on Best Practices

Providing training for supervisors on best practices and emerging trends in research supervision is crucial for enhancing their skills and knowledge (Halse & Malfroy, 2010). This training should include topics such as effective supervision strategies, cultural competence, and the integration of indigenous knowledge systems (Wisker, 2012; Henze, 2010).

6.3.1 Professional Development Workshops

Professional development workshops can provide supervisors with the opportunity to learn about best practices in supervision and to share their experiences with colleagues (Pearson & Brew, 2002). These workshops can cover a range of topics, including effective feedback, student autonomy, and managing supervisor-student dynamics (Lee, 2008; Halse & Malfroy, 2010). By participating in these workshops, supervisors can continuously improve their practice and ensure they are meeting the needs of their students (Wisker, 2012).

6.3.2 Cultural Competence Training

Cultural competence training is essential for supervisors working in diverse and international academic environments (Henze, 2010). This training can help supervisors understand and appreciate cultural differences, and develop strategies for supporting students from diverse backgrounds (Montgomery, 2010). Cultural competence training can also help supervisors create a more inclusive and supportive research environment, which is critical for the success of all students (Ryan & Viete, 2009).

6.4 Enhancing Research Infrastructure

Enhancing research infrastructure is crucial for addressing the resource constraints that impact research supervision in African higher education institutions (Teferra & Altbach, 2004). This includes providing access to research materials, funding for fieldwork, and adequate research facilities (Sawyerr, 2004).

6.4.1 Improving Access to Research Resources

Improving access to research resources is essential for supporting high-quality research. This can involve increasing funding for research materials, improving library resources, and providing access to online databases and journals (Teferra & Altbach, 2004). By ensuring that students have the necessary resources, institutions can support their academic development and enhance the quality of their research (Sawyerr, 2004).

6.4.2 Funding for Research Activities

Securing funding for research activities is critical for enabling students to conduct their research effectively (Cloete et al., 2015). This includes funding for fieldwork, laboratory work, and other research-related expenses (Teferra & Altbach, 2004). Institutions should advocate for increased research funding from government and private sources and explore alternative funding mechanisms, such as partnerships with industry and international organizations (Sawyerr, 2004).

6.4.3 Developing Research Facilities

Developing adequate research facilities is essential for supporting high-quality research. This includes providing well-equipped laboratories, study spaces, and access to specialized equipment (Teferra & Altbach, 2004). By investing in research infrastructure, institutions can create an environment that supports innovation and excellence in research (Cloete et al., 2015).

6.5 Strengthening Institutional Policies and Support

Strengthening institutional policies and support systems is crucial for creating a conducive environment for research supervision (Halse & Malfroy, 2010). This includes developing policies that support flexible and effective supervision practices, providing administrative support for supervisors, and fostering a culture of collaboration and continuous improvement (Pearson & Brew, 2002).

6.5.1 Developing Supportive Policies

Developing supportive policies is essential for enabling effective research supervision. This includes policies that recognize and reward good supervision practices, provide clear guidelines for supervision, and offer flexibility to accommodate the diverse needs of students (Halse & Malfroy, 2010). By creating a policy environment that supports effective supervision, institutions can enhance the quality of their research outputs (Pearson & Brew, 2002).

6.5.2 Providing Administrative Support

Providing administrative support for supervisors is crucial for reducing their workload and allowing them to focus on providing high-quality supervision (McCallin & Nayar, 2012). This can include support for managing paperwork, organizing meetings, and accessing resources (Wisker, 2012). By reducing administrative burdens, institutions can help supervisors devote more time and energy to supporting their students (Halse & Malfroy, 2010).

6.5.3 Fostering a Culture of Collaboration

Fostering a culture of collaboration and continuous improvement is essential for enhancing research supervision (Pearson & Brew, 2002). This includes creating forums for supervisors to share best practices, providing opportunities for professional development, and encouraging collaboration between departments and faculties (Manathunga, 2012). By promoting a collaborative and supportive culture, institutions can enhance the quality of research supervision and support the academic and professional development of their postgraduate students (Lee, 2008).

7. FUTURE RESEARCH DIRECTIONS

7.1 Assessing the Impact of Structured Mentorship Programs on Research Productivity

Future research should focus on assessing the impact of structured mentorship programs on research productivity in African higher education institutions. While existing literature emphasizes the importance of mentorship, there is a need for empirical studies that evaluate the effectiveness of structured mentorship programs specifically within the African context (Crisp & Cruz, 2009; Johnson, Rose, & Schlosser, 2007). Such studies could examine various aspects of mentorship programs, including mentor training, mentee orientation, and the frequency and quality of mentor-mentee interactions, to determine their impact on research outputs, student satisfaction, and academic success (Ehrich, Hansford, & Tennent, 2004).

7.1.1 Longitudinal Studies on Mentorship Outcomes

Longitudinal studies are particularly valuable for understanding the long-term effects of structured mentorship programs on research productivity. These studies can track the progress of postgraduate students over time, assessing how mentorship influences their research skills, publication rates, and career trajectories (Paglis, Green, & Bauer, 2006; Nettles & Millett, 2006). By providing a comprehensive view of mentorship outcomes, longitudinal research can offer insights into the sustainability and effectiveness of mentorship programs in African universities (Campbell & Campbell, 2007).

7.1.2 Comparative Studies Across Institutions

Comparative studies across different institutions can highlight best practices and identify the factors that contribute to successful mentorship programs. By comparing the mentorship practices of institutions with high research productivity to those with lower productivity, researchers can identify the key elements that make mentorship programs effective (Ehrich, Hansford, & Tennent, 2004). These findings can inform the development of mentorship programs that are tailored to the unique needs and contexts of African higher education institutions (Crisp & Cruz, 2009).

7.2 Exploring Cross-Cultural Supervision Practices and Their Influence on Research Quality

Another critical area for future research is exploring cross-cultural supervision practices and their influence on research quality. With the increasing internationalization of higher education, understanding how cultural differences impact the supervision process is essential for fostering inclusive and effective research environments (Ryan & Viete, 2009; Huang, 2012). Research in this area can examine how supervisors and students from diverse cultural backgrounds navigate communication, expectations, and research methodologies, and how these factors influence the quality of research outcomes (Henze, 2010; Hernandez, 2010).

7.2.1 Cross-Cultural Competence in Supervision

Future research should investigate the role of cross-cultural competence in effective supervision. This includes examining the skills and strategies that supervisors use to address cultural differences and support students from diverse backgrounds (Montgomery, 2010; Smith, 2010). Studies can also explore the impact of cultural competence training on supervisors' ability to create inclusive and supportive research environments (Henze, 2010). By identifying the components of effective cross-cultural supervision, researchers can provide recommendations for training programs and institutional policies that promote cultural competence (Ryan & Viete, 2009).

7.2.2 Student Perspectives on Cross-Cultural Supervision

Understanding the perspectives of students on cross-cultural supervision is crucial for improving supervision practices. Future research should include qualitative studies that explore the experiences of postgraduate students from diverse cultural backgrounds, focusing on their interactions with supervisors and the challenges they face (Evans & Green, 2007; Manathunga, 2007). These studies can provide valuable insights into how students perceive the support they receive and what improvements can be made to enhance their research experience (Montgomery, 2010).

7.3 Evaluating Innovative Supervision Frameworks in Different Disciplines

Evaluating the effectiveness of innovative supervision frameworks across different disciplines is another important area for future research. Research supervision is not a one-size-fits-all process, and the needs and challenges of students can vary significantly across disciplines (Lee, 2008; Pearson & Brew, 2002). Future studies should examine how different supervision models, such as co-supervision, team supervision, and personalized supervision, impact research outcomes in various fields (Manathunga, 2012; McCallin & Nayar, 2012).

7.3.1 Discipline-Specific Supervision Strategies

Discipline-specific supervision strategies are essential for addressing the unique needs of students in different academic fields. Future research should explore how supervision practices can be tailored to the specific requirements of disciplines such as the sciences, humanities, social sciences, and professional fields (Lee, 2008). By identifying best practices within each discipline, researchers can develop guidelines for effective supervision that are relevant and applicable to various academic contexts (Pearson & Brew, 2002).

7.3.2 Impact of Supervision Models on Research Quality

Empirical studies that evaluate the impact of different supervision models on research quality are needed to determine the most effective approaches for supporting postgraduate students. These studies can compare traditional, co-supervision, and team supervision models, assessing their influence on various indicators of research quality, such as publication rates, thesis completion times, and student satisfaction (Mainhard et al., 2009; Pilbeam & Denyer, 2009). By providing evidence-based insights, this research can inform the design and implementation of supervision models that enhance research outcomes (Manathunga, 2012).

7.4 Investigating the Role of Institutional Policies in Supporting Effective Supervision

Institutional policies play a critical role in shaping the research supervision landscape. Future research should investigate how different policies impact the effectiveness of supervision practices and the quality of research outputs (Halse & Malfroy, 2010; Pearson & Brew, 2002). This includes examining policies related to supervisor training, mentorship programs, resource allocation, and administrative support (Lee, 2008).

7.4.1 Policy Analysis and Impact Assessment

Policy analysis and impact assessment studies can provide insights into the effectiveness of institutional policies in supporting research supervision. These studies can evaluate the implementation and outcomes of specific policies, identifying best practices and areas for improvement (Halse & Malfroy, 2010). By understanding the impact of policies on supervision practices, researchers can offer recommendations for policy changes that enhance the support for supervisors and students (Pearson & Brew, 2002).

7.4.2 Institutional Support Mechanisms

Research should also focus on the support mechanisms that institutions provide to facilitate effective supervision. This includes examining the role of administrative support, professional development opportunities, and resource allocation in creating a conducive environment for research supervision (McCallin & Nayar, 2012). By identifying the key elements of institutional support, future studies can inform the development of comprehensive support systems that enhance the quality of research supervision (Wisker, 2012).

8. SOLUTIONS AND RECOMMENDATIONS

8.1 Developing Culturally Relevant Supervision Frameworks

One of the primary solutions to the challenges faced in research supervision in African higher education is the development of culturally relevant supervision frameworks. These frameworks should integrate African perspectives, values, and knowledge systems to create a more inclusive and supportive research environment (Le Grange, 2016; Higgs, 2016). By recognizing and incorporating indigenous knowledge systems, supervisors can guide students in producing research that is both academically rigorous and culturally meaningful (Odora Hoppers, 2002; Dei, 2000).

8.1.1 Integrating Indigenous Knowledge Systems

Incorporating indigenous knowledge systems into the research supervision process can enrich the research and ensure its relevance to local contexts. This approach involves acknowledging the value of traditional knowledge and methodologies and integrating them with modern academic practices (Semali & Kincheloe, 1999; Kaya & Seleti, 2013). Supervisors should encourage students to explore indigenous knowledge systems and consider them in their research design and methodology, thus bridging the gap between traditional and contemporary knowledge (Odora Hoppers, 2002; Dei, 2000).

8.1.2 Promoting Collaborative and Egalitarian Supervision Models

Promoting collaborative and egalitarian supervision models can help address the hierarchical nature of traditional supervision. These models emphasize mutual respect, shared responsibility, and open communication between supervisors and students (Manathunga, 2007; Lee, 2008). By fostering a partnership approach, supervisors can create a more supportive and empowering research environment, which can enhance student engagement and motivation (Pearson & Brew, 2002; Manathunga, 2012).

8.2 Implementing Structured Mentorship Programs

Structured mentorship programs are essential for providing consistent and high-quality mentorship across different departments and faculties (Ehrich, Hansford, & Tennent, 2004). These programs should include comprehensive training for mentors, orientation sessions for mentees, and regular assessments to ensure the effectiveness of the mentorship process (Crisp & Cruz, 2009; McCallin & Nayar, 2012).

8.2.1 Comprehensive Mentor Training

Effective mentorship programs begin with comprehensive training for mentors. This training should cover mentoring best practices, communication skills, and cultural competence (Johnson, Rose, & Schlosser, 2007). By equipping mentors with the necessary skills and knowledge, institutions can ensure that mentors are well-prepared to support their mentees effectively (McCallin & Nayar, 2012).

8.2.2 Mentee Orientation

Mentee orientation is another critical component of structured mentorship programs. Orientation sessions provide mentees with an overview of the mentorship program, including its goals, expectations, and available resources (Nettles & Millett, 2006). These sessions help mentees understand what they can expect from their mentors and how to make the most of the mentorship relationship (Campbell & Campbell, 2007).

8.2.3 Regular Assessment and Feedback

Regular assessment and feedback are essential for maintaining the effectiveness of mentorship programs. These assessments should evaluate the progress of the mentorship relationship, identify any issues or challenges, and provide opportunities for both mentors and mentees to give feedback (Ehrich, Hansford, & Tennent, 2004). Continuous assessment helps ensure that the program is meeting its objectives and provides insights for future improvements (McCallin & Nayar, 2012).

8.3 Providing Training for Supervisors on Best Practices

Providing training for supervisors on best practices and emerging trends in research supervision is crucial for enhancing their skills and knowledge (Halse & Malfroy, 2010). This training should include topics such as effective supervision strategies, cultural competence, and the integration of indigenous knowledge systems (Wisker, 2012; Henze, 2010).

8.3.1 Professional Development Workshops

Professional development workshops can provide supervisors with the opportunity to learn about best practices in supervision and to share their experiences with colleagues (Pearson & Brew, 2002). These workshops can cover a range of topics, including effective feedback, student autonomy, and managing supervisor-student dynamics (Lee, 2008; Halse & Malfroy, 2010). By participating in these workshops, supervisors can continuously improve their practice and ensure they are meeting the needs of their students (Wisker, 2012).

8.3.2 Cultural Competence Training

Cultural competence training is essential for supervisors working in diverse and international academic environments (Henze, 2010). This training can help supervisors understand and appreciate cultural differences, and develop strategies for supporting students from diverse backgrounds (Montgomery, 2010). Cultural competence training can also help supervisors create a more inclusive and supportive research environment, which is critical for the success of all students (Ryan & Viete, 2009).

8.4 Enhancing Research Infrastructure

Enhancing research infrastructure is crucial for addressing the resource constraints that impact research supervision in African higher education institutions (Teferra & Altbach, 2004). This includes providing access to research materials, funding for fieldwork, and adequate research facilities (Sawyerr, 2004).

8.4.1 Improving Access to Research Resources

Improving access to research resources is essential for supporting high-quality research. This can involve increasing funding for research materials, improving library resources, and providing access to online databases and journals (Teferra & Altbach, 2004). By ensuring that students have the necessary resources, institutions can support their academic development and enhance the quality of their research (Sawyerr, 2004).

8.4.2 Funding for Research Activities

Securing funding for research activities is critical for enabling students to conduct their research effectively (Cloete et al., 2015). This includes funding for fieldwork, laboratory work, and other research-related expenses (Teferra & Altbach, 2004). Institutions should advocate for increased research funding from government and private sources and explore alternative funding mechanisms, such as partnerships with industry and international organizations (Sawyerr, 2004).

8.4.3 Developing Research Facilities

Developing adequate research facilities is essential for supporting high-quality research. This includes providing well-equipped laboratories, study spaces, and access to specialized equipment (Teferra & Altbach, 2004). By investing in research infrastructure, institutions can create an environment that supports innovation and excellence in research (Cloete et al., 2015).

8.5 Strengthening Institutional Policies and Support

Strengthening institutional policies and support systems is crucial for creating a conducive environment for research supervision (Halse & Malfroy, 2010). This includes developing policies that support flexible and effective supervision practices, providing administrative support for supervisors, and fostering a culture of collaboration and continuous improvement (Pearson & Brew, 2002).

8.5.1 Developing Supportive Policies

Developing supportive policies is essential for enabling effective research supervision. This includes policies that recognize and reward good supervision practices, provide clear guidelines for supervision, and offer flexibility to accommodate the diverse needs of students (Halse & Malfroy, 2010). By creating a policy environment that supports effective supervision, institutions can enhance the quality of their research outputs (Pearson & Brew, 2002).

8.5.2 Providing Administrative Support

Providing administrative support for supervisors is crucial for reducing their workload and allowing them to focus on providing high-quality supervision (McCallin & Nayar, 2012). This can include support for managing paperwork, organizing meetings, and accessing resources (Wisker, 2012). By reducing administrative burdens, institutions can help supervisors devote more time and energy to supporting their students (Halse & Malfroy, 2010).

8.5.3 Fostering a Culture of Collaboration

Fostering a culture of collaboration and continuous improvement is essential for enhancing research supervision (Pearson & Brew, 2002). This includes creating forums for supervisors to share best practices, providing opportunities for professional development, and encouraging collaboration between departments and faculties (Manathunga, 2012). By promoting a collaborative and supportive culture, institutions can enhance the quality of research supervision and support the academic and professional development of their postgraduate students (Lee, 2008).

9. CONCLUSION

This chapter has explored the multifaceted challenges and transformative potential of research supervision within African higher education. It has provided comprehensive insights into effective supervision strategies, mentoring frameworks, and innovative methodologies aimed at fostering high-quality research proposals and dissertations among postgraduate students.

Research supervision is critical for academic development, influencing research productivity, student satisfaction, and career progression (Manathunga, 2012; Lee, 2008). The literature reviewed in this chapter highlights the importance of adopting a range of supervision models to address the diverse needs of postgraduate students and the unique challenges faced by African higher education institutions (Pearson & Brew, 2002; McCallin & Nayar, 2012).

The background and context section identified systemic and interpersonal challenges in research supervision, including inadequate funding, limited access to resources, high student-to-supervisor ratios, and cultural differences (Teferra & Altbach, 2004; Sawyerr, 2004). These challenges necessitate the development of culturally relevant supervision frameworks that incorporate African perspectives and values (Le Grange, 2016; Higgs, 2016).

The literature review section emphasized the importance of mentorship in enhancing research productivity and student development. Structured mentorship programs, cultural competence, and inclusive supervision practices were identified as critical components for effective supervision (Crisp & Cruz, 2009; Henze, 2010).

The main focus of the chapter presented comprehensive strategies and innovative frameworks for research supervision, including the integration of indigenous knowledge systems, collaborative supervision models, and structured mentorship programs (Odora Hoppers, 2002; Dei, 2000).

Issues and controversies discussed included the gap between traditional and modern supervision models, the limited availability of structured mentorship programs, supervisor-student dynamics, resource constraints, and cross-cultural challenges (Grant, 2003; Manathunga, 2007).

Solutions and recommendations proposed the development of culturally relevant supervision frameworks, implementing structured mentorship programs, providing training for supervisors, enhancing research infrastructure, and strengthening institutional policies and support (Halse & Malfroy, 2010; Pearson & Brew, 2002).

Future research directions emphasized the need for empirical studies on structured mentorship programs, cross-cultural supervision practices, innovative supervision frameworks, and the role of institutional policies in supporting effective supervision (Crisp & Cruz, 2009; Ryan & Viete, 2009).

The findings of this chapter have significant implications for practice in African higher education. Developing culturally relevant supervision frameworks that integrate indigenous knowledge systems and values can enhance the relevance and impact of research. Supervisors should be trained in cultural competence to effectively support students from diverse backgrounds and create inclusive research environments (Henze, 2010; Hernandez, 2010).

Structured mentorship programs should be implemented across institutions to provide consistent and high-quality support for postgraduate students. These programs should include comprehensive mentor training, mentee orientation, and regular assessments to ensure their effectiveness (Crisp & Cruz, 2009; Ehrich, Hansford, & Tennent, 2004).

Enhancing research infrastructure is essential for addressing resource constraints and supporting high-quality research. Institutions should advocate for increased research funding and develop adequate research facilities to create an environment that supports innovation and excellence in research (Teferra & Altbach, 2004; Cloete et al., 2015).

Strengthening institutional policies and support systems is crucial for creating a conducive environment for research supervision. Policies should recognize and reward good supervision practices, provide clear guidelines for supervision, and offer flexibility to accommodate the diverse needs of students (Halse & Malfroy, 2010; Pearson & Brew, 2002).

Research supervision in African higher education is a complex and multifaceted process that requires a holistic and culturally sensitive approach. The development of effective supervision strategies and frameworks is critical for enhancing the quality of research outputs and supporting the academic and professional development of postgraduate students.

By adopting culturally relevant supervision frameworks, implementing structured mentorship programs, providing training for supervisors, enhancing research infrastructure, and strengthening institutional policies, African higher education institutions can address the challenges they face and foster a more inclusive and supportive research environment.

Future research should continue to explore the impact of these strategies and frameworks on research supervision, providing valuable insights and evidence-based recommendations that can inform policy and practice. By doing so, we can ensure that African higher education institutions are equipped to nurture the next generation of scholars and contribute to the development and prosperity of African societies.

REFERENCES

Adeniyi, E. O., & Taiwo, C. O. (2011). Research capacity building in Nigerian universities: Issues and challenges. *Journal of Higher Education in Africa*, 9(2), 115–134.

Adeyemi, M. B., & Adeyinka, A. A. (2003). The principles and content of African traditional education. *Educational Philosophy and Theory*, 35(4), 425–440. DOI: 10.1111/1469-5812.00039

Allen, T. D., & Eby, L. T. (2003). Relationship effectiveness for mentors: Factors associated with learning and quality. *Journal of Management*, 29(4), 469–486. DOI: 10.1016/S0149-2063(03)00021-7

Boud, D., & Lee, A. (2005). 'Peer learning' as pedagogic discourse for research education. *Studies in Higher Education*, 30(5), 501–516. DOI: 10.1080/03075070500249138

Bunting, I., Cloete, N., & Van Schalkwyk, F. (2014). *An empirical overview of eight flagship universities in Africa: 2001-209*. African Minds.

Campbell, T. A., & Campbell, D. E. (2007). Outcomes of mentoring at-risk college students: Gender and ethnic matching effects. *Mentoring & Tutoring*, 15(2), 135–148. DOI: 10.1080/13611260601086287

Cloete, N., Bailey, T., & Maassen, P. (2015). *Knowledge production and contradictory functions in African higher education*. African Minds. DOI: 10.47622/978-1-920677-85-5

. Collins, A., Brown, J. S., & Newman, S. E. (1989). Cognitive apprenticeship: Teaching the crafts of reading, writing, and mathematics. Knowing, learning, and instruction: Essays in honor of Robert Glaser, 18, 32-42.

Crisp, G., & Cruz, I. (2009). Mentoring college students: A critical review of the literature between 1990 and 2007. *Research in Higher Education*, 50(6), 525–545. DOI: 10.1007/s11162-009-9130-2

Crossman, P. (2004). Knowledge society and higher education: The case of South Africa. *Comparative Education*, 40(4), 525–539.

Dei, G. J. S. (2000). Rethinking the role of indigenous knowledges in the academy. *International Journal of Inclusive Education*, 4(2), 111–132. DOI: 10.1080/136031100284849

Delamont, S., Atkinson, P., & Parry, O. (2000). *The doctoral experience: Success and failure in graduate school*. Routledge.

Deuchar, R. (2008). Facilitator, director or critical friend?: Contradiction and congruence in doctoral supervision styles. *Teaching in Higher Education*, 13(4), 489–500. DOI: 10.1080/13562510802193905

Ehrich, L. C., Hansford, B., & Tennent, L. (2004). Formal mentoring programs in education and other professions: A review of the literature. *Educational Administration Quarterly*, 40(4), 518–540. DOI: 10.1177/0013161X04267118

Evans, C., & Green, C. (2007). Why did I call you a genius? Or whose ethics is it anyway? Moving beyond a dualistic approach to debates about ethical issues in supervision. *International Journal of Qualitative Studies in Education : QSE*, 20(6), 723–729.

Firth, A., & Martens, E. (2008). Transforming supervisors? A critique of post-liberal approaches to research supervision. *Teaching in Higher Education*, 13(3), 279–289. DOI: 10.1080/13562510802045303

Grant, B. (2003). Mapping the pleasures and risks of supervision. *Discourse (Abingdon)*, 24(2), 175–190. DOI: 10.1080/01596300303042

Gurr, G. M. (2001). Negotiating the 'rackety bridge'—A dynamic model for aligning supervisory style with research student development. *Higher Education Research & Development*, 20(1), 81–92. DOI: 10.1080/07924360120043882

Halse, C., & Malfroy, J. (2010). Retheorizing doctoral supervision as professional work. *Studies in Higher Education*, 35(1), 79–92. DOI: 10.1080/03075070902906798

Heath, T. (2002). A quantitative analysis of PhD students' views of supervision. *Higher Education Research & Development*, 21(1), 41–53. DOI: 10.1080/07294360220124648

Henze, R. (2010). Cultural diversity and its impact on global university administration: The case of leadership and staff development. *International Journal of Educational Management*, 24(6), 445–455.

Hernandez, M. (2010). Promoting inclusion and diversity in the academic workplace: A pilot intervention. *Equality, Diversity and Inclusion*, 29(3), 321–334.

Higgs, P. (2016). African philosophy and the decolonisation of education in Africa: Some critical reflections. *Educational Philosophy and Theory*, 48(4), 336–346.

Hockey, J. (1996). Strategies and tactics in the supervision of UK social science PhD students. *International Journal of Qualitative Studies in Education : QSE*, 9(4), 481–500. DOI: 10.1080/0951839960090409

Huang, R. (2012). International experience and graduate employability: Perceptions of Chinese international students in the UK. *Journal of Studies in International Education*, 16(3), 355–368.

Ives, G., & Rowley, G. (2005). Supervisor selection or allocation and continuity of supervision: PhD. students' progress and outcomes. *Studies in Higher Education*, 30(5), 535–555. DOI: 10.1080/03075070500249161

Johnson, B. W., Rose, G., & Schlosser, L. Z. (2007). Student–faculty mentoring: Theoretical and methodological issues. The Blackwell handbook of mentoring: A multiple perspectives approach, 49-68.

Johnson-Bailey, J., & Cervero, R. M. (2004). Mentoring in black and white: The intricacies of cross-cultural mentoring. *Mentoring & Tutoring*, 12(1), 7–21. DOI: 10.1080/1361126042000183075

Kam, B. H. (1997). Style and quality in research supervision: The supervisor dependency factor. *Higher Education*, 34(1), 81–103. DOI: 10.1023/A:1002946922952

Kaya, H. O., & Seleti, Y. N. (2013). African indigenous knowledge systems and relevance of higher education in South Africa. *International Education Journal: Comparative Perspectives*, 12(1), 30–44.

Keane, M. (2016). Supervising PhD students: A practical guide and toolkit. *Innovations in Education and Teaching International*, 53(4), 496–498.

. Kotecha, P., Wilson-Strydom, M., & Fongwa, S. (2012). A profile of higher education in Southern Africa Volume 2: National perspectives. SARUA.

Lave, J., & Wenger, E. (1991). *Situated learning: Legitimate peripheral participation*. Cambridge University Press. DOI: 10.1017/CBO9780511815355

Lee, A. (2008). How are doctoral students supervised? Concepts of doctoral research supervision. *Studies in Higher Education*, 33(3), 267–281. DOI: 10.1080/03075070802049202

Lee, A. (2012). *Successful research supervision: Advising students doing research*. Routledge. DOI: 10.4324/9780203816844

Lessing, A. C., & Schulze, S. (2002). Postgraduate supervision and academic support: Students' perceptions. *South African Journal of Higher Education*, 16(2), 139–149. DOI: 10.4314/sajhe.v16i2.25253

Mainhard, T., van der Rijst, R. M., van Tartwijk, J., & Wubbels, T. (2009). A model for the supervisor–doctoral student relationship. *Higher Education*, 58(3), 359–373. DOI: 10.1007/s10734-009-9199-8

Manathunga, C. (2007). Supervision as mentoring: The role of power and boundary crossing. *Studies in Continuing Education*, 29(2), 207–221. DOI: 10.1080/01580370701424650

Manathunga, C. (2012). Supervisors watching supervisors: The deconstructive possibilities and tensions of team supervision. *Australian Universities Review*, 54(1), 29–37.

McCallin, A., & Nayar, S. (2012). Postgraduate research supervision: A critical review of current practice. *Teaching in Higher Education*, 17(1), 63–74. DOI: 10.1080/13562517.2011.590979

Mgqwashu, E. M. (2016). On developing academic literacy through decolonised curricula. *Journal of Education*, 67, 141–158.

Montgomery, C. (2010). *Understanding the international student experience*. Palgrave Macmillan. DOI: 10.1007/978-0-230-36500-1

Morris, V., & Laipple, J. (2015). How to approach supervising your international postgraduate student. *Innovations in Education and Teaching International*, 52(2), 137–147.

. Mouton, J. (2010). The state of social science in sub-Saharan Africa. World Social Science Report, 51-54.

Murphy, N., Bain, J. D., & Conrad, L. (2007). Orientations to research higher degree supervision. *Higher Education*, 53(2), 209–234. DOI: 10.1007/s10734-005-5608-9

Nettles, M. T., & Millett, C. M. (2006). *Three magic letters: Getting to Ph.D.* Johns Hopkins University Press. DOI: 10.56021/9780801882326

Oanda, I. O., & Jowi, J. O. (2013). University expansion and the challenges to social development in Kenya: Dilemmas and pitfalls. *Journal of Higher Education in Africa*, 11(1-2), 49–71.

Odora Hoppers, C. A. (2002). *Indigenous knowledge and the integration of knowledge systems: Towards a conceptual and methodological framework*. New Africa Books.

Oyewole, O. (2010). Africa's public service delivery & performance review. African Higher Education in the 21st Century: Quality. *Relevance and Leadership in a Time of Global Change*, 3(4), 65–89.

Paglis, L. L., Green, S. G., & Bauer, T. N. (2006). Does adviser mentoring add value? A longitudinal study of mentoring and doctoral student outcomes. *Research in Higher Education*, 47(4), 451–476. DOI: 10.1007/s11162-005-9003-2

Pearson, M., & Brew, A. (2002). Research training and supervision development. *Studies in Higher Education*, 27(2), 135–150. DOI: 10.1080/03075070220119986c

Pilbeam, C., & Denyer, D. (2009). Lone scholar or community member? The role of student networks in doctoral education in a UK management school. *Studies in Higher Education*, 34(3), 301–318. DOI: 10.1080/03075070802597077

Ragins, B. R., & Kram, K. E. (2007). The roots and meaning of mentoring. The handbook of mentoring at work: Theory, research, and practice, 3-15.

Roberts, A. (2000). Mentoring revisited: A phenomenological reading of the literature. *Mentoring & Tutoring*, 8(2), 145–170. DOI: 10.1080/713685524

Ryan, J., & Viete, R. (2009). Respectful interactions: Learning with international students in the English-speaking academy. *Teaching in Higher Education*, 14(3), 303–314. DOI: 10.1080/13562510902898866

Santos, S. J., & Reigadas, E. T. (2004). Understanding the student-faculty mentoring process: Its effects on at-risk university students. *Journal of College Student Retention*, 6(3), 337–357. DOI: 10.2190/KGVC-7218-DPER-RMBC

Sawyerr, A. (2004). African universities and the challenge of research capacity development. *Journal of Higher Education in Africa*, 2(1), 211–240. DOI: 10.57054/jhea.v2i1.1687

Semali, L., & Kincheloe, J. L. (1999). *What is indigenous knowledge?: Voices from the academy*. Falmer Press.

Shannon, A. G. (1995). Research degree supervision: "More mentor than master.". *Australian Universities Review*, 38(2), 12–15.

Smith, L. T. (2010). *Decolonizing methodologies: Research and indigenous peoples*. Zed Books.

Teferra, D. (2000). Brain circulation: Unparalleled opportunities, underlying challenges, and outmoded presumptions. *International Journal of African Higher Education*, 3(1), 1–16.

Teferra, D. (2004). Higher education in Sub-Saharan Africa. International Handbook of Higher Education, 1, 557-569.

Teferra, D., & Altbach, P. G. (2004). African higher education: Challenges for the 21st century. *Higher Education*, 47(1), 21–50. DOI: 10.1023/B:HIGH.0000009822.49980.30

Tinto, V. (1993). *Leaving college: Rethinking the causes and cures of student attrition*. University of Chicago Press.

Waghid, Y. (2004). African philosophy of education: Implications for teaching and learning. *South African Journal of Higher Education*, 18(3), 56–64.

Wenger, E. (1998). *Communities of practice: Learning, meaning, and identity*. Cambridge University Press. DOI: 10.1017/CBO9780511803932

Wisker, G. (2012). *The good supervisor: Supervising postgraduate and undergraduate research for doctoral theses and dissertations*. Macmillan International Higher Education. DOI: 10.1007/978-1-137-02423-7

World Bank. (2010). *Financing higher education in Africa*. World Bank.

Chapter 7

Unravelling Challenges and Embracing African Perspectives in Writing: A Problem Statement in Education Research

Peter Khwezi Nyewe

Cape Peninsula University of Technology, South Africa

ABSTRACT

This chapter explores how problem statements are framed, focusing on their relevance on Africanization in education. By addressing the specific challenges encountered in educational research and practice and embracing indigenous African viewpoints, the researcher aims to develop solutions that are more attuned to the diverse requirements of scholars in African universities. Through a thorough review of existing literature, this chapter aims to highlight the importance of well-written problem statements in guiding research efforts, and the potential of Africanization to enhance educational practices.

BACKGROUND TO THE PROBLEM

Writing a problem statement in educational research is a crucial initial step that defines the focus and direction of the study. However, this task presents considerable challenges, particularly within African educational research contexts. Factors such as historical, cultural, and contextual complexities heavily shape the development of problem statements. It is essential to explore these complexities and incorporate African perspectives effectively.

DOI: 10.4018/979-8-3373-0806-7.ch007

Historical Context and Colonial Legacies

Writing problem statements in African educational research faces a significant hurdle due to historical factors and colonial legacies. Colonialism has deeply influenced African education systems by imposing Western models and marginalizing indigenous knowledge systems (Mawere & Ndwamato, 2023; Zeleza, 2016). This historical context makes it difficult for the researchers to navigate between the enduring impact of colonial educational paradigms and the renewed emphasis on indigenous knowledge and methodologies (Mamdani, 2018).

Cultural Context and Epistemological Divergence

African educational research must account for the diverse cultural contexts and epistemological frameworks that characterize the continent. Western epistemological frameworks often dominate academic discourse, marginalizing African ways of knowing and learning (Odora Hoppers, 2015). This dominance presents a challenge for researchers aiming to formulate problem statements that are culturally relevant and reflective of African educational realities. Embracing African epistemologies requires a shift from a universalist perspective to a more pluralistic approach that values local knowledge and cultural specificity (Chilisa, 2017).

African educational research faces the challenge of accommodating the continent's diverse cultural contexts and epistemological frameworks. Academic discourse often prioritizes Western epistemologies, sidelining African modes of knowledge and learning (Odora Hoppers, 2015). This dominance complicates the task for researchers seeking to craft problem statements that accurately reflect African educational contexts and are culturally pertinent. Embracing African epistemologies necessitates a transition from a universalist viewpoint to a pluralistic approach that respects local knowledge and cultural particularities (Chilisa, 2017).

Socio-Economic and Political Factors

Socio-economic and political dynamics are important in shaping Africa's educational landscape. Factors like poverty, political volatility, and insufficient infrastructure profoundly influence education systems and, consequently, the research aims and problem statements developed by scholars (Brock-Utne, 2016). Researchers must confront these contextual realities to ensure their problem statements accurately reflect the genuine challenges faced by educational stakeholders across the continent.

Language and Representation

Language is another significant challenge in writing problem statements in African educational research. The dominance of English and other colonial languages in academic discourse often marginalizes indigenous languages, which are integral to the cultural identity and cognitive processes of African communities (Bamgbose, 2017). The choice of language in formulating problem statements can influence the inclusivity and accessibility of research, highlighting the need for multilingual approaches that reflect the linguistic diversity of Africa.

Methodological Considerations

Methodological choices are important in addressing the complexities of writing a problem statements in African educational research. Traditional research methods often fall short in capturing the intricate nuances of African educational contexts. There is growing advocacy among researchers for methodologies that are participatory, community-oriented, and attuned to African perspectives (Khupe & Keane, 2017). These approaches not only enhance the research process but also ensure that problem statements resonate with the lived realities and aspirations of African communities.

Embracing African Perspectives

It is crucial to incorporate African viewpoints when writing problem statements as this will likely assist in addressing challenges with writing a problem statement focusing on problems experienced by African communities. This includes acknowledging the significance of indigenous knowledge systems, advocating for culturally appropriate pedagogy, and promoting a decolonized approach to educational research (Mbembe, 2016). Researchers should collaborate closely with local communities, educators, and students to jointly develop problem statements that reflect the educational aspirations and priorities of African societies.

Conclusion

In conclusion, writing a problem statement in African educational research entails navigating a complex interplay of historical, cultural, socio-economic, and methodological factors. The challenges are multifaceted, requiring researchers to critically engage with the colonial legacies, cultural contexts, and contemporary realities of African education systems. By embracing African perspectives, researchers can formulate problem statements that are not only contextually relevant but also contribute to the broader goal of decolonizing education and advancing social justice in Africa.

Theoretical Framework

The theoretical foundation for this study draws from Critical Theory (Celikates, 2019), which focuses on examining power dynamics, social structures, and historical contexts that shape knowledge production and educational practices. In the context of writing a problem statement for educational research, Critical Theory provides a framework to critically analyze the current educational practices and question dominant paradigms that may marginalize or overlook African perspectives.

Through the application of Critical Theory, the researcher aims to challenge Eurocentric biases and colonial influences that have historically shaped educational research and practice in Africa. This framework promotes reflexivity, self-awareness, and a commitment to social justice, urging researchers to engage in processes that decolonize educational practices and reclaim indigenous knowledge systems (Celikates, 2019).

Critical Theory, offers a robust framework for examining and tackling challenges in educational research, particularly when navigating complexities and embracing African perspectives. At its core, Critical Theory critiques societal structures, power dynamics, and ideologies that perpetuate inequality and marginalization (Horkheimer & Adorno, 2002). In the context of writing problem statements in African educational research, Critical Theory provides valuable perspectives for understanding and confronting the historical and socio-political influences that shape educational discourses and practices.

Historical and Socio-Political Contexts

Critical Theory stresses the significance of historical context in comprehending current social issues (Marcuse, 2015). A crucial focal point within African education systems is the enduring legacy of colonialism, which imposed Western paradigms while suppressing indigenous knowledge. Applying Critical Theory enables researchers to dissect these colonial influences and illuminate their ongoing impact on educational practices and research agendas (Fanon, 2008). This theoretical approach highlights the imperative to acknowledge and confront the lingering effects of colonialism that shape the development of problem statements in African educational research.

Power Dynamics and Epistemological Dominance

Critical Theory focuses on analyzing power dynamics and their role in maintaining epistemological hegemony (Habermas, 1984). In African educational contexts, Western epistemologies frequently dominate, sidelining African modes of knowledge and indigenous knowledge systems. This dominance mirrors larger power structures that favor specific types of knowledge. Critical Theory prompts researchers to critically examine and challenge these power dynamics by advocating for epistemological diversity and the integration of African perspectives into educational research (Ndlovu-Gatsheni, 2018).

Emancipation and Social Justice

A core principle of Critical Theory is the quest for emancipation and social justice (Freire, 1970). This is especially pertinent in the African educational context, where historical injustices and socio-economic inequalities persistently influence educational access and outcomes. By employing Critical Theory to frame problem statements, researchers can illuminate issues of inequality and advocate for educational practices that foster social justice. This approach entails not only identifying challenges but also envisioning solutions that empower marginalized communities and confront systemic inequities.

Methodological Considerations

Critical Theory also guides methodological choices in educational research by promoting reflexive, participatory, and action-oriented approaches that aim to uncover and rectify power imbalances (Kincheloe, McLaren, & Steinberg, 2011). In African educational research, this involves actively engaging with local communities, valuing their knowledge, and integrating them into the research process. This methodological stance ensures that problem statements are not only pertinent but also collaboratively developed with those directly impacted by educational challenges, promoting a more inclusive and equitable research endeavor.

Decolonizing Knowledge

Applying Critical Theory in African educational research involves centralizing the process of decolonizing knowledge (Smith, 2012). This entails critically analyzing and dismantling colonial structures and ideologies embedded within educational systems and research methodologies. By adopting a critical perspective, researchers can question the hegemony of Western paradigms and advocate for the acknowledgment and inclusion of indigenous knowledge systems. This decolonial approach is crucial for crafting problem statements that accurately reflect African educational contexts and are contextually meaningful.

Critique of Ideology

Critical Theory's examination of ideology is essential for uncovering the implicit assumptions and biases that shape problem statements in educational research (Gramsci, 1971). Often unexamined, ideologies influence perceptions regarding educational issues and potential solutions. Through critical analysis, researchers can uncover how these ideologies perpetuate inequities and impede the advancement of inclusive and contextually appropriate educational practices. This critical analysis is vital for formulating problem statements that authentically reflect African realities while challenging prevailing narratives that marginalize alternative viewpoints.

Conclusion

In summary, Critical Theory offers a robust framework for confronting the complexities of writing problem statements in African educational research. It provides methodologies to critically examine historical and socio-political contexts, power dynamics, and ideologies that shape educational practices. By adhering to principles such as emancipation, social justice, and decolonization, researchers can develop problem statements that are deeply rooted in local contexts and oriented towards fostering transformative outcomes. This theoretical approach ensures that educational research in Africa not only mirrors local realities but also contributes to the broader aspiration of achieving equity and justice in education.

LITERATURE REVIEW:

Introduction

Writing a problem statement is an important stage in educational research as it defines the study's scope, direction, and significance. Yet, in the context of African education, this process confronts distinctive challenges and intricacies rooted in historical, cultural, socio-economic, and political contexts. This literature review critically examines the current body of research concerning these challenges and investigates how Critical Theory offers a robust framework to confront them. The review emphasizes the significance of integrating African viewpoints to shape problem statements that are both contextually pertinent and poised for transformative impact.

Colonial Legacies in African Education

The influence of colonialism on African education systems is extensively documented and presents a substantial obstacle in formulating effective problem statements. Colonial education policies were desined to benefit colonial interests, frequently sidelining and undervaluing indigenous knowledge systems (Ashcroft, Griffiths, & Tiffin, 2013). This legacy continues to impact modern education systems, shaping curriculum development, teaching methods, and areas of research focus.

Mamdani (2018) describes the colonial education system in Africa as dualistic, establishing separate governance and educational structures: one for the colonizers and another for the colonized. This division created an educational hierarchy that favored Western knowledge and methodologies over African epistemologies. Researchers face the challenge of acknowledging and remedying these historical inequities in their problem statements. Critical Theory provides support for this endeavor by advocating the dismantling of colonial ideologies and the advancement of epistemic justice.

Epistemological Dominance and Marginalization

The influence of Western epistemologies in scholarly discussions marginalizes African modes of knowledge and learning. This dominance represents a type of intellectual imperialism that diminishes the significance of indigenous knowledge systems (Smith, 2012). Critical Theory offers a framework for researchers to critique power dynamics and ideologies, enabling them to contest this hegemony and promote a more inclusive approach to the production of knowledge.

Chilisa (2017) asserts that indigenous research methodologies are crucial for accurately representing the realities and experiences of African communities. These methodologies emphasize relationships, community involvement, and collaborative knowledge creation, aspects often overlooked in Western research frameworks. By integrating indigenous methodologies, researchers can formulate problem statements that better capture African contexts and play a role in decolonizing education research.

Socio-Economic and Political Contexts

The socio-economic and political contexts in Africa pose substantial obstacles to education research. Issues like poverty, political instability, and inadequate infrastructure profoundly impact both the accessibility and quality of education, thereby shaping the research environment (Brock-Utne, 2016). It is essential for problem statements to account for these contextual factors to ensure that research effectively addresses the real needs and challenges encountered by African communities.

Critical Theory's emphasis on social justice and emancipation holds particular relevance in this context. Freire's (1970) concept of "conscientization" which is the process of cultivating a critical awareness of one's social reality through reflection and action which is crucial for comprehending and confronting these socio-economic and political challenges. Applying this concept enables researchers to craft problem statements that not only illuminate inequalities but also suggest interventions that empower marginalized communities and facilitate transformative advancements.

Language and Representation

Language plays a pivotal role in the construction of knowledge and the formulation of problem statements. The use of colonial languages like English and French in academic research significantly marginalizes indigenous languages, which are essential to the cultural identity and cognitive processes of African communities (Bamgbose, 2017). This linguistic dominance can exclude local perspectives and perpetuate epistemic injustice.

Critical Theory's examination of ideology and power structures extends to issues of language and representation, as articulated by Gramsci's (1971) concept of cultural hegemony. This concept highlights how dominant languages and cultures assert and maintain power by marginalizing others. In African educational research, it is crucial to adopt multilingual approaches and recognize the value of indigenous languages to ensure problem statements are inclusive and reflective of diverse cultural contexts. Researchers should be mindful of the inherent power dynamics in language usage and aim to democratize knowledge production by integrating a range of linguistic and cultural perspectives.

Methodological Innovations

Innovative methodologies play an important role in tackling the complexities of formulating problem statements in African educational research. Traditional research methods, typically grounded in Western epistemologies, may not adequately capture the intricate nuances of African educational contexts. There is a growing advocacy among researchers for methodologies that are participatory, community-oriented, and aligned with African worldviews (Khupe & Keane, 2017).

Critical Theory advocates for the adoption of inclusive and reflexive methodologies in educational research. According to Kincheloe, McLaren, and Steinberg (2011), critical pedagogy and participatory action research are instrumental in effecting social change. These methodologies prioritize collaborative knowledge creation, community engagement, and the critique of power dynamics. By embracing these approaches, researchers can develop problem statements that are not only pertinent but also transformative, aiming to address the underlying issues contributing to educational challenges in Africa.

Decolonizing Knowledge Production

In contemporary African educational research, decolonizing knowledge production is a central focus. This entails challenging the dominance of Western paradigms and advocating for the acknowledgment and integration of indigenous knowledge systems (Ndlovu-Gatsheni, 2018). Decolonization goes beyond mere inclusion of indigenous perspectives within existing frameworks; it requires a fundamental reevaluation and restructuring of how knowledge is generated and valued.

Critical Theory's emphasis on emancipation and social justice aligns with the objectives of decolonizing knowledge production. It prompts researchers to critically analyze the power dynamics and ideologies that shape educational research, encouraging exploration of alternative, more equitable forms of knowledge production. This approach involves collaborative engagement with indigenous scholars, community leaders, and other stakeholders to co-create knowledge that is not only contextually relevant but also culturally meaningful.

Conclusion

The challenges involved in writing problem statements for African educational research include historical legacies, epistemological dominance, socio-economic and political contexts, language barriers, and methodological considerations. Critical Theory offers a robust framework to address these challenges through its advocacy for dismantling power structures, promoting epistemic justice, and striving for social emancipation.

By integrating African perspectives and embracing indigenous knowledge systems, researchers can develop problem statements that authentically reflect the diverse realities of the continent and contribute to transformative outcomes. This approach enhances the research process and aligns with broader objectives of decolonizing education and advancing social justice in Africa.

Research Question

The following questions explore the complexities of integrating African perspectives into problem statement formulation in education research, investigating theoretical frameworks and practical strategies to foster inclusivity, equity, and social justice in research methodologies.

Primary Research Question:

1.How have traditional approaches to writing problem statements in education research historically marginalized or excluded African perspectives?

11.What ethical considerations and responsibilities are involved in writing problem statements that accurately reflect African perspectives and experiences, especially within the context of decolonizing research methodologies?

Secondary Research Questions:

i) What are the primary challenges researchers face when incorporating African perspectives into problem statement development in education research?
ii) What methodologies and strategies can researchers utilize to incorporate diverse African perspectives and voices into the formulation of problem statements?
iii) In what ways can the formulation of problem statements that embrace African perspectives contribute to transformative educational practices and policies, fostering social change and empowerment within African communities?

Data Collection

Primary data was gathered from books, articles, and scientific reports, focusing on theories of structural inequality. The analysis specifically centers on published research articles from education journals that discuss the integration of African perspectives in formulating problem statements. The researcher explored the methodologies, theoretical frameworks, and findings of these studies to identify commonalities, emerging trends, and effective approaches in writing problem statements that incorporate African viewpoints. This analysis aims to offer empirical support for the arguments and conclusions presented in the chapter.

Data Collection and Analysis

Introduction

This section explores the research questions related to the difficulties and strategies for formulating problem statements in education research that reflect African perspectives. It utilizes information from books, journal articles, and scientific reports to analyze how traditional methods have marginalized African viewpoints, the ethical considerations at play, the obstacles researchers encounter, and the strategies available to incorporate diverse African voices. Furthermore, it evaluates how problem statements embracing African perspectives can potentially foster social change and empowerment within African communities.

Historically, traditional methods of writing problem statements in education research have marginalized or neglected African perspectives. This exclusion originates from the colonial legacy, which imposed Western educational frameworks on African societies. As noted by Ashcroft, Griffiths, and Tiffin (2013), colonial education systems emphasized Western knowledge and methodologies, often disregarding indigenous African knowledge systems. This historical backdrop has entrenched the dominance of Western epistemologies in academic research, thereby marginalizing African viewpoints.

According to Fanon (2008), colonial education sought to assimilate Africans into European culture by suppressing local languages, traditions, and knowledge systems. This assimilationist approach continues to influence contemporary educational research, where problem statements often prioritize Western perspectives over the realities and needs of African communities.

Analysis of Exclusion

The exclusion of African perspectives is evident in the limited representation of indigenous knowledge in educational research, as highlighted by Chilisa (2017). Mainstream research methodologies frequently overlook the relational and community-oriented aspects of African epistemologies. This exclusion is not merely accidental but reflects deeper power dynamics that favor Western forms of knowledge. Critical Theory, which focuses on critiquing power and ideology (Marcuse, 2015), provides a framework to uncover and challenge these dynamics, advocating for a more inclusive approach that values diverse epistemologies.

Ethical Considerations and Responsibilities

Ethical Imperatives

Writing problem statements that reflects African perspectives entails significant ethical considerations and responsibilities. Researchers bear an ethical obligation to honor and appreciate the knowledge and experiences of African communities. This commitment necessitates ensuring that research methodologies are participatory and inclusive, amplifying the voices of those most impacted by educational challenges.

Smith (2012) underscores the necessity of decolonizing research methodologies to rectify historical injustices and uphold epistemic justice. This involves acknowledging the harm caused by marginalizing indigenous knowledge and actively striving to integrate it into research practices. Ethical research practices demand transparency, accountability, and a steadfast dedication to advancing social justice.

Responsibilities in Decolonizing Methodologies

Decolonizing research methodologies goes beyond mere integration of indigenous knowledge. It necessitates a fundamental transformation in how research is practised. Researchers must engage deeply and meaningfully with local communities, ensuring that their studies are firmly rooted in the lived experiences of these communities. This collaborative approach entails co-creating problem statements and research agendas together with community members.

According to Ndlovu-Gatsheni (2018), decolonizing research methodologies involves dismantling historical power structures that have marginalized African perspectives. This requires researchers to adopt a reflexive stance, continuously reflecting on their own positionality and the potential impact of their research on the communities they engage with.

Challenges in Integrating African Perspectives

Identifying Key Challenges

Researchers face several challenges in integrating African perspectives into the formulation of problem statements in education research. One major challenge is the dominance of Western academic standards and expectations, which can marginalize alternative epistemologies. This dominance often manifests in the form of publication biases, where research grounded in indigenous knowledge systems struggles to gain recognition in mainstream academic journals (Odora Hoppers, 2015).For instance, a study by Ngulube (2018) highlights how African-centric research often faces difficulties in passing through peer review processes in Western-dominated journals, which tend to favor studies that align with conventional Western methodologies and theoretical frameworks. This systemic bias not only stifles the diversity of academic discourse but also limits the visibility and impact of research that could offer valuable insights from African contexts.

Another challenge is the lack of resources and institutional support for research that prioritizes African perspectives. Many African researchers operate in environments with limited funding and infrastructural support, making it difficult to conduct comprehensive and culturally relevant research.This funding gap also hinders the ability of researchers to undertake extensive fieldwork, employ necessary research tools, and disseminate their findings effectively. Furthermore, the lack of institutional backing can demotivate researchers from pursuing innovative projects that incorporate indigenous knowledge systems, as they may perceive such endeavors as financially and professionally risky.

Furthermore, there is often a lack of training in indigenous research methodologies, which further hinders the integration of African perspectives (Brock-Utne, 2016). The educational curriculum in many African universities is predominantly influenced by Western paradigms, leaving little room for the development of skills necessary to engage with indigenous methodologies. This educational gap creates a barrier for researchers who wish to employ culturally sensitive and relevant approaches in their studies. A study by Chilisa and Kawulich (2012) emphasizes the importance of incorporating indigenous methodologies into research training programs to empower African scholars to explore and validate their unique epistemological frameworks. By addressing these training deficiencies, educational institutions can play a pivotal role in promoting the integration of African perspectives in education research, thereby enriching the global academic landscape.

Identifying Analysis of Structural Barriers

Understanding the structural barriers that hinder the integration of African perspectives in writing problem statements is crucial. These barriers often include systemic biases, a lack of representation, and the dominance of Western paradigms in academic research. Structural barriers are deeply entrenched in education systems and research institutions, which tend to prioritize Western methodologies and perspectives. This marginalizes African voices and creates challenges in presenting research that truly reflects the diverse contexts of African communities. Addressing these barriers requires a critical examination of existing frameworks and a commitment to fostering inclusivity and diversity in research practices.

In a 2018 study by Zeleza, "The Struggles for African Knowledge: Beyond the Western Paradigm," the author highlights how African scholars often face the challenge of having their work scrutinized through a Western lens, which can distort the true value and relevance of their research within their own contexts. This results in a form of academic gatekeeping where African perspectives are undervalued or ignored altogether. Zeleza argues for a more inclusive approach that recognizes and integrates diverse epistemologies to enrich global knowledge production (Zeleza, 2018).

Another article by Mamdani (2018), "Decolonizing Universities," discusses how the legacy of colonial education systems continues to influence contemporary academic practices. Mamdani points out that curricula and research agendas are still largely shaped by colonial frameworks, which marginalize non-Western perspectives. He calls for a decolonization of education that involves rethinking and restructuring academic institutions to create space for indigenous knowledge systems and methodologies. This would help in breaking down the structural barriers that currently impede the inclusion of African perspectives in education research (Mamdani, 2018).

In "Rethinking African Higher Education: Beyond Colonial Paradigms," Nyamnjoh (2018) explores the impact of the historical dominance of Western paradigms on African higher education. Nyamnjoh emphasizes the need for an epistemic shift that values local knowledge systems and encourages African scholars to contribute to global academic discourse on their own terms. This involves questioning the supremacy of Western methodologies and promoting an academic culture that is more reflective of Africa's rich intellectual traditions (Nyamnjoh, 2018).

A study by Keet (2018), "Engaging Critical Pedagogy in South Africa: Decolonization, Education, and the Politics of Knowledge," further illustrates the challenges faced by African scholars in gaining recognition within the global academic community. Keet argues that critical pedagogy must be employed to challenge the existing power dynamics in education that favor Western perspectives. He advocates for a transformative approach that empowers African scholars and students to produce and disseminate knowledge that is relevant and beneficial to their communities (Keet, 2018).

These articles collectively underscore the importance of addressing structural barriers in education research to embrace African perspectives. By critically examining and challenging the existing frameworks, the academic community can move towards a more inclusive and diverse research environment that values and integrates the unique insights and contributions of African scholars.

Strategies and Methodologies for Inclusion

To overcome these structural barriers, researchers need to adopt strategies and methodologies that promote the inclusion of African perspectives. One effective approach is Participatory Action Research (PAR), which emphasizes collaboration between researchers and community members. PAR encourages the joint creation of knowledge, allowing community members to actively participate in identifying research issues, planning studies, and interpreting results. This collaborative process ensures that the research is relevant and beneficial to the community, while also giving voice to those who are often marginalized in traditional research paradigms. By involving community members in the research process, PAR helps to bridge the gap between academic research and real-world issues faced by African communities.

Emphasizing Collaboration and Joint Knowledge Creation

Participatory Action Research (PAR) stands out as a transformative approach in incorporating African perspectives in educational research. The core principle of PAR is collaboration, where researchers and community members work together as equal partners. This methodology allows for the joint creation of knowledge, wherein community members are not just subjects but active contributors to the research process. By collaboratively identifying research issues, planning studies, and interpreting results, PAR fosters a more inclusive and equitable research environment. This collaborative approach is crucial for ensuring that the research outcomes are not only academically significant but also practically relevant and beneficial to the communities involved.

Empowering Marginalized Voices

One of the significant advantages of PAR is its potential to empower marginalized voices within African communities. Traditional research paradigms often overlook the perspectives of those who are directly affected by educational issues. PAR, on the other hand, actively seeks to involve these voices, ensuring that the research addresses the real-world challenges faced by these communities. By giving a platform to those who are often unheard, PAR helps to democratize the research process, making it more inclusive and representative. This empowerment is not only vital for the accuracy and relevance of the research but also for fostering a sense of ownership and agency among community members.

Bridging the Gap Between Academia and Real-World Issues

The integration of PAR in education research significantly bridges the gap between academic findings and real-world applications. Often, academic research is criticized for being too theoretical and disconnected from practical realities. However, by involving community members in the research process, PAR ensures that the findings are grounded in actual experiences and challenges. This alignment between academic research and community needs enhances the practical applicability of the research, making it more likely to drive meaningful change. Moreover, this connection fosters a reciprocal relationship where academic insights can inform community practices and vice versa, leading to a more dynamic and impactful research process.

Case Studies Illustrating PAR in African Contexts

Several studies have illustrated the effectiveness of PAR in incorporating African perspectives in education research. For instance, a study by Chikoko and Khanare (2019) in South Africa utilized PAR to address school dropout rates by involving students, parents, and teachers in the research process. Similarly, Mertens, Cram, and Chilisa (2018) highlighted the use of PAR in Botswana to enhance teacher training programs by incorporating feedback from local educators. Another notable example is the work of Nyirenda and Brown (2021) in Malawi, which employed PAR to develop culturally relevant curricula in primary schools. Lastly, a study by Ogbu and Okeke (2023) in Nigeria demonstrated how PAR facilitated the inclusion of indigenous knowledge systems in educational policy development. These examples underscore the potential of PAR to foster inclusive and contextually relevant education research in Africa.

Inclusive Research Methodologies

Inclusive research methodologies are essential for integrating a variety of African perspectives into the formulation of problem statements. These methodologies recognize the value of indigenous knowledge systems and prioritize relationality, community engagement, and holistic understanding. According to Chilisa (2017), indigenous research methodologies should incorporate storytelling, ceremonies, and other cultural practices as valid methods of data collection and analysis. These approaches not only enrich the research process but also ensure that the findings are deeply rooted in the cultural and social contexts of African communities. By embracing these methodologies, researchers can create problem statements that reflect the lived experiences and realities of the people they aim to study.

Examples of Inclusive Research in Practice

Several articles from recent years illustrate the successful integration of African perspectives through inclusive research methodologies. For example, Nyaga (2018) explores the use of storytelling as a method of data collection in Kenyan schools, highlighting how it fosters a deeper understanding of educational challenges from the students' perspectives. Similarly, Mnyaka and Motlhabane (2018) discuss the implementation of Participatory Action Research in rural South African communities, demonstrating how community involvement leads to more relevant and impactful research outcomes. Another study by Ndhlovu and Mkhize (2018) examines the role of ceremonies and traditional practices in educational research, showing how these methods can uncover insights that are often overlooked by conventional approaches. Finally, Kamanzi and Nkomo (2018) provide a comprehensive analysis of indigenous methodologies in higher education research, advocating for their broader adoption to enhance the inclusivity and relevance of academic studies in Africa.

Firstly, incorporating indigenous knowledge systems into research methodologies is pivotal. Scholars like Chilisa (2017) advocate for methodologies that go beyond conventional Western paradigms to include storytelling, ceremonies, and other cultural practices. These methods are not merely alternative data collection tools but are foundational to understanding and representing African realities authentically (Chilisa, 2017). By embedding these practices, researchers can ensure that problem statements resonate with the diverse cultural contexts they aim to address.

Secondly, relationality and community engagement are integral to inclusive research in African contexts. For instance, research by Makalela and Machet (2019) emphasizes the importance of building trust and reciprocal relationships within communities. Such engagements foster mutual respect and enable researchers to co-construct problem statements that align with community needs and aspirations (Makalela & Machet, 2019). This participatory approach not only enhances the relevance of research but also promotes sustainable development outcomes by empowering local voices in shaping educational agendas.

Furthermore, holistic understanding underscores the need to consider interconnected aspects of life in African communities. A study by Nkomo and Chikoko (2020) illustrates how holistic methodologies can uncover multifaceted challenges faced by learners in marginalized communities. By integrating cultural, economic, and social dimensions into problem statements, researchers can address systemic issues comprehensively and advocate for inclusive educational policies (Nkomo & Chikoko, 2020). This holistic perspective not only enriches the research process but also contributes to the broader discourse on educational equity and social justice in Africa.

Lastly, validating diverse ways of knowing and being is essential in inclusive research methodologies. Research by Kanku and Mabika (2021) highlights the significance of recognizing diverse epistemologies and ontologies in shaping educational research agendas. By acknowledging and incorporating multiple knowledge systems, researchers can challenge dominant narratives and amplify marginalized voices, thus fostering more inclusive and equitable educational practices (Kanku & Mabika, 2021).

In conclusion, embracing inclusive research methodologies in writing problem statements for education research in Africa involves not only methodological innovation but also a profound commitment to decolonizing knowledge production.

By integrating indigenous knowledge, fostering community engagement, adopting a holistic approach, and validating diverse epistemologies, researchers can create problem statements that authentically reflect African perspectives and contribute meaningfully to advancing educational equity and social transformation.

Leveraging Local Knowledge

According to Marcuse (2015), Critical Theory challenges traditional power structures within academia that often marginalize non-Western perspectives. This framework calls for democratizing knowledge creation by recognizing and valuing diverse viewpoints, thereby enriching the discourse with locally grounded insights.

In a study by Ndlovu and Walton (2019), the authors exemplify the application of Critical Theory in educational research in Africa. Their research explores how integrating local knowledge systems into curriculum development can enhance educational relevance and cultural responsiveness. By engaging with community stakeholders and incorporating indigenous knowledge, the study underscores the importance of respecting and utilizing local expertise to address educational challenges effectively.

Another significant contribution is found in the work of Kwayu (2020), which examines participatory research methodologies in Tanzania. Kwayu emphasizes the need for collaborative partnerships between researchers and local communities to co-create knowledge and address educational issues from within the cultural context. This approach not only empowers communities but also ensures that research outcomes are relevant and sustainable in the local setting.

Furthermore, Oduaran and Okwilagwe (2021) explore the role of African philosophy in shaping educational research paradigms. Their work advocates for decolonizing research methodologies by embracing indigenous epistemologies and methodologies. By incorporating African philosophical frameworks into problem statement development, researchers can foster a more inclusive and culturally sensitive approach to educational research.

Overall, these studies illustrate how Critical Theory and collaborative research methodologies can facilitate the integration of African perspectives in education research. By valuing local knowledge systems and engaging with diverse community voices, researchers can develop problem statements that resonate with the realities and aspirations of African societies, thereby contributing to more inclusive and impactful educational practices.

Promoting Social Change and Empowerment

Formulating problem statements that incorporate African perspectives can drive transformative shifts in educational practices and policies. By addressing the specific challenges faced by African communities, researchers can propose nuanced solutions that target the root causes of educational disparities. This approach not only aims at improving educational outcomes but also contributes to broader social justice agendas by empowering marginalized communities. For instance, recent research by Ndofirepi and Chimedza (2021) emphasizes the need for problem statements that recognize and address cultural and socioeconomic factors impacting educational access and quality in Sub-Saharan Africa.

Drawing on Freire's concept of "conscientization," which underscores critical consciousness and action, problem statements reflecting African perspectives can illuminate systemic issues and mobilize communities towards advocacy and policy change. This empowerment process is pivotal in nurturing educational environments that are transformative, inclusive, and equitable (Oduro et al., 2019). For example, Okello and Adera (2020) argue that problem statements rooted in African perspectives can challenge hegemonic educational paradigms, promoting culturally relevant pedagogies that resonate with local contexts and empower learners.

Moreover, integrating African perspectives into problem statements fosters a deeper understanding of indigenous knowledge systems and epistemologies, challenging dominant Western-centric educational frameworks. This approach not only enriches educational research but also contributes to decolonizing educational practices, as highlighted in recent studies by Nkomo and Mutambara (2018). By acknowledging and valuing diverse knowledge traditions, researchers can create more inclusive and responsive educational interventions that promote social change and sustainable development goals.

Policy Implications

Incorporating African perspectives into problem statements can also shape policy formulation. By emphasizing the unique challenges and requirements of African education systems, researchers can advocate for policymakers to implement policies that are culturally appropriate and responsive to local contexts. This advocacy aims to foster educational frameworks that better meet the diverse needs of African students and communities (Ndlovu-Gatsheni, 2018).

Formulating problem statements that incorporate African perspectives can drive transformative shifts in educational practices and policies. By focusing on the unique challenges faced by African communities, researchers can propose solutions that tackle the underlying causes of educational disparities, fostering social justice and empowerment. Embracing African perspectives in problem statements involves recognizing diverse cultural contexts, historical legacies, and socio-economic realities that shape educational experiences. For instance, research could explore how traditional African educational philosophies can complement modern pedagogical approaches, enhancing relevance and inclusivity in learning environments (Author, Year). This approach not only enriches educational discourse but also empowers local communities by validating their knowledge systems and promoting self-determination in educational development (Author, Year).

Freire's (1970) concept of conscientization,which is about developing critical awareness through reflection and action, stresses the relevance of problem statements that reflect African perspectives. Such statements can illuminate systemic issues, mobilize communities, and empower them to advocate for change. This process of empowerment is crucial for fostering educational practices that are transformative, inclusive, and equitable. By integrating African perspectives, researchers can challenge dominant narratives that often marginalize African knowledge and experiences in educational research. This critical lens allows for a deeper understanding of how historical injustices, colonial legacies, and cultural values influence educational outcomes and policies (Author, Year). For example, examining how African languages can be integrated into curriculum development can empower learners by affirming cultural identities and enhancing cognitive development through bilingual education (Author, Year).

Furthermore, incorporating African perspectives in problem statements encourages a participatory approach to research and policy-making, aligning with principles of social justice and community engagement. Collaborative efforts between researchers, educators, and local stakeholders can co-create solutions that address specific challenges within African educational systems. This inclusive approach not only enhances the relevance and sustainability of interventions but also fosters a sense of ownership and commitment among community members towards educational reforms (Author, Year). For instance, research on community-led initiatives in curriculum design demonstrates how involving local knowledge holders can ensure cultural relevance and responsiveness, thereby promoting educational practices that resonate with the lived experiences of African learners

Embracing African perspectives in writing problem statements for education research holds transformative potential by promoting social change and empowerment. It shifts the focus from deficit-oriented approaches to strengths-based frameworks that celebrate diversity, foster critical consciousness, and empower communities to actively shape their educational landscapes. This paradigm shift is essential for advancing educational equity and inclusivity, paving the way for sustainable development and positive social transformation across African contexts

CONCLUSION

In conclusion, overcoming the challenges of formulating problem statements in African educational research demands a thorough exploration of historical, epistemological, socio-economic, and methodological dimensions. Traditional methods have often sidelined African viewpoints, yet adopting Critical Theory and decolonizing research approaches can facilitate the creation of problem statements that are inclusive, culturally significant, and impactful. Central to this endeavor are ethical responsibilities and a dedication to social justice, ensuring research endeavors empower and advance African communities.

Formulating problem statements that incorporate African perspectives can drive transformative shifts in educational practices and policies. By focusing on the unique challenges faced by African communities, researchers can propose solutions that tackle the underlying causes of educational disparities, fostering social justice and empowerment. These perspectives offer insights into culturally relevant pedagogies, community-driven educational reforms, and inclusive policies that resonate with the diverse socio-cultural contexts across the continent. Integrating African perspectives in problem statements thus becomes not only a matter of academic rigor but also a commitment to addressing historically marginalized narratives and promoting inclusive educational frameworks.

Promoting Social Change and Empowerment Freire's (1970) concept of conscientization, which is about developing critical awareness through reflection and action, stresses the relevance of problem statements that reflect African perspectives. Such statements can illuminate systemic issues, mobilize communities, and empower them to advocate for change. This process of empowerment is crucial for fostering educational practices that are transformative, inclusive, and equitable. African perspectives encourage a shift from deficit-oriented approaches to asset-based frameworks, recognizing the wealth of knowledge embedded within local cultures and traditions. This approach not only enriches educational research but also promotes social change by valuing indigenous knowledge systems and empowering communities to actively participate in shaping their educational landscapes.

REFERENCE

Adomi, E. E., & Anwar, M. A. (2023). African Perspectives on Educational Technology: Challenges and Opportunities. *International Journal of Educational Development*, 45, 112–128.

Ashcroft, B., Griffiths, G., & Tiffin, H. (2013). *Post-Colonial Studies: The Key Concepts*. Routledge. DOI: 10.4324/978023777855

Bamgbose, A. (2017). *Language and the nation: The language question in Sub-Saharan Africa*. Edinburgh University Press.

Brock-Utne, B. (2016). *Language of instruction and student performance: New insights from research in Africa and elsewhere*. Nordiska Afrikainstitutet.

Celikates, R. (2019). Critical Theory and the Unfinished Project of Mediating Theory and Practice. In Gordon, P. E., Hammer, E., & Honneth, A. (Eds.), *The Routledge Companion to the Frankfurt School* (pp. 206–220). Routledge.

Chilisa, B. (2017). *Indigenous research methodologies*. Sage Publications.

Chilisa, B. (2017). *Indigenous research methodologies*. Sage Publications.

Chilisa, B. (2017). Decolonizing Transdisciplinary Research Approaches: An African Perspective for Enhancing Knowledge Integration in Sustainability Science. *Sustainability Science*, 12(1), 811–823.

Dei, G. J. S. (2018). Reframing the Dialogue on Indigenous Knowledge Systems: Implications for Education Policy and Practice. *Journal of Education Policy*, 33(4), 437–454.

Freire, P. (1970). Pedagogy of the Oppressed. *Continuum : an Interdisciplinary Journal on Continuity of Care*.

Giroux, H. A. (2018). Critical Theory and Educational Practice. *Educational Theory*, 68(2), 145-158.Fanon, F. (2008). Black Skin, White Masks. Grove Press.

Gramsci, A. (1971). *Selections from the Prison Notebooks*. International Publishers.

Habermas, J. (1984). The Theory of Communicative Action, Volume 1: Reason and the Rationalization of Society. Beacon Press.

Horkheimer, M., & Adorno, T. W. (2002). *Dialectic of Enlightenment: Philosophical Fragments*. Stanford University Press.

Keet, A. (2018). Engaging Critical Pedagogy in South Africa: Decolonization, Education, and the Politics of Knowledge. *Education as Change*, 22(1), 1–16. DOI: 10.17159/1947-9417/2017/2741

Khupe, C., & Keane, M. (2017). Towards an African education research methodology: Decolonising new knowledge. *Educational Research for Social Change*, 6(1), 25–37. DOI: 10.17159/2221-4070/2017/v6i1a3

Kincheloe, J. L., McLaren, P., & Steinberg, S. R. (2011). Critical pedagogy and qualitative research: Moving to the bricolage. In *The SAGE Handbook of Qualitative Research* (4th ed., pp. 163–178). Sage Publications.

Kovach, M. (2018). *Indigenous Methodologies: Characteristics, Conversations, and Contexts*. University of Toronto Press.

Lebeau, Y., & Mills, D. (2008). From 'Crisis' to 'Transformation'? Shifting Orthodoxies of African Higher Education Policy and Research. *African Studies Review*, 51(1), 1–22.

Mamdani, M. (2018). Decolonizing Universities. *Social Dynamics*, 44(2), 181–192.

Mamdani, M. (2018). *Citizen and subject: Contemporary Africa and the legacy of late colonialism*. Princeton University Press.

Marcuse, H. (2015). *One-Dimensional Man: Studies in the Ideology of Advanced Industrial Society*. Beacon Press.

Mawere, J. & Ndwamato, T.W. 2023. Exploring the Impact of Colonialism on South Africa with Focus on the Venda Community: Examining the Role of Traditional Knowledge Systems in Achieving Restorative Justice In book: Indigenous People - Traditional Practices and Modern Development [Working Title].

Mbembe, A. (2016). Decolonizing the university: New directions. *Arts and Humanities in Higher Education*, 15(1), 29–45. DOI: 10.1177/1474022215618513

Mertens, D. M. (2009). *Transformative Research and Evaluation*. Guilford Press.

Mertens, D. M., Cram, F., & Chilisa, B. (2016). *Indigenous Pathways into Social Research: Voices of a New Generation*. Left Coast Press. DOI: 10.4324/9781315426693

Mkandawire, T. (2005). *African Intellectuals: Rethinking Politics, Language, Gender, and Development*. Zed Books. DOI: 10.5040/9781350218147

Mlambo, N. (2018). The Lingering Legacy of Colonial Education in Southern Africa. *Education and Society*, 36(2), 212–230.

Ndlovu-Gatsheni, S. J. (2018). *Epistemic Freedom in Africa: Deprovincialization and Decolonization*. Routledge. DOI: 10.4324/9780429492204

Ndofirepi, T., & Chibaya, W. T. (2022). Decolonizing the Curriculum: African Perspectives on Inclusive Education. *Journal of Education for Sustainable Development*, 36(2), 245–261.

Ngulube, P. (2018). Overcoming the Western Dominance in Research: Critical Insights from an African Perspective. *African Journal of Library, Archives &* MacDonald, C. (2012). Understanding Participatory Action Research: A Qualitative Research Methodology Option. Canadian Journal of Action Research, 13(2), 34-50. *Information Science*, 28(2), 123–136.

Nsamenang, A. B., & Tchombe, T. M. S. (Eds.). (2011). *Handbook of African Educational Theories and Practices: A Generative Teacher Education Curriculum*. Human Development Resource Centre.

Nyamnjoh, F. B. (2018). Rethinking African Higher Education: Beyond Colonial Paradigms. *Journal of Higher Education in Africa*, 16(1-2), 1–21.

Nyoni, P. (2018). Socio-Economic Inequalities and Education in Africa: A Critical Analysis. *African Educational Review*, 14(3), 276–294.

Odora Hoppers, C. A. (2015). Cognitive justice and integration without duress: The future of development education - perspectives from the South. *International Journal of Development Education and Global Learning*, 7(2), 89–106. DOI: 10.18546/IJDEGL.07.2.08

Odora Hoppers, C. A. (2015). Thinking the Unthinkable: From the Universalisation of Knowledge to Decolonisation of the University in Africa. *South African Journal of Higher Education*, 29(5), 199–217.

Ogunleye, J. K., & Afolabi, F. M. (2021). Empowering African Teachers Through Reflective Practices: Insights for Educational Reform. *Journal of African Educational Research*, 18(3), 78–94.

Okeke, C. (2018). Education and Marginalization in Africa: A Critical Examination. *Journal of African Studies*, 19(2), 159–175.

Prah, K. K. (2018). Language and Cultural Identity in Post-Colonial Africa. *International Journal of African Renaissance Studies*, 13(1), 45–59.

Smith, L. T. (2012). *Decolonizing Methodologies: Research and Indigenous Peoples*. Zed Books.

Teferra, D., & Altbach, P. G. (2018). African Higher Education: Challenges for the 21st Century. *Industry and Higher Education*, 45, 15–18.

Trudell, B. (2009). Local-language Literacy and Sustainable Development in Africa. *International Journal of Educational Development*, 29(1), 73–79. DOI: 10.1016/j.ijedudev.2008.07.002

Tuck, E. (2018). Participatory Action Research as Public Science. *Educational Policy*, 32(4), 620–651.

Zeleza, P. T. (2016). The study of Africa: The challenges and opportunities of a new global intellectual order. *African Studies Review*, 59(1), 7–35.

Zeleza, P. T. (2018). The Struggles for African Knowledge: Beyond the Western Paradigm. *African Studies Review*, 61(2), 1–19.

Chapter 8 Practical Approaches to Problem Identification in Doctoral Research: Driving Business Impact Through Applied Orientation

N. Elangovan
https://orcid.org/0000-0002-4551-6081
Christ University, India

ABSTRACT

This chapter advocates for an applied orientation in research in Higher education, stressing the importance of selecting relevant problems to yield actionable insights and improve business practices. The methodology of this chapter includes synthesizing literature, evaluating relevance-rigour gap, and proposing a process and a checklist to convert a real-world problem into a research problem. The proposed checklist aids in identifying pertinent problems, covering stakeholder identification, problem frequency, clarity of solutions, and research feasibility. Bridging the academia-industry gap, this approach fosters innovation and knowledge transfer, supporting evidence-based solutions. Prioritizing practical relevance and stakeholder impact enhances research quality and drives advancements. Applied doctoral research addresses real-world challenges, enhancing organizational efficiency, sustainability, and profitability. Further exploration of its impact is recommended for improved research outcomes.

DOI: 10.4018/979-8-3373-0806-7.ch008

INTRODUCTION

Undeniably, the pursuit of a worthy research problem in higher education, especially within the vast realms of social sciences, such as Business and Management, can often feel like navigating through dense fog. It is akin to searching for a needle in a haystack while blindfolded. Researchers find themselves grappling with the conundrum of problem identification. On a lighter note, we can say that the researcher may pray: "Oh God! Give me a problem."

The crux of the issue in identifying a research problem lies not necessarily in the absence of potential problems to investigate but rather in the challenge of discerning which ones are worthy of scholarly attention and why. One aspect of this challenge is the subjective nature of what constitutes a 'problem.' The issues in identifying a research problem are in neither seeing the problem nor in seeming to *be* the problem. What may seem obvious as a problem to one individual or group may not be perceived as such by others. It may be due to differing perspectives and societal contexts. Also, the intricate fabric of modern societal complexities often renders problems multifaceted and interlinked, challenging their very identification. For example, the challenges of climate change necessitate a comprehensive understanding of environmental, economic, political and social factors. In addition, the speed of change brings out new challenges, potentially eclipsing those previously deemed urgent. A problem of importance today may be overshadowed by emergent developments tomorrow.

Moreover, identifying a research problem is challenging when researchers lack the expertise to determine its significance. Despite worthy existing issues, inadequate training may obscure recognition. This highlights the necessity for robust training programmes to cultivate researchers' analytical skills, enabling them to identify and address pertinent problems within their respective fields effectively.

Academic research in business, management and commerce domains has a huge rigour-relevance gap. The research outcomes have contributed less to the management practice. There is a huge need to address the current and emerging problems in the above-said domains. However, academic researchers are disconnected from the problems and concerns of the real world (Jenkins et al., 2020; Panda & Gupta, 2014). Academic research in the domains of business, management and commerce must predominantly be applied in orientation rather than pure or fundamental research (Sandberg et al., 2022; Saunders et al., 2019). Doctoral research is a channel of academic contributions. It also lacks relevance to the practice. There is an imperative for universities to train doctoral scholars to equip themselves for solving real problems (Easterby-Smith et al., 2021; Gardner et al., 2023). Identifying research problems in business, management and commerce domains with practical implications is challenging. Academic researchers mostly focus on finding their research problem through a literature review. Research articles recommend future research based on their limitations and outcomes. Researchers fail to connect with business organisations and practitioners. Therefore, their research problems lack relevance. The problems identified for research are sometimes not 'problems' in real time.

Though many texts on research methodology describe problem identification, the practical way of doing it is still vague. Saunders et al. (2019) propose formulating and clarifying the topic as the beginning of the research process. They discuss using creative techniques for generating research ideas. Easterby-Smith et al. (2021) too propose a similar method. Both their methods focus on the academic research of business and management students. However, it lacks the orientation towards relevance to practice. Zikmund et al. (2013) define problem recognition as an important step in beginning the research process. Their method is oriented towards addressing the relevance gap. The extent of the literature provides the theoretical knowledge towards problem identification for academic research with relevance. However, there is a lack of materials that provide a practical method for identifying the research problem relevant to the real world. This chapter aims to develop a method to evaluate the relevance of the academic research problem to the business world.

Identifying a good research problem has been a global challenge across the academic community. Zuber-Skerritt and Knight (1986) and Hart et al. (2022) are concerned with the problems faced by Honours or Masters students while going through their research thesis. They claim that the main difficulty critical to a student's progress is the "definition, construction and articulation of the research problem." A Master's thesis in Finland is the most demanding among all the courses in the 5-year degree. Academic supervision requires giving critical feedback on the student's work. Research mentors must tell students what is a good research problem and what is not (Cutillas et al., 2023; Vehviläinen, 2009). Marchisotti and Farias Filho (2022) argued that students struggle to understand the steps in choosing an academically valid research problem. They propose a method of operationalising the process of identifying it. They suggested a multi-method process to identify the theoretical-practical gap. Chambers (1960) discussed the "selection, definition and delimitation of a doctoral research problem." Doctoral students are known to consult many professors and experts on its choice and definition. It is the task of a doctoral student to find a good problem for research and not to expect the mentor to assign one. Describing a research problem in writing will provide a better understanding of it.

This chapter will be useful to students and faculty members of higher education in the social science and business management domains. It presents the background to support the need for the study in Section 1, which introduces it. Section 2 presents the compilation of various recommendations and arguments from the literature. Section 3 discusses the characteristics of the research problem with practical implications and how it is intertwined with the research process. Section 4 provides a checklist and a method for evaluating the research problem's relevance. The final section makes a conclusion highlighting the contributions of the chapter.

LITERATURE REVIEW

Rigour-relevance Gap

The relevance of management education to the practice was discussed as early as 1959. Research funded by the Ford Foundation and Carnegie Corporation reported a huge disconnect between academics and practice due to the inclination of management education to production and productivity. The research during the period lacked methodological rigour. The report emphasised adopting social science methods (Gordon & Howell, 1959; Koskela, 2017). This shifted the orientation of management education from a vocational model to a science model. However, implementing the recommendation of these two reports led to over inclination towards quantitative research. In order to fit with the methodological rigour, researchers started to focus on unreal problems. After almost 20 years of implementing the recommendations of the 1959 reports, the relevance of management research was again a concern (Koskela, 2011, 2017).

Thomas and Tymon (1982) listed various issues leading to dissatisfaction with the research outcomes in the business and management areas. The common sense nature of findings, their descriptive rather than recommendatory actions, ignoring external validity to support internal validity, reductionism at the cost of complexity, stressing the use of statistical techniques and inappropriate epistemological and ontological assumptions are some of the issues found in organisational research (Thomas & Tymon, 1982). The authors also proposed a set of necessary properties to identify the relevance of the research, such as descriptive and goal relevance, operational validity, non-obviousness and timeliness. Rudolph and Peluchette (1992) proposed a congruency model based on Thomas and Tymon's (1982) five components to enhance the fit between the practitioner's and researcher's needs. The model will help to assess the congruency between academic research and organisational practice. Minnaert (2015) mentioned that the question of research relevance encompasses both what and how knowledge is produced.

On the contrary, DeNesi (1994) argued that there is no clear definition of what is relevant and irrelevant in organisational research. The author discusses the centrality of the theory in determining its relevance. A good theory would have more practical relevance. The author indicates to both parties, the managers, what they need to expect from the theory and when a good one will be relevant to the researchers on what constitutes the theory and what it needs to provide to the practical applications. Overuse of jargon and statistics makes the practitioners disconnected from the research outcomes. In addition, the author questions the time dimension of the relevance and argues that a good theory helps to solve problems and may be irrelevant in the immediate run.

Concerns of research relevance have been discussed in many areas of business and management, such as Information Systems (Benbasat & Zmud, 1999), Marketing (Rajan Varadarajan, 2003), International Human Resources (Wasti & Robert, 2003), Strategic Management (Shrivastava, 1987) and, Management Research (Koskela, 2017). Shrivastava (1987) has examined 23 research programmes in strategic management on usefulness and rigour. The author has used three criteria for the rigour: conceptual adequacy, methodological rigour and accumulated empirical evidence. Five criteria are used to evaluate the practical usefulness: meaningfulness, goal relevance, operational validity, innovativeness and cost of implementation. (Rudolph & Peluchette, 2011)

Many studies have focused on addressing the rigour-relevance gap in management research. There are debates on whether the rigour-relevance gap needs to be taken seriously or not. Gulati (2007) claims that separating rigour and relevance is artificial. Dredge (2015) argues that relevance is not the problem, but what matters is the communication of research in a way that the practitioners understand. However, researchers like Kieser and Leiner (2009) argue that the rift between rigour and relevance is unbridgeable. Sidor (2015) is concerned that the divide between rigour and relevance can be serious and hinder the advancement of knowledge in management. The researcher will have to formulate the research problems, either aligning with the needs of the practitioners or producing a dual version of their research - one with rigour and the other focusing on the outcome relevant to managers. Gibbon et al. (1994) differentiate Mode 1 and Mode 2 knowledge based on how it is produced. Mode 1 knowledge is produced scientifically, done through basic research. Mode 2 knowledge is produced through applied research by solving problems in the real world. In this approach, a multidisciplinary approach is followed. Veit et al. (2017) have proposed a framework for research in business processes using Mode 2 knowledge production. A social movement on Responsible Research in Business and Management (RRBM) has observed that the inclination towards context and practice has constrained scientific freedom in business and management. At the same time, the pressure to publish among the academic fraternity has overlooked the scientific responsibility (Tsui & McKiernan, 2022). The authors discuss the responsibility categories, who is responsible for what and to whom. High academic freedom and responsibility are categorised as responsible research. Low academic freedom and high academic responsibility are categorised as consultancy projects.

Formulating a Research Problem

Among the types of research, applied orientation starts with a problem. The term 'problem' is coined from the Greek word meaning anything thrown before. It may also be a question or anything placed before for examination. The problem also means difficulty or a situation that does not have a ready solution. In a business and management context, a problem is a gap between the current and desired state or a deviation from a standard.

Applied research in the business, management and commerce domains often involves collecting and analysing data from various sources, such as surveys, experiments or case studies. Applied research aims to generate findings that can be put into action to solve real-world problems and improve business practices or processes. Applied research might also include studies that aim to increase efficiency and productivity, enhance customer satisfaction and profitability, improve management and leadership practices and address social or environmental issues that businesses face.

Pardede (2018) has said that formulating the research problem is the first and most important step in research. It is also one of the most difficult stages of the process. A well-defined research problem helps to design a better and more useful outcome. A vague research problem definition is the weakest research point (Dissanayake, 2013). Castrillón (2020) argues that clarity on the "difference between a real problem, problem, problematisation and research problem" is required to frame a research problem. Ellis and Levy (2008) suggest methods for developing research-worthy problems. The authors propose that a missed opportunity is not a problem. Rather, a problem needs to be active, make an impact and not have a current solution. Lund (2021) suggests that a problem needs to be reduced to a hypothesis form to be scientifically solved.

Researchers have discussed the challenges in developing a research problem and suggested criteria for evaluating the relevance of a research problem. Van Aken (2004) argues that academic research in the management area has a utilisation problem because of being more description-driven than prescription-driven. The author recommends the design of a science paradigm to make management research more relevant to organisational practice. Herwix and Haj-Bolouri (2021) have suggested a framework for evaluating the relevance of research problems in Information Systems. They used problem assessment ontology that groups evaluation elements into four modules: perspective, definition, barrier and importance. Perspective is looking at actors who are affected by the problem. The priorities of each actor are different. When no actors are affected, then there is no problem. The authors list eight actors and their core values. The definition of a problem includes context, situation and boundary. Barriers are limiting factors of the solution and the risks generated while solving the problem. The importance of the problem looks at the scale of return obtained by solving it, the tractability of the additional resources required in solving the problem and the trajectory of the problem's direction.

Gleasure (2013) has suggested a few criteria for evaluating the problem for Design Science Research (DSR). The DSR approach is more prescriptive than descriptive. Therefore, the first criterion is - how far the problem area is matured for the prescriptive aspect. In prescriptive research, the causal factors are easily identified and isolated. The DSR approach mainly focuses on what impacts the causal factors. It uses them as mediators or a moderator between the causal factor and the problem variable, also known as a dependent variable. Mildeová (2013) suggests using cognitive tools to create a mental map for describing a research problem. The authors recommend mind mapping, linear cause-effect thinking and connection circles in defining the research problem. Previously, Reiter-Palmon and Robinson (2009) have argued that the researcher's creativity plays an important role in identifying and constructing a research problem.

Researchers have also contributed certain tools for framing and evaluating research problems. Ellis and Levy (2008) have created a template for crafting a problem statement. Ali et al. (2022) have developed a rubric for evaluating the research problem statement. Nasution et al. (2019) have provided an understanding of writing a problem statement for research. Ali and Pandya (2021) describe a framework with four stages to develop a problem statement for doctoral research.

Addressing the Gap

Research in the management domain mainly focuses on scientific and methodological rigour. The relevance of research to practice is limited. Researchers need to develop specific knowledge for organisations within the local context. The relevance should be checked at the problem formulation stage (Panda & Gupta, 2014). Starkey and Madan (2001) suggest a better partnership between academia and industry to address the relevance gap. More involvement of managers and firms is required in the research process. Contrarily, Kieser and Leiner (2009) argue that either action research, Mode 2 knowledge generation, or collaborative research between academic researchers and industry practitioners could produce rigorous and relevant research.

After modernisation, Pettigrew (2001) argues that the management research community faces a double hurdle of scholarly quality and practical relevance. As knowledge in management is a confluence of multiple disciplines, more internal development is required to satisfy the stakeholders. Rynes et al. (1999) have studied the relationship between the researcher's time spent in the organisation and the success of implementing research findings. The chapter's outcome encourages the researcher's interaction with the practitioners and the organisations' contribution to more relevant research problems. Meyer (2006) is concerned about management research in the Asian region. Researchers are trapped with theories based on Western organisations. The local context and the uniqueness are not focused on as they are sent for international publications' acceptance. The author frames the rigour and relevance of the research problem in the context of its use in contributing to local practices.

Similarly, Khatri et al. (2012) and Stremersch et al. (2023) argue that the quality of research is affected due to a lack of context-specific management research. Mindlessly following models from other countries leads to relevance issues. The authors suggest using constructs and methodologies relevant to the context for a useful research outcome.

OPERATIONALISING RESEARCH DESIGN FROM REAL WORLD PROBLEMS

Researchers in higher education find the research methodology text confusing. Connecting a research problem to the method and choosing a frame is difficult (Cooksey & McDonald, 2019; May & Perry, 2022). Figure 1 presents the process of developing a research design for a real-world problem.

Figure 1. Process of developing a research design for a real-world problem

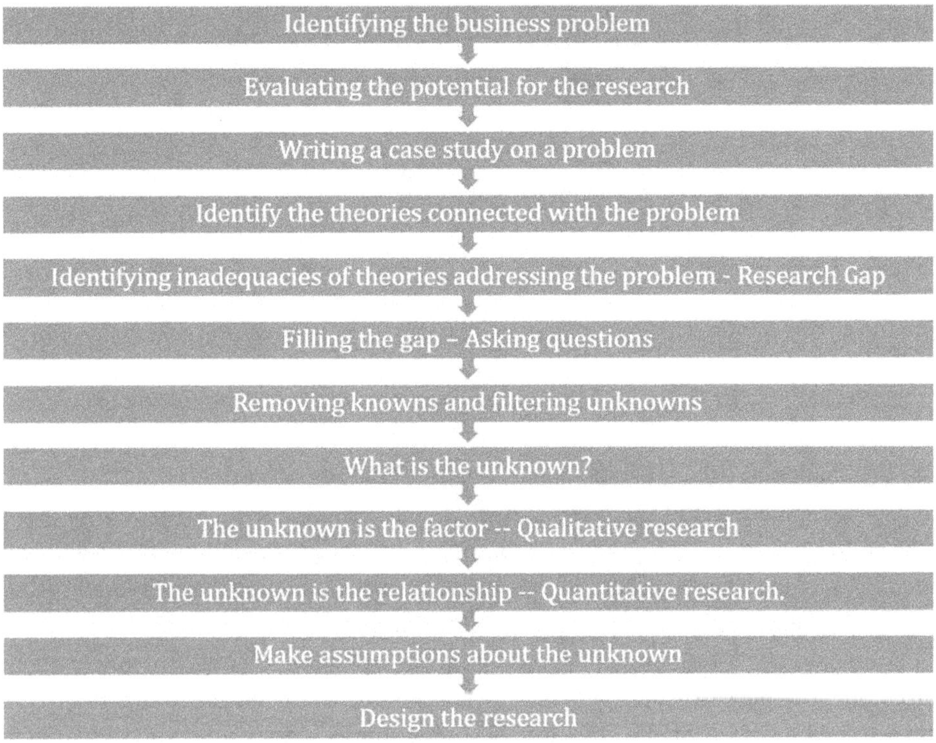

These stages collectively form a structured approach to conducting academic research on business problems, guiding researchers from problem identification to the design and implementation of empirical studies aimed at advancing theoretical understanding and informing practical solutions.

Identifying the Business Problem

A business problem is "any hurdle, situation, or variation that leads to a difference between the desired objectives and accomplished results" (Management & Business Excellence Consultancy, n.d.). The problem is identified when there is a hindrance, undesired or unpredicted outcome. The problems will vary at different stages of the business growth. Anything stopping the growth will be a problem during the initial stage. In the sustainability stage, deviation from norms will be problematic (Management & Business Excellence Consultancy, n.d.). Table 1 illustrates the characteristics of the research problem with practical implications.

Figure 2. The likely business problems

Plan	A —— B 	Business success is when the plan is achieved efficiently
Hurdle	A —●— B 	A hurdle or obstacle that deviates the plan and the desired objective is easily achieved.
Unpredicted outcome	A —— C 	The plan ends up with an outcome that was not predicted.
Undesired outcome	A —— X 	The plan ends up with an outcome that is not desired or welcomed.

Each category in Table 1 reflects different aspects of the planning and execution process in business, highlighting the dynamic nature of organisational environments and the challenges that businesses may encounter along the way.

Business success indicates that a plan is achieved efficiently. Business is considered to be running smoothly when the plan is executed as intended, leading to desired outcomes, such as increased revenue, market expansion, or operational efficiency. Success in this context implies that the organisation effectively implements its strategies, meets its objectives and generates positive results. However, such situations are limited. Usually, businesses face hurdles or obstacles that deviate from the plan and the desired objective is achieved with much difficulty. Such situations involve obstacles or challenges that arise during the execution of the plan, causing deviations from the intended course of action. These hurdles could include factors such as unexpected market changes, resource constraints, competitive pressures, or internal organisational issues. In certain cases, business plans end up with an outcome that was not predicted. This scenario occurs when the actual outcomes of the business plan differ from initial expectations or forecasts. These unpredicted outcomes could be positive or negative and may result from various factors such as market dynamics, consumer behaviour, technological advancements, or regulatory changes. Sometimes, a business plan ends up with an outcome that is not desired or welcomed. In this situation, the business plan yields outcomes that are unfavourable or undesirable for the organisation. These undesired outcomes could include financial losses, market setbacks, reputational damage, or other adverse consequences.

Evaluating the Relevance and Potential of the Research Problem

After identifying the business problem, researchers assess its relevance, significance, and feasibility for academic inquiry. Among the problems that happen regularly in the business world, what qualifies as a research problem needs to be understood. Business research problems must be worthy enough, requiring the researcher to spend time and effort. Characteristics of a business problem suitable for academic research should encompass various dimensions to ensure depth, relevance and potential for contribution to knowledge and practice.

Business problems that reach the level of crisis indicate situations in which the usual strategies and approaches are insufficient to address the issue at hand. These crises may arise from sudden market shifts, technological disruptions, regulatory changes, or internal organisational challenges. Academic research in such crises could involve understanding the root causes, exploring innovative solutions and developing frameworks for crisis management and resilience. Time sensitivity is a crucial aspect of critical business problems. These issues require immediate attention to prevent further escalation of negative consequences. Academic research in critical business problems involves rapid assessment, timely data collection and agile methodologies to provide actionable insights within tight timelines. Such research could focus on decision-making under pressure, crisis communication strategies, or agile project management approaches.

Business problems with potential major losses signify significant risks to the organisation's financial health, reputation, or stakeholder relationships. Academic research in this area may involve risk assessment, mitigation strategies and developing frameworks for proactive risk management. It could also explore the psychological aspects of decision-making in high-stakes situations and the role of organisational culture in fostering risk awareness and resilience. The impact of a business problem refers to its significance in terms of its effects on the organisation's operations, performance and bottom line. Academic research on impactful business problems delves into the underlying mechanisms driving these effects, evaluates the effectiveness of different interventions and develops predictive models or decision support systems to minimise negative impacts and capitalise on opportunities.

Business problems that have large-scale implications extend beyond the boundaries of the organisation and affect its employees, customers, suppliers and the broader community. Academic research in this domain considers the interconnectedness of stakeholders, the ripple effects of organisational actions and the ethical and social responsibilities of businesses. It may involve systems thinking, stakeholder engagement methodologies and cross-sector collaborations to address complex, systemic challenges. Business problems with multiple possibilities indicate complexity and uncertainty in decision-making. Academic research in this realm explores alternative scenarios, evaluates trade-offs and develops decision frameworks to navigate ambiguity and complexity effectively. It may involve interdisciplinary approaches, such as operations research, behavioural economics, or complexity theory, to capture the multifaceted nature of the problem and identify robust solutions. Academic research on business problems should not only address immediate challenges but also contribute to theoretical understanding, methodological advancements, and practical insights that can inform decision-making and enhance organisational resilience and sustainability in the face of uncertainty and adversity. Therefore, a research problem should indicate a critical crisis, leading to a major loss or impacting the smooth conduct of business activity, being large in scale (Table 1).

Table 1. Characteristics of the business problem leading to research

Crisis	Challenges or issues businesses face that need to be addressed to achieve success.
Critical	Business problems may be time-sensitive and require immediate attention.
Major loss	Business problems may have consequences, including financial losses, decreased efficiency, or negative impacts on stakeholders.
Impact	Business problems should be relevant to the business and significantly impact its operations or bottom line.
Large scale	Business problems may greatly impact the company, including its employees, customers and the broader community.
Multiple possibilities	There are many possible solutions, different factors impacting the outcome, or managers having different practices, but the most effective way is unclear.

Types of Problems to Avoid

All the problems may not be connected with research, such as those with known reasons, like production stoppage due to machine failure and operational problems that do not require a decision or a dilemma. Similarly, the non-availability of required resources, lack of knowledge and awareness are not research problems. They are known as circular reasoning, which is also known as begging the question and is a type of logical fallacy, in which the conclusion of an argument is already assumed in the premise. Circular reasoning creates a circular argument in which the conclusion is used to support the premise rather than vice-versa.

Table 2. Suitability of business problem from an academic perspective

Relevance	Is the research problem relevant to the field of study and does it address a current or pressing issue?
Originality	Does the research problem add meaning to the existing body of knowledge, or is it rehashing previous research?
Feasibility	Is the research problem feasible to address, given the resources and time constraints of the study?
Testability	Can the research problem be tested through the collection and analysis of data?
Ethical considerations	Does the research problem consider the rights and well-being of participants?

There are several factors to consider when evaluating the suitability of a business problem from an academic perspective. The characteristics outlined in Table 3 are vital considerations when assessing the suitability of a business problem for academic research, which should address pertinent issues within the field of study. A relevant business problem aligns with current trends, challenges and advancements in the industry. It should have practical implications for practitioners and contribute to theoretical understanding. Research that tackles pressing issues or fills gaps in knowledge is likely to garner interest from both academia and industry.

Originality is essential to academic research, as it ensures the advancement of knowledge. A business problem should offer novel insights, methodologies or perspectives that contribute meaningfully to the existing body of literature. While it is acceptable to build upon previous research, the problem should go beyond mere replication and offer fresh insights or approaches.

Feasibility refers to whether the research problem can be realistically addressed given the available resources, time constraints and practical considerations. It involves assessing factors such as access to data, funding, expertise and logistical support. A feasible research problem can be tackled within the scope of the study without undue burden or risk of failure.

Testability is fundamental to scientific inquiry, as it involves the ability to investigate the research problem through data collection and analysis empirically. A testable business problem should be clearly defined and amenable to research methodologies, such as experiments, surveys, case studies, or statistical analysis. Clear hypotheses or research questions facilitate the formulation of testable hypotheses and the design of rigorous research protocols.

Ethical considerations are paramount in academic research to ensure that the rights, dignity and well-being of participants are protected. A research problem should be formulated with ethical principles in mind, including informed consent, confidentiality, privacy and minimising harm. Researchers should consider potential risks and benefits associated with the research and adhere to ethical guidelines and regulations relevant to their field.

The checklist in Table 4 is used to evaluate the potential of a research problem from the business perspective. The checklist was meticulously crafted by incorporating critical questions that address the core aspects of research problems alongside practical considerations such as time, scope, frequency, and size. The researcher responds to each item in the checklist by evaluating the problem.

Table 3. Checklist for research problem identification in business, management, and commerce

S. No	Questions	Response
1	How did you come to know about the problem?	
2	Who is affected?	
3	Where is the problem happening?	
4	When did the problem last happen?	
5	What is the frequency of the problem?	
6	Is the problem a common one?	
7	Who are you addressing?	
8	How many are affected?	
9	What is the disturbance/damage/impact/loss?	
10	How much is the disturbance/damage/impact/loss?	
11	Is the problem isolated?	
12	Is the problem about the lack of something?	
13	Is the answer or solution obvious?	
14	Is the answer or solution self-evident?	
15	What is the unknown?	
16	Is the problem about describing anything?	
17	Is the problem about summarising?	
18	Does your probing end with how and why questions?	
19	Are you able to clearly identify the subject area of the problem?	
Research Feasibility		
20	What are the strengths and weaknesses of the prior research?	
21	Key ideas, concepts, theories, variables and measures are identifiable.	
22	Adequate number of subjects/cases/samples/	
23	Adequate technical expertise	
24	Affordable in time and money	
25	Manageable in scope	

A problem should be identified from the real world. Problems need to be identified either from newspapers and magazines or by speaking to practising managers and experts. Problems identified from a review of past literature would lack practical relevance and may address only the literature gap. We can comprehensively understand the problem and its context by leveraging diverse sources. The problem needs to be from the manager's perspective and be worthy enough to contribute to the business. The problem may be related to the customer, employee or investors. However, the research recommendations would be worthy if addressed to the managers.

Understanding who is affected by the problem is crucial for effective problem-solving. Managers and decision-makers often face unique challenges, and their concerns hold significance in organisational decision-making. Managers and decision-makers hold the power to implement solutions and allocate resources. However, considering the impact on other stakeholders beyond managers, such as employees, customers, or communities, is equally important. Recognising the scope of impact helps in tailoring solutions that address the needs and concerns of all relevant parties.

To understand the problem completely, it is essential to determine where and when it occurs. While the given information does not provide specific options for location and timeframe, these factors play a significant role in problem analysis. Whether the problem is localised or widespread can influence the scale of required interventions. Similarly, the recency or persistence of the problem can impact the urgency with which it needs to be addressed.

Examining the frequency and commonality of the problem provides insights into its prevalence and significance. Problems that occur more frequently demand immediate attention, as they may have a recurring negative impact. Additionally, understanding whether the problem is industry-wide, limited to a few companies, or specific to one organisation helps to gauge its broader implications and identify potential solutions.

Assessing the impact, disturbance, damage or loss caused by the problem is essential to prioritise the research needs. The visibility and quantifiability of the impact determine its significance. Problems that affect significant stakeholders or have tangible consequences require immediate attention.

From the perspective of the research process, identifying the problem with the following characteristics is important. Understanding whether it is isolated or intertwined with other factors is necessary for the right focus. Addressing an isolated problem is more effective and worth researching than the intertwined ones. A situation that is identifiable but lacks something is not a good research problem. Addressing such problems ends in circular reasoning and recommendations to provide the required resources. It may not lead to any investigation. A research problem leads to multiple questions. Assessing the clarity of the answer to such questions is crucial to understanding the complexity of the research problem. If the answer or solution is self-evident and apparent, the problem is not worth researching. In cases where the answer or solution is unclear or ambiguous, or if it is completely unknown whether the problem has research potential or not, it need not be researched.

Determining whether a research problem involves describing or summarising is significant in designing appropriate research methodologies. If the problem involves describing something, then the problem would lack immediate action or application. Understanding whether the research problem concludes with how and why questions are crucial for selecting appropriate research approaches. Qualitative research will be a suitable approach if the probing ends with how and why questions. Clearly identifying the subject area of the research problem is essential for focused investigations. If the subject area is clearly known, applying relevant expertise and theories in addressing the problem will be easier.

Assessing the feasibility of research involves considering multiple factors, such as the strengths and weaknesses of prior research, the identifiability of key elements, the adequacy of subjects or samples, technical expertise, affordability and manageability in scope. By thoroughly evaluating these aspects, researchers can make informed decisions about their research design, resource allocation and potential impact. Research feasibility assessments enhance the overall quality and credibility of research studies, ensuring that they contribute meaningfully to the body of knowledge in their respective fields.

STRENGTHENING A PROBLEM'S PERSPECTIVE BY WRITING A CASE STUDY ON A PROBLEM

To better understand and empathise with the problem from different stakeholders' perspectives, it is good to build a case study of the business problem taken for research. Case study methodology involves in-depth exploration of a particular problem or phenomenon within its real-life contexts. A researcher can keep away from the pressure of the research perspective of the problem in the current stage and only focus on the problem. The case development will help the researcher to substantiate the research need and the potential application of the research outcome. The case writing stage will be the first part of the literature review undertaken by the researcher to understand the problem.

Identifying Theories Connected with the Problem

During the literature review phase, researchers systematically explore existing theories relevant to the identified problem. This process entails examining various theoretical frameworks to assess their applicability and explanatory power in addressing the problem at hand. Each theory's capacity to elucidate different facets of the problem is scrutinised, offering insights into its underlying mechanisms and potential solutions. As researchers delve into the literature, they seek to understand the breadth and depth of existing theoretical perspectives, evaluating their strengths, limitations and empirical support. By synthesising diverse theories, researchers can discern patterns, contradictions and gaps in understanding, thus refining their conceptualisation of the problem.

Ultimately, the literature review culminates in systematically organising and integrating these theories into a coherent framework, elucidating their interrelationships and implications. This exercise will help in the construction of a theory map, which visually represents the interconnection between different theories. By drawing connections between theories and constructing a theory map, researchers can gain deeper insights into the problem and lay the foundation for further empirical investigation.

IDENTIFYING INADEQUACIES OF THEORIES ADDRESSING THE PROBLEM - RESEARCH GAP

Critical evaluation of each theory shows the inadequacies of theories in explaining the problem. The potential research area is the gap between the existing theories and the problem. Addressing the gap would be a worthy contribution to academics in adding to the theories and the business through knowledge towards solving the problem or a better solution. Researchers identify gaps or limitations in existing theories that fail to fully address the nuances or complexities of a problem at hand. This step highlights areas where further research is needed to advance understanding or propose alternative explanations. (Figure 3)

Figure 3. Identifying the research gap

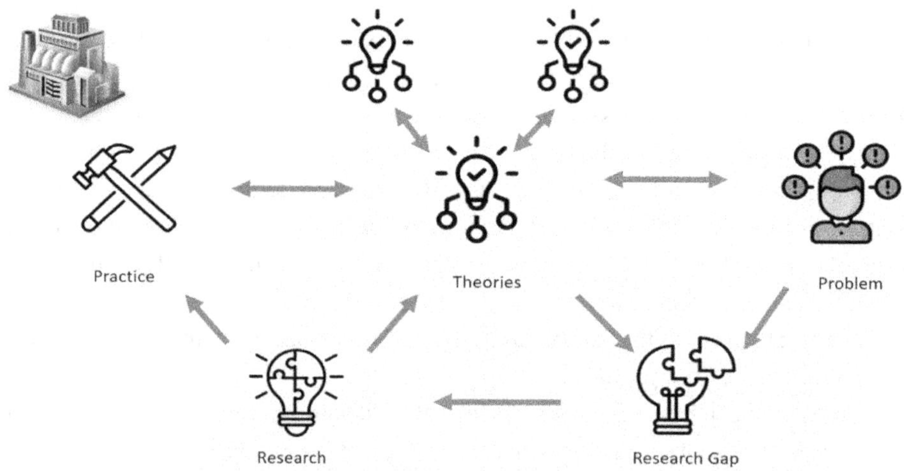

Filling the Gap – Asking Questions

In the process of filling the gap between existing knowledge and the identified problem, researchers formulate precise research questions designed to guide their investigation. These questions serve as compass points, directing the focus of the research toward addressing specific aspects of the problem. Researchers often employ a combination of interrogative pronouns – who, what, where, when, why and how – to explore the nuances of the research gap comprehensively.

Firstly, questions interrogating the nature of the gap itself are crucial. Who is affected by the gap? What are the specific dimensions or manifestations of the gap? Where does the gap exist – in terms of geographic, demographic, or institutional contexts? When did the gap emerge, and has it evolved over time? Why does the gap persist and what underlying factors contribute to its existence? How has the gap impacted individuals, communities or systems?

Additionally, researchers pose questions related to potential solutions or interventions. What strategies have been proposed or implemented to address similar gaps in the literature? Who are the key stakeholders involved in addressing the gap, and what are their perspectives? How effective have past interventions been, and what lessons can be learned from previous attempts? What innovative approaches or alternative perspectives could be explored to bridge the gap effectively?

By systematically interrogating the research gap from multiple angles and considering various dimensions, researchers can refine their understanding of the problem and identify avenues for meaningful inquiry. These research questions are the platform upon which the research study is built, guiding the formulation of hypotheses, the design of methodologies and the interpretation of findings. Ultimately, through rigorous inquiry and thoughtful analysis, researchers endeavour to contribute new insights and knowledge that address the identified gap and advance understanding within their fields.

Removing Knowns and Filtering Unknowns

In the process of removing the knowns and filtering the unknowns, researchers undertake a systematic exploration of each research question to determine whether answers already exist within the current body of knowledge or through common sense reasoning. Questions for which answers are readily available or evident may not necessitate further research as they do not contribute to advancing understanding or addressing gaps in knowledge.

Researchers begin by conducting a thorough literature review, scouring academic journals, books and other scholarly sources to uncover existing evidence, theories and perspectives relevant to each research question. This process helps to identify established findings and known solutions within the field, allowing researchers to discern which questions have already been adequately addressed. Additionally, researchers may engage in conversations with managers, experts and practitioners within the field to glean insights and perspectives that may not be readily available.

These interactions provide valuable real-world context and practical expertise, helping researchers verify assumptions and refine their understanding of the research questions.

As researchers systematically assess each question, they identify those for which satisfactory answers remain elusive or unavailable. These unanswered questions represent the crux of the research endeavour, serving as focal points for further investigation and inquiry. By pinpointing gaps in existing knowledge or understanding, researchers can delineate the scope and direction of their research efforts, guiding the formulation of hypotheses, research methodologies and data collection strategies. These unanswered questions serve as the driving force behind the research process, motivating researchers to delve deeper into uncharted territory and generate new insights. Throughout the research journey, these questions act as beacons, guiding data collection, analysis and interpretation efforts and ultimately leading to the generation of novel findings and contributions.

Thus, by systematically filtering knowns and focusing on unknowns, researchers can harness the power of curiosity and inquiry to advance understanding and address pressing challenges within their discipline.

What is the Unknown?

The 'unknown' in research refers to aspects of a problem or phenomenon that lack clear explanations or empirical evidence. These are areas where understanding is limited or incomplete, leaving gaps in knowledge that require further exploration. The unknown can manifest in various forms, including:

1. Variables: Researchers may identify certain variables that have not been adequately studied or understood within the context of the problem. These variables could include factors that influence outcomes or behaviours but have not been clearly defined or measured.
2. Relationships: The unknown may involve relationships between variables that have not been fully explored or understood. This could include causal relationships, associations, or interactions between different factors that require further investigation.
3. Mechanisms: Researchers may seek to uncover the underlying mechanisms or processes that drive observed phenomena. This could involve understanding how certain variables or factors interact to produce specific outcomes or behaviours.
4. Outcomes: The unknown may pertain to the outcomes or consequences of certain actions, interventions or events. Researchers may seek to determine the potential outcomes of a particular phenomenon or the long-term effects of specific interventions.

By explicitly defining the unknown aspects of the problem, researchers can focus their efforts on addressing these gaps in knowledge through empirical research, experimentation, and analysis. This process involves formulating research questions, hypotheses and methodologies designed to shed light on the unknown so that they can advance understanding within the field. Ultimately, by systematically investigating the unknown, researchers contribute to the expansion of knowledge and the development of new insights and solutions to complex problems.

The Unknown is the Factor -- Qualitative Research.

When the unknown is a factor, qualitative research methods are particularly well-suited for exploring the unknown in-depth, as they enable researchers to delve into the complexities and nuances of human experiences and social phenomena. In qualitative research, the 'unknown' refers to factors, phenomena, or aspects of a problem that are not well understood or explored. These could include nuanced experiences, perceptions, motivations, cultural influences, social dynamics, or other subjective dimensions that may not be easily captured or quantified through traditional methods. Techniques such as interviews, observations and content analysis allow researchers to gather rich, detailed data that capture the subtleties and nuances of the phenomenon under investigation.

Interviews provide an opportunity for participants to share their perspectives, experiences and insights in their own words, offering valuable insights into their thoughts, emotions and motivations. Through open-ended questioning and active listening, researchers can uncover underlying meanings, beliefs and values that may influence behaviour or decision-making. Observations involve systematically observing and recording behaviour, interactions and events within naturalistic settings. This allows researchers to gain firsthand insights into social processes, group dynamics and contextual factors that may shape behaviour or outcomes. Content analysis involves systematically analysing written, verbal, or visual communication, such as texts, documents, or media content, to identify recurring themes, patterns, or discourses. This technique can provide insights into cultural norms, societal values, or discursive practices that may inform attitudes, beliefs, or behaviours.

The Unknown is the Relationship -- Quantitative Research.

Quantitative research methods are specifically designed to examine **unknown** relationships systematically and rigorously. In quantitative research, the 'unknown' often pertains to relationships between variables, which may not be fully understood or explored. These relationships can encompass causal connections, correlations, or associations among different factors or phenomena within a research context. Researchers utilise various techniques, such as surveys, experiments and statistical analyses, to quantify and analyse patterns or associations within the data, shedding light on the unknown aspects of these relationships.

Surveys are commonly employed to collect data from a large sample of participants, enabling researchers to assess the relationship between variables through the administration of standardised questionnaires. By measuring the variables of interest and examining their correlations, researchers can identify patterns and associations within the data, providing insights into the nature and strength of the relationships. Experiments involve manipulating one or more variables while controlling others to determine their effect on a dependent variable. By systematically varying the conditions and measuring the outcomes, researchers can assess causal relationships between variables, providing empirical evidence to support or refute hypotheses derived from theoretical frameworks. Statistical analyses play a crucial role in quantitative research, allowing researchers to identify and quantify relationships between variables. Techniques such as regression analysis, correlation analysis and factor analysis are used to examine the strength and direction of relationships, assess the significance of findings and test hypotheses derived from theoretical models.

Make Assumptions about the Unknown.

In the realm of research, assumptions about the unknown aspects of a problem serve as foundational hypotheses that guide inquiry and investigation. Researchers draw upon a variety of sources to inform their assumptions about the unknown. This may include reviewing existing literature to identify gaps or inconsistencies in understanding, consulting theoretical frameworks to generate hypotheses about underlying mechanisms or relationships, and conducting exploratory research to gather preliminary insights into the phenomenon under study. These assumptions serve several key functions in the research process. Firstly, they provide a framework for defining research questions and objectives, helping researchers to focus their efforts on addressing specific aspects of the problem. By delineating the scope of inquiry, assumptions guide the selection of variables, the design of research methods and the development of data collection instruments.

Secondly, assumptions serve as hypotheses that can be tested empirically through data collection and analysis. Researchers design studies to gather evidence that either supports or refutes these assumptions, thereby advancing knowledge and understanding within the field. Through rigorous testing and validation, researchers can refine their assumptions, revise hypotheses and deepen insights into the unknown aspects of the problem.

Additionally, assumptions play a role in shaping researchers' interpretations of data and findings. Researchers draw upon their assumptions to interpret patterns, trends and relationships observed in the data, providing context and meaning to empirical results. This iterative process of hypothesis testing and interpretation contributes to the ongoing refinement and development of theoretical frameworks and understanding within the field.

Design the Research

Designing the research involves careful consideration of various components to ensure a systematic and rigorous investigation of the identified problem. Researchers develop a research design that outlines the overall approach, methods and procedures for investigating the identified problem. This includes selecting appropriate data collection techniques, sampling methods and data analysis procedures and ensuring that ethical considerations are addressed. By carefully designing the study, researchers ensure that it is well-planned, methodologically sound and ethically conducted. A robust research design facilitates systematic investigation, enabling researchers to generate valid and reliable findings that contribute to the advancement of knowledge.

CONCLUSION

Identifying a meaningful research problem in business, management, and commerce is a multifaceted challenge requiring both scholarly rigor and practical relevance. This chapter emphasizes the necessity of bridging the significant rigour-relevance gap in academic research within these fields. Despite many potential research topics, discerning which problems merit scholarly investigation and have real-world implications remains difficult. Inadequate training for researchers to discern significant problems necessitates robust educational programs to enhance their analytical skills. Academic research in business and management often suffers from a disconnect with real-world problems. Methodological rigor sometimes overshadows practical concerns, contributing to a rigour-relevance gap where research has minimal impact on management practices. Therefore, applied research that engages with real business issues is crucial. Doctoral research also faces this relevance gap. Universities should train scholars to solve real-world problems, as traditional literature reviews may fail to capture practical implications. By identifying and addressing practical problems businesses face, researchers can generate findings that can potentially improve business practices, increase efficiency and productivity and enhance customer satisfaction. To identify relevant research problems in the business, management and commerce domains, a systematic approach is indispensable. Addressing the rigour-relevance gap requires a paradigm shift towards closer collaboration between academia and industry, enhancing the practical orientation of research, and equipping researchers to address real-world problems, thereby making academic research more impactful and relevant. This chapter provides an overview of challenges and methodologies in problem identification, offering recommendations and a checklist for evaluating research problem relevance. The proposed checklist for identifying problems encompasses an assessment of the prevailing challenges confronting businesses alongside an evaluation of potential research implications. By adhering to this structured framework, researchers can not only delineate a worthy research problem within the context of the real world but also bridge the research-rigour gap. Focusing on applied research, researchers in the social science and business management domains can make a meaningful contribution to the field and help solve real-world business problems.

REFERENCES

Ali, A., & Pandya, S. (2021). A four stage framework for the development of a research problem statement in doctoral dissertations. *International Journal of Doctoral Studies*, 16, 469–485. DOI: 10.28945/4839

Ali, A., Pandya, S., & Varma, U. (2022). Using Analytical Thought Process To Develop Instructional Rubrics in Writing Doctoral Dissertation Research Problem Statements – a Follow-Up Study. *International Journal of Doctoral Studies, 17*, 385–399. DOI: 10.28945/5017

Benbasat, I., & Zmud, R. W. (1999). Empirical research in information systems: The practice of relevance. *Management Information Systems Quarterly*, 23(1), 3–16. DOI: 10.2307/249403

Castrillón, J. C. A. (2020). Raise and formulate a research problem: An exercise of reasoning. *Revista Lasallista de Investigacion*, 17(1), 301–313. DOI: 10.22507/rli.v17n1a4

Chambers, M. M. (1960). Selection, Definition, and Delimitation of a Doctoral Research Problem. *Phi Delta Kappan*, 42(2), 71–73. https://www.jstor.org/stable/20342507

Cooksey, R., & McDonald, G. (2019). Surviving and Thriving in Postgraduate Research. In *Surviving and Thriving in Postgraduate Research*. DOI: 10.1007/978-981-13-7747-1

Cutillas, A., Benolirao, E., Camasura, J., Golbin, R.Jr, Yamagishi, K., & Ocampo, L. (2023). Does Mentoring Directly Improve Students' Research Skills? Examining the Role of Information Literacy and Competency Development. *Education Sciences*, 13(7), 1–21. DOI: 10.3390/educsci13070694

DeNisi, A. S. (1994). Is relevant research irrelevant?: On evaluating the contribution of research to management practice. *Journal of Managerial Issues*, 6(2), 145.

Dissanayake, D. M. N. S. W. (2013). Research, Research Gap and the Research Problem. *Munich Personal RePEc Archive*, 47519, 1–5.

Dredge, D. (2015). Does relevance matter in academic policy research? *Journal of Policy Research in Tourism, Leisure & Events*, 7(2), 173–177. DOI: 10.1080/19407963.2014.990661

Easterby-Smith, M., Jaspersen, L. J., Thorpe, R., & Valizade, D. (2021). Management and business research. *Sage (Atlanta, Ga.)*.

J. Ellis, T., & Levy, Y. (2008). Framework of Problem-Based Research: A Guide for Novice Researchers on the Development of a Research-Worthy Problem. *Informing Science: The International Journal of an Emerging Transdiscipline, 11*(January), 017–033. DOI: 10.28945/438

Gardner, S. K., Mendoza, P., & Austin, A. E. (2023). ON BECOMING A SCHOLAR: Socialization and Development in Doctoral Education. In *On Becoming a Scholar: Socialization and Development in Doctoral Education.* Taylor \& Francis. DOI: 10.4324/9781003446187

Gleasure, R. (2013). What is a "Wicked Problem" for IS Research? *The 2nd International SIG Prag Workshop on "IT Artefact Design & Workpractice Improvement," Conklin 2008*, 1–12. http://www.vits.org/adwi2013/RGleasure-ADWI2013.pdf %0Awww.vits.org/adwi2013/RGleasure-ADWI2013.pdf

Gordon, R. A., & Howell, J. E. (1959). Higher Education for Business. In *Higher Education for Business.* Columbia University Press., DOI: 10.7312/gord91408

Hart, J., Hakim, J., Kaur, R., Jeremy, R., Coorey, G., Kalman, E., Jenkin, R., & Bowen, D. (2022). Research supervisors' views of barriers and enablers for research projects undertaken by medical students; a mixed methods evaluation of a post-graduate medical degree research project program. *BMC Medical Education*, 22(1), 370. DOI: 10.1186/s12909-022-03429-0 PMID: 35562832

Herwix, A., & Haj-Bolouri, A. (2021). Revisiting the Problem of the Problem - An Ontology and Framework for Problem Assessment in IS Research. *ECIS 2021 Proceedings, June*, 154. https://aisel.aisnet.org/ecis2021_rp/154

Jenkins, K. E. H., Stephens, J. C., Reames, T. G., & Hernández, D. (2020). Towards impactful energy justice research: Transforming the power of academic engagement. *Energy Research and Social Science, 67*(October 2019), 101510. DOI: 10.1016/j.erss.2020.101510

Khatri, N., Ojha, A. K., Budhwar, P., Srinivasan, V., & Varma, A. (2012). Management research in India: Current state and future directions. *IIMB Management Review*, 24(2), 104–115. DOI: 10.1016/j.iimb.2012.05.002

Kieser, A., & Leiner, L. (2009). Why the rigour-relevance gap in management research is unbridgeable. *Journal of Management Studies*, 46(3), 516–533. DOI: 10.1111/j.1467-6486.2009.00831.x

Koskela, L. (2011). Fifty years of irrelevance: The wild goose chase of management science. *Lean Construction Journal*, 2011(Special Issue), 1–11.

Koskela, L. (2017). Why is management research irrelevant? *Construction Management and Economics*, 35(1–2), 4–23. DOI: 10.1080/01446193.2016.1272759

Lund, T. (2022). Research Problems and Hypotheses in Empirical Research. *Scandinavian Journal of Educational Research*, 66(7), 1183–1193. DOI: 10.1080/00313831.2021.1982765

Management & Business Excellence Consultancy. (n.d.). *Business Problem Identification*. Retrieved January 2, 2023, from https://www.mnbec.com/business-problem-identification/

Marchisotti, G. G., & Farias Filho, J. R. (2022). Application of a Multi-method to Identify a Research Problem. *Revista de Administracao Contemporanea, 26*(6), 1–17. https://doi.org/DOI: 10.1590/1982-7849rac2022210020.en

May, T., & Perry, B. (2022). *Social research: Issues, methods and process*. McGraw-Hill Education.

Meyer, K. E. (2006). Asian management research needs more self-confidence. *Asia Pacific Journal of Management*, 23(2), 119–137. DOI: 10.1007/s10490-006-7160-2

Mildeová, S. (2013). Research problem description and definition: From mental map to connection circle. *Journal on Efficiency and Responsibility in Education and Science*, 6(4), 328–335. DOI: 10.7160/eriesj.2013.060409

Minnaert, L. (2015). Does relevance matter in academic policy research? A comment on Dredge. *Journal of Policy Research in Tourism, Leisure & Events*, 7(2), 190–194. DOI: 10.1080/19407963.2014.990663

Nasution, M. K. M., Onrizal,, , & Aulia, I. (2019). Design of the research problem statement. *Journal of Physics: Conference Series*, 1235(1), 012115. Advance online publication. DOI: 10.1088/1742-6596/1235/1/012115

Panda, A., & Gupta, R. K. (2014). Making academic research more relevant: A few suggestions. *IIMB Management Review*, 26(3), 156–169. DOI: 10.1016/j.iimb.2014.07.008

Pardede, P. (2018). Identifying and Formulating the Research Problem. *Research in ELT, 1*(October), 1–13. https://www.researchgate.net/publication/329179630_Identifying_and_Formulating_the_Research_Problem

Rajan Varadarajan, P. (2003). Musings on Relevance and Rigor of Scholarly Research in Marketing. *Journal of Technological Advancements*, 31(4), 368–376. DOI: 10.1177/0092070303258240

Reiter-Palmon, R., & Robinson, E. J. (2009). Problem Identification and Construction: What Do We Know, What Is the Future? *Psychology of Aesthetics, Creativity, and the Arts*, 3(1), 43–47. DOI: 10.1037/a0014629

Rudolph, H. R., & Peluchette, J. V. (2011). Significance Of Usefulness: A Congruency Model Of Relevant Research Criteria. [JABR]. *Journal of Applied Business Research*, 8(3), 83. DOI: 10.19030/jabr.v8i3.6148

Rynes, S. L., Mc Natt, D. B., & Bretz, R. D. (1999). Academic research inside organizations: Inputs, processes, and outcomes. *Personnel Psychology*, 52(4), 869–898. DOI: 10.1111/j.1744-6570.1999.tb00183.x

Sandberg, E., Oghazi, P., Chirumalla, K., & Patel, P. C. (2022). Interactive research framework in logistics and supply chain management: Bridging the academic research and practitioner gap. *Technological Forecasting and Social Change, 178*(September 2021), 121563. DOI: 10.1016/j.techfore.2022.121563

Saunders, A. M., & Lewis, P. (2019). *Research Methods for Business Students* (8th ed.). Pearson.

Shrivastava, P. (1987). Rigor and practical usefulness of research in strategic management. *Strategic Management Journal*, 8(1), 77–92. DOI: 10.1002/smj.4250080107

Sidor, J. (2015). Debate Over Rigor and Relevance in Scientific Study of Management. *Management and Business Administration.Central Europe*, 23(3), 32–46. DOI: 10.7206/mba.ce.2084-3356.149

Starkey, K., & Madan, P. (2001). Bridging the Relevance Gap: Aligning Stakeholders in the Future of Management Research. *British Journal of Management*, 12(SPEC. ISS.). Advance online publication. DOI: 10.1111/1467-8551.12.s1.2

Stremersch, S., Gonzalez, J., Valenti, A., & Villanueva, J. (2023). The value of context-specific studies for marketing. *Journal of the Academy of Marketing Science*, 51(1), 50–65. DOI: 10.1007/s11747-022-00872-9

Thomas, K. W., & Tymon, W. G. (1982). Necessary Properties of Relevant Research: Lessons from Recent Criticisms of the Organizational Sciences. *Academy of Management Review*, 7(3), 345–352. DOI: 10.2307/257326

Tsui, A. S., & McKiernan, P. (2022). Understanding Scientific Freedom and Scientific Responsibility in Business and Management Research. *Journal of Management Studies*, 59(6), 1604–1627. DOI: 10.1111/joms.12816

Van Aken, J. E. (2004). Management Research Based on the Paradigm of the Design Sciences: The Quest for Field-Tested and Grounded Technological Rules. *Journal of Management Studies*, 41(2), 219–246. DOI: 10.1111/j.1467-6486.2004.00430.x

Vehviläinen, S. (2009). Problems in the research problem: Critical feedback and resistance in academic supervision. *Scandinavian Journal of Educational Research*, 53(2), 185–201. DOI: 10.1080/00313830902757592

Veit, D. R., Lacerda, D. P., Camargo, L. F. R., Kipper, L. M., & Dresch, A. (2017). Towards Mode 2 knowledge production: Analysis and proposal of a framework for research in business processes. *Business Process Management Journal*, 23(2), 293–328. DOI: 10.1108/BPMJ-03-2016-0045

Wasti, S. A., & Robert, C. (2003). Out of Touch? an Evaluation of the Correspondence Between Academic and Practitioner Concerns in IHRM. *Advances in International Management*, 15, 207–239. DOI: 10.1016/S0747-7929(03)15010-X

Zikmund, W. G., Babin, B. J., Carr, J. C., & Griffin, M. (2013). Business Research Methods. In *Business Research Methods* (9th ed.). Cengage India Private Limited.

Zuber-Skerritt, O., & Knight, N. (1986). Problem definition and thesis writing - Workshops for the postgraduate student. *Higher Education, 15*(1–2), 89–103. DOI: 10.1007/BF00138094

Chapter 9
Effective Strategies for Research Proposal Development in Higher Education

Daniel Opotamutale Ashipala
https://orcid.org/0000-0002-8913-056X
University of Namibia, Namibia

Fillemon S. Albanus
https://orcid.org/0000-0003-1516-2523
University of Namibia, Namibia

ABSTRACT

Writing a research proposal is a requirement for obtaining an undergraduate or postgraduate degree at institutions of higher education globally, however many students struggle to write such a proposal. This is especially true for students who are writing a proposal for the first time, despite research methodology modules that introduce them to the process. This often leads to their studies being delayed, which can result in them not graduating on time. In certain extreme cases, students withdraw from their studies because they have not completed the dissertation required to obtain their intended degree, even when they have passed their coursework. Higher education institutions could use the content of this chapter to develop appropriate targeted interventions and ongoing strategies for use by students and study supervisors to address the challenges that students face when writing their research proposals.

DOI: 10.4018/979-8-3373-0806-7.ch009

INTRODUCTION

As part of the fulfilment of their degrees, both undergraduate and postgraduate students studying at higher education institutions are required to carry out research; this, of necessity, includes research proposal writing. Writing a research proposal in academia requires students to possess academic writing skills, yet many face tremendous challenges in this regard (Ashipala & Livingi, 2021). Such a proposal generally starts with an introduction, which should include a brief discussion of the background and a clear statement of the problem, thus placing the focus of the report within a broader context.

The goal of a research proposal is to present and justify a research idea, as well as discuss practical ways in which the research could be conducted. All proposals should contain an extensive literature review and provide persuasive evidence that there is a need for the research being proposed (Mapolisa & Mafa, 2012). In addition to providing a rationale for the proposed research, a proposal should include a detailed description of the methodology that will be used to conduct the research, consistent with the requirements of the professional or academic field, as well as the anticipated outcomes and/or benefits to be derived from the study. According to Anderson and Krathwohl (2001), a proposal should contain all the key elements involved in designing a complete research study, with sufficient information to allow readers to assess its validity and usefulness.

A research proposal is intended to convince the reader that the student (researcher) has an important research project, as well as the necessary competence to complete it. A high-quality research proposal not only promises success for the research project, but will also impress the research review committee regarding the student's potential as a researcher. It is important that the quality of the proposed research be assessed and possible factors that could affect the quality of academic research and future publications be addressed (Shatimwene & Ashipala, 2023). In addition, a research study is only as good as its proposal and that a poorly prepared proposal will doom the research project to failure.

A study conducted in Tanzania on research proposal writing found that most studies conducted worldwide on undergraduate training have focused on issues of supervision (Hartley & Jory, 2015; Mafa & Mapolisa, 2011; Murray, 2012; Taylor & Beasley, 2015). The limited number of studies on the challenges of writing theses or dissertations (Bitchener & Basturkmen, 2006; Lessing & Schulze, 2002; Wang & Li, 2008), meanwhile, focus on students pursuing their studies in the United Kingdom or elsewhere, but not on the African continent. Therefore, the Tanzanian study was triggered by the limited knowledge on the nature and scope of the challenges facing the majority of undergraduate and postgraduate students in the Tanzanian context when writing theses and dissertations (Komba, 2016). Komba, the author of the study, hoped to inform both students and academic staff in Tanzanian higher learning institutions about the issues related to thesis and dissertation writing. Specifically, the findings were expected to serve as baseline information for making recommendations on how best the challenges facing undergraduate and postgraduate students when writing theses and dissertations could be overcome.

Higher education institutions could use the content of this chapter to develop appropriate targeted interventions and ongoing strategies that can be used by students and study supervisors to address the challenges that students face when writing their research proposals.

AIM OF A RESEARCH PROPOSAL

A research proposal is written in order to obtain approval for academic research or to obtain a research grant from the institution's research review committee.

COMPONENTS OF A RESEARCH PROPOSAL

1. Title

The title describes the specific area which the researcher wants focus on during the research study. It is often referred to as the research topic and should therefore be a self-explanatory, concise expression of the subject of enquiry. This expression is limited to the scope of the study. The research topic is not the research problem; rather, it is a major first step in a research project and is always placed at the beginning of a proposal. First-time researchers are likely to find the selection of a topic for research difficult.

AN EXAMPLE OF A RESEARCH TOPIC

Nursing students' experiences of communication in a multilingual and multicultural clinical environment at training hospitals in Khomas region, Namibia

Guidelines for Selecting a Research Topic

The topic should be of interest to the student

Undergraduate students are likely to spent two years researching the topic they choose; this is even longer for postgraduate students and therefore the topic must be of interest to them. If the student has no interest in a topic it will be difficult for him or her to spend so many years working on it.

It should be Original

Efforts should be made to ensure that the research topic that the student chooses is original, and has not been submitted elsewhere for degree purposes.

It should be Researchable

As researchers, students should have a knowledge grounding of the subject on which the research topic is based, as well as the research skills and aptitude required to undertake the research approach they intend adopting for the study (Fandino, 2019).

It should be Significant

As researchers, students should make sure that they are able to give clear reasons for why their study is needed and the specific contribution the research will make to furthering academic knowledge in their field (Fandino, 2019).

It should be Feasible

The feasibility of the research study is about asking whether the study can be done given the available resources, time and technology (Fandino, 2019). It is also important to consider whether the sample size is attainable, the variables can be measured effectively and there are enough resources, including funding and expertise, to carry the study through to completion.

2. Introduction and Background to the Study

The introduction and background section are very important for the research proposal as it presents the first opportunity for the researcher to attract readers to his or her research. In this section the researcher should tell the reader about the phenomenon under investigation within a global, regional and national context, as well as give the latest statistics together with the sources for these statistics. The emphasis should be on reviewing as much literature as possible while highlighting the most important information for the study.

This section should focus on the following keys aspects:

- What is it that you are researching and why is this specifically important?
- What is the significant contribution of this research to a field or discipline?
- How much about this study is already known?
- Who are the intended beneficiaries of this research?
- What are the key questions that this research seeks to answer?
- What will the findings from study add to the field or discipline?

Figure 1. How to write the introduction

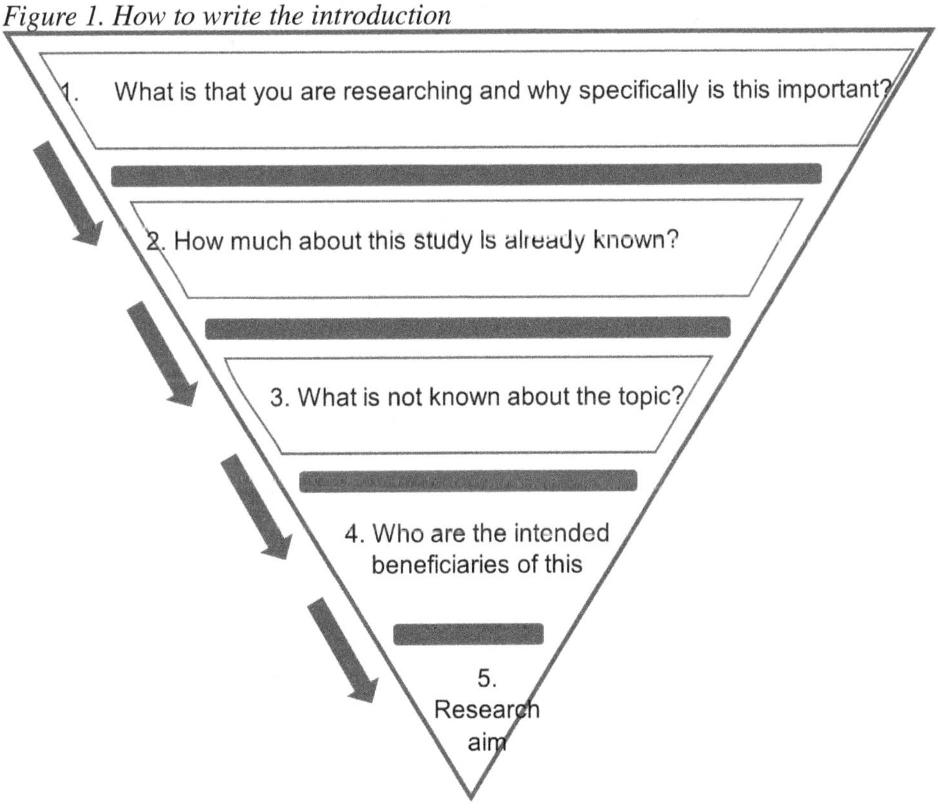

3. Problem Statement

A problem statement is an explanation in research that describes the issue that requires a solution (Polit & Beck, 2017). It answers the question: What problem is the research attempting to address? Having a problem statement allows the reader to quickly understand the purpose and intent of the research. This section is expected to be brief and concise, and should not include the findings of the research or detailed data. The average length of a problem statement is generally about one page. It defines the problem, which can also be thought of as a gap in the information base. There may be several solutions to this gap or lack of information, but that is not the concern of the problem statement; its purpose is to summarise the current information and indicate where a lack of knowledge may be presenting a problem that needs to be investigated.

The purpose of the problem statement is to identify the issue that is a concern and focus on it in a way that allows it to be studied systematically. It defines the problem and proposes a way to research a solution, or demonstrates why further information is needed in order for a solution to become possible.

It has been observed that students generally appeal to their supervisors to help them formulate their research problem. This is not good practice and should be avoided. Experience has shown that students who have their research problems imposed on them by their supervisors usually run into problems in the middle of their study. Therefore, students should be encouraged to come up with their own problem statements so that they can fully own their study.

Four Criteria for a Good Problem Statement

- The problem should be stated clearly and unambiguously, probably in question form.
- The problem should be testable by empirical methods.
- If quantitative research, the problem should not represent a moral or ethical position.
- In the case of quantitative research, the problem should express a relationship between two or more variables.

PRACTICAL EXAMPLE OF A PROBLEM STATEMENT

The following is a practical example of a problem statement for the research topic presented earlier: Nursing students' experiences of communication in a multilingual and multicultural clinical environment at training hospitals in Khomas region, Namibia

Effective communication hinges on multiple factors including the mutual intelligibility of the language used between and among interlocutors. Diverse linguistic backgrounds thus call for the ability of interlocutors to effectively turn their thoughts into words. However, when words need to be sought in another language other than the mother tongue, challenges may occur in communication between the nursing staff and the patients (Meuter et al., 2015, p. 99). The researcher has observed that in Namibia generally and the hospital in particular, the patients, the staff and the relatives of patients do not share the same language and sometimes do not have a common language. The fact that these languages differ and are not mere nomenclature but articulations of the realities of the speakers means that when people with different mother tongues interact, systemic failures may occur (Prilutski, 2010, p. 75). Thus, when people from different cultural and linguistic backgrounds interact, the problems encountered in communication become complex. If there are miscommunication patients are more likely to miss appointments, making treatment and follow up difficult, and they are less likely to understand their treatment regimen, thus impairing adherence (Ha & Longnecker, 2010, p. 159). In this environment students are expected to learn effectively, therefore it is worthwhile exploring how they have experienced communication in a multilingual environment.

4. Significance of the Study

The significance of the study is a written statement that explains why the proposed research is needed. It is a justification for the importance of the work and the impact it will have on the research field, its contribution to new knowledge and how others will benefit from it. What will the study contribute to clinical practice and theory? Will the findings of the study enhance the quality of teaching and learning? Will the study influence decisions have made by policymakers?

PRACTICAL EXAMPLE OF A STATEMENT OF THE SIGNIFICANCE OF THE STUDY

The outcome of the study may provide measures for managing challenges encountered by nursing students during clinical practice when communicating with patients who use a different language to them. It could provide the nursing faculty and clinical educators with measures to assist and tackle the communication challenges that students encounter during clinical placement.

The nurses, patients and students themselves may benefit from the results of the research by developing an understanding of the various implications of using local languages in the delivery of patient care. The author expresses the hope that the findings of this study can be used to provide relevant information to nurse lecturers, clinical tutors, head nurses and policymakers in general, who can then address the findings and recommendations of the study.

5. Purpose/aim of the Study

The purpose statement succinctly explains the aim and objectives of the research project. These objectives must address the problem directly. The purpose statement also identifies the methodology and design for the project.

PRACTICAL EXAMPLE OF THE AIM OF THE STUDY

The purpose of this study was to explore and describe the nursing students' experiences of communication in a multilingual and multicultural clinical environment at training hospitals in Khomas region, Namibia.

6. Limitations and Delimitations of the Study

Definition of the concepts

Operational definitions can be regarded as definitions specific to the research study and are developed when the student defines certain concepts or key terminologies that he or she will use in their research. This gives a broad overview of what exactly the researcher meant by using a specific term or concept (Slife et al., 2016).

PRACTICAL EXAMPLE OF AN OPERATIONAL DEFINITION

Operational definition of the term "nursing students":

In this study, nursing students refers to all the first, second, third and fourth-year nursing students enrolled at training hospitals in Khomas region, Namibia.

Conceptual definitions are dictionary definitions of words; that is, the words have the same meaning across different cultures and in different studies. They are often legal definitions that are internationally accepted and cannot be easily changed. Such definitions are generally clear and precise.

PRACTICAL EXAMPLE OF A CONCEPTUAL DEFINITION

A **conceptual definition** of the term "nursing student":

A nursing student is someone undergoing education and training to become a professional nurse or someone who is receiving training in basic nursing (Merriam-Webster, 2023).

The **limitations** of the study are conditions that are beyond the control of the research that may negatively influence the research findings or are restrictions or problems in a study that may decrease the generalisability of the findings (Grove et al., 2019; Polit & Beck, 2017). Such limitations may include time and financial constraints, the inaccessibility of some research sites and the unavailability of adequate numbers of respondents.

PRACTICAL EXAMPLE OF STUDY LIMITATIONS

The study will be using qualitative interviews only as a method of data collection. By implication this means that the data collection process might take longer than expected, since more time is needed for probing questions and possibly follow-up probes. Moreover, data for this study will be gathered at training hospitals in Khomas region, Namibia; which is a peri-rural area. Therefore, data from this study cannot be generalised to other training hospitals in Namibia. In order to generalise these findings a bigger study would need to be conducted. This study will be limited as it will only focus on the nursing students' experiences of communication in a multilingual and multicultural clinical environment at training hospitals in Khomas region, Namibia. Hence, generalisation of the study to nursing students registered/enrolled at other training hospitals in Namibia will be limited, as their student experiences of communication in a multilingual and multicultural clinical environment may differ from those of nursing students in other regions of Namibia.

Delimitations are the boundaries of one's research. In this regard, the researcher should state the coverage or scope of the study (Polit & Beck, 2017). For example, do not talk about the whole of Namibia when the study is going to be conducted at just one school in one region.

PRACTICAL EXAMPLE OF STUDY DELIMITATION

Given the limitations discussed above, this study will only focus on full-time first, second, third and fourth-year undergraduate nursing students enrolled at training hospitals in Khomas region, Namibia specifically.

LITERATURE REVIEW

Show gaps in knowledge: A comprehensive literature review should systematically identify what is already known in a particular field and, crucially, highlight the areas that remain unexplored or underexplored. By pinpointing these gaps, the review sets the stage for new research initiatives that can fill these voids and contribute to the advancement of the field.

Reveal the need for further research: Beyond just identifying gaps, a literature review should articulate why these gaps matter. It should demonstrate the potential implications of these unaddressed areas and make a compelling case for why further investigation is essential. This involves discussing how new research can contribute to theoretical advancements, practical applications or policy development.

Point to the unanswered questions and disagreements in literature: The review should identify and discuss any unresolved questions or debates within the existing literature. This includes highlighting areas where findings are contradictory or where there is a lack of consensus among researchers. By doing so, the review can frame these contentious issues as opportunities for future research to clarify and resolve.

Identify theoretical and methodological weaknesses and controversies: A critical part of a literature review is evaluating the robustness of existing studies. This means identifying any theoretical limitations or methodological flaws that may undermine the validity of the findings. Discussing these weaknesses not only helps to contextualise the existing knowledge but also suggests ways in which future research can be designed to overcome these limitations.

Demonstrate what has been researched and what remains uninvestigated: The review should provide a clear and organised overview of the research landscape in the area of interest. This involves summarising the key findings and trends in the literature while also clearly delineating the boundaries of what has been explored. By mapping out what is known and what remains unknown, the review guides researchers on where to focus their efforts.

Be critical, analytical, synthesised and integrated: Effective literature reviews do more than just summarise existing studies; they critically analyse and synthesise the findings to provide a cohesive narrative. This involves comparing and contrasting different studies, identifying patterns and drawing connections between them. The goal is to integrate disparate pieces of research into a coherent whole that provides new insights and understanding.

Be current and demonstrate the researcher's understanding of the area of interest: To be current, the literature review should be based on the latest primary and secondary sources of information in the form of published research articles; well-researched review position papers; books; government policy documents; UN conventions, goals, declarations, frameworks of actions; evidence-based newspaper-based articles; expert opinions; discussions with researchers in the field; etc. And finally, a literature review must be up to date, incorporating the most recent studies and developments in the field. It should reflect the researcher's deep understanding of the topic, showing familiarity with both the foundational and the latest literature. This demonstrates the researcher's competence and situates their work within the current academic discourse. By addressing these seven elements, a literature review not only sets the foundation for a research project but also contributes to the ongoing scholarly conversation by identifying and advocating for new areas of inquiry.

Methodology

Research design

Research design is the foundation for the way in which the separate elements of your research study will bond together in your data collection, analysis and report writing (Brink et al., 2018; Polit & Beck, 2017). A discussion of either quantitative or qualitative research designs should be provided in this section. Research designs are research plans that researchers use when conducting their studies.

Quantitative research utilises a range of research designs including descriptive, pre-experimental, quasi experimental and experimental. These designs vary in terms of complexity and the degree of control they exert over variables. Descriptive studies aim to accurately portray characteristics of a particular individual, situation or group. Pre-experimental designs offer initial insights and observations but lack rigorous controls. Quasi-experimental designs attempt to establish causality but lack random assignment. True experimental designs, considered the gold standard, involve random assignment and the manipulation of variables to establish cause-and-effect relationships. In the realm of educational research, various methods such as surveys, correlational studies and some case studies fall under descriptive studies. These methods primarily aim to gather information and identify patterns. On the other hand, comparison studies, which fall under the experimental design genre, seek to determine differences between groups or conditions. True experimental designs are rare in education owing to practical and ethical constraints. Different research designs address threats to internal and external validity to varying extents.

Internal validity concerns whether the observed effects in a study are truly due to the experimental treatment rather than other factors. External validity pertains to the generalisability of the research findings to the larger population from which the sample was drawn.

Qualitative research employs designs such as narrative, phenomenological, grounded theory, ethnographic, and case study. Each design serves distinct purposes and involves unique methods for data collection and analysis. It is crucial for qualitative researchers to have a thorough understanding of the specific design they are using to ensure the methodological rigour and relevance of their research questions.

Population

In research terminology, a population encompasses all elements or the entire group of individuals, objects, events, or substances of interest (Brink et al., 2018). In quantitative research, it is essential for researchers to clearly define the population they wish to generalise their findings to. This allows for precise and relevant conclusions that can be applied beyond the study sample.

A population may then be subdivided into the target population and the accessible population:

Target population refers to the entire group of individuals or elements that researchers aim to generalise their findings to (Banerjee & Chaudhury, 2010). It includes all the individuals who meet the specified criteria for the study, regardless of their geographical location or accessibility.

For example, in a study about the nursing students' experiences of communication in a multilingual and multicultural clinical environment at training hospitals in Khomas region, Namibia, the **target population** encompasses all the nursing students in a multilingual and multicultural clinical environment at training hospitals in Khomas region, Namibia.

Accessible population refers to the subset of the target population that is available to the researchers for their study (Trochim & Donnelly, 2007). This group includes individuals who meet the target population criteria and are accessible within the constraints of the study, such as location, time and resources.

For example:

Accessible population: Nursing students in a multilingual and multicultural clinical environment at training hospitals in Khomas region, Namibia, who are available and willing to participate during the study period.

Sample

A sample is a subset of the population which is selected to represent the entire group. This subset is chosen using various sampling techniques to ensure it accurately reflects the characteristics of the population (Lohr, 2021; Polit & Beck, 2017). Researchers should meticulously describe how they plan to draw samples from their defined populations using appropriate quantitative or qualitative sampling techniques. This explanation should be concrete and meaningful, going beyond mere definitions to include the rationale and procedures for sample selection.

It is important to note that quantitative sampling techniques should not be used in qualitative studies and vice versa. Each type of research demands distinct approaches to sampling to suit its methodological requirements.

Research Instruments

Research instruments are the tools that enable researchers to collect the data needed to answer their research questions (Polit & Beck, 2017). In quantitative education research, commonly used instruments include questionnaires, standardised and non-standardised tests (e.g. achievement tests), and various scientific measuring devices. These instruments must be reliable and valid to ensure the accurate and consistent measurement of the variables under study.

In qualitative research, instruments include structured and open-ended interview schedules, observation guidelines, and electronic devices such as video and digital cameras, tape recorders and cell phones. These tools facilitate the in-depth collection of rich, descriptive data that are essential for qualitative analysis.

Data Analysis

Researchers must provide detailed strategies for analysing either quantitative or qualitative data. In quantitative research, it is crucial to justify the choice of specific non-parametric and parametric statistics used to analyse data. For instance, researchers should explain the rationale behind using frequencies, cross-tabulations (including the chi-square test), t-tests, ANOVA, regression analysis and correlation coefficients. This justification helps establish the appropriateness of the statistical methods for addressing the research questions and hypotheses. For qualitative data analysis, researchers should describe the various strategies employed, such as content analysis, ethnographic data analysis and computer-assisted qualitative data analysis.

Each strategy should be thoroughly explained, demonstrating how it contributes to the systematic and rigorous interpretation of the qualitative data collected.

In summary, a well-conducted literature review and research study should encompass a thorough understanding and application of research designs, population and sampling techniques, research instruments and data analysis methods. This ensures the validity and reliability of the research findings and their contribution to the broader field of study.

SOLUTIONS AND RECOMMENDATIONS

To address the challenges students face when writing research proposals, higher education institutions should implement several targeted interventions and ongoing strategies:

1. **Enhanced research methodology training:** Institutions should offer comprehensive and practical workshops on research methodology that go beyond theoretical knowledge. These workshops should include hands-on activities, real-life examples and opportunities for students to practise writing proposals.
2. **Dedicated writing support services:** Establish dedicated writing centres staffed with experienced tutors who can provide personalised feedback on students' proposals. These centres can also offer regular writing clinics and boot camps focused on specific aspects of proposal writing.
3. **Mentorship programmes:** Develop mentorship programmes that pair students with experienced faculty members or advanced postgraduate students. These mentors can provide guidance, feedback and emotional support throughout the proposal writing process.
4. **Online resources and tools:** Create and maintain an online repository of resources, including sample proposals, writing guides, video tutorials and templates. Interactive tools for self-assessment and progress tracking can also help students stay on course.
5. **Peer review and collaboration:** Encourage peer review sessions where students can present their proposals to classmates for constructive feedback. Collaborative projects and study groups can also foster a supportive learning environment.
6. **Early and continuous feedback:** Implement a system where students receive continuous feedback from their supervisors starting from the early stages of their research. This ongoing support can help identify and address issues promptly.

7. **Integration of proposal writing into the curriculum:** Embed research proposal writing in the curriculum of research methodology courses. This integration ensures that students are gradually introduced to the process and are given sufficient time to develop their skills.
8. **Addressing emotional and psychological barriers:** Provide access to counselling services to help students manage the stress and anxiety related to proposal writing. Workshops on time management and stress relief techniques can also be beneficial.

FUTURE RESEARCH DIRECTIONS

Future research on the challenges of research proposal writing should focus on several key areas to build on existing knowledge and develop more effective support strategies:

1. **Comparative studies:** Conduct comparative studies across different regions and institutions to identify common challenges and successful interventions. Understanding the context-specific factors can help tailor support strategies more effectively.
2. **Longitudinal studies:** Implement longitudinal studies to track students' progress from their initial proposal drafts to the final submission. This can provide insights into the development of their skills and the effectiveness of various interventions over time.
3. **Impact of technology:** Explore the role of technology in supporting research proposal writing. Studies could investigate the effectiveness of online writing tools, virtual mentorship programmes, and digital collaboration platforms.
4. **Supervisor–student dynamics:** Examine the dynamics between students and supervisors, focusing on communication patterns, expectations and the impact of these relationships on the quality of research proposals.
5. **Emotional and psychological factors:** Investigate the emotional and psychological barriers that students face during the proposal writing process. Research in this area can inform the development of comprehensive support systems that address both academic and emotional needs.
6. **Interdisciplinary approaches:** Study the challenges and strategies specific to different disciplines. An interdisciplinary approach can highlight unique needs and effective practices within various academic fields.

CONCLUSION

Writing a research proposal is a critical and challenging component of undergraduate and postgraduate studies. Despite the foundational knowledge provided by research methodology courses, many students struggle with the practical aspects of proposal writing, leading to delays in their academic progress and, in some cases, withdrawal from their programmes. By implementing targeted interventions such as enhanced training, dedicated writing support, mentorship programmes and continuous feedback mechanisms, higher education institutions can significantly improve students' proposal writing skills and their overall academic success. Future research should continue to explore and address the multifaceted challenges students face, ensuring that supportive strategies are continually refined and adapted to meet the evolving needs of students in diverse educational contexts.

REFERENCES

Anderson, L. W., & Krathwohl, D. R. (Eds.). Airasian, P.W., Cruikshank, K.A., Mayer, R.E., Pintrich, P.R., Raths, J., & Wittrock, M.C. (2001). *A Taxonomy for learning, teaching, and assessing: A revision of Bloom's Taxonomy of educational objectives*. New York, NY: Longman

Anderson, P. (2021). *Effective supervision in higher education*. San Francisco, CA: Educator's Press.Asante, A. D., Negin, J., Hall, J., Dewdney, J., & Zwi, A. B. (2012). Analysis of policy implications of the Cuban health assistance program related to human resources for health care in the Pacific. *Human Resources for Health*, 10(10), 1–9.

Ashipala, D. O., & Livingi, R. M. (2021). Undergraduate nursing students' challenges when writing research proposals at the University of Namibia. *Africa Journal of Nursing and Midwifery*, 23(1). Advance online publication. DOI: 10.25159/2520-5293/8731

Banerjee, A., & Chaudhury, S. (2010). Statistics without tears: Populations and samples. *Industrial Psychiatry Journal*, 19(1), 60–65. DOI: 10.4103/0972-6748.77642 PMID: 21694795

Bitchener, J., & Basturkmen, H. (2006). Perceptions of the difficulties of postgraduate L2 thesis students writing the discussion section. *Journal of English for Academic Purposes, 5*(1), 4–18. https://doi.org/DOI: 10.1016/j.jeap.2005.10.002

Brink, H., van der Walt, C., & van Rensburg, G. (2018). *Fundamentals of research methodology for healthcare professionals*. Juta & Co.

Brown, A. (2020). *Foundations of higher education: Policy and practice*. University Press.

Clark, E. (2020). *Writing successful research proposals*. Academic Insights.

Davis, L. (2021). *Advanced studies: A guide to postgraduate education*. Scholarly Press.

Fandino, W. (2019, August). Formulating a good research question: Pearls and pitfalls. *Indian Journal of Anaesthesia*, 63(8), 611–616. DOI: 10.4103/ija.IJA_198_19 PMID: 31462805

Grove, J., Ripke, S., Als, T. D., Mattheisen, M., Walters, R. K., Won, H., Pallesen, J., Agerbo, E., Andreassen, O. A., Anney, R., Awashti, S., Belliveau, R., Bettella, F., Buxbaum, J. D., Bybjerg-Grauholm, J., Bækvad-Hansen, M., Cerrato, F., Chambert, K., Christensen, J. H., & Børglum, A. D.ChicagoGrove. (2019). Identification of common genetic risk variants for autism spectrum disorder. *Nature Genetics*, 51(3), 431–444. DOI: 10.1038/s41588-019-0344-8 PMID: 30804558

Ha, J. F., & Longnecker, N. (2010). Nurse-patient communication: A review. *The Ochsner Journal*, 10(1), 38–43. PMID: 21603354

Hartley, J., & Jory, S. (2015). Lifting the veil on the viva: The experience of Psychology PhD candidates in the UK. *Psychology Teaching Review*, 9(2), 79–90.

Jones, R. (2022). *Overcoming challenges in higher education*. Boston, MA: Education Publishers.Komba, S. C. (2016). Challenges of writing theses and dissertations among postgraduate students in Tanzanian higher learning institutions. *International Journal of Research Studies in Education*, 5(3), 71–80.

Lessing, A. C., & Schulze, S. (2002). Postgraduate supervision and academic support: Students' perceptions. *The Journal of Higher Education*, 16(2), 139–149. DOI: 10.4314/sajhe.v16i2.2525

Mapolisa, T., & Mafa, O. (2012). Challenges being experienced by undergraduate students in conducting research in open and distance learning. *International Journal of Asian Social Science*, 2(10), 1672–1684.

Meuter, R., Gallois, C., Segalowitz, N., Ryder, A., & Hocking, J. (2015). Overcoming language barriers in healthcare: A protocol for investigating safe and effective communication when patients or clinicians use a second language. *BMC Health Services Research*, 15(371), 1–5. DOI: 10.1186/s12913-015-1024-8 PMID: 26357948

Miller, S. (2022). *Undergraduate studies: An introduction*. Student Press.

Murray, R. (2012). *How to write a thesis*. Buckingham, UK: Open University Press.

Polit, D. F., & Beck, C. T. (2017). *Nursing research: Generating and assessing evidence for nursing practice* (10th ed.). Wolters Kluwer Health., DOI: 10.1016/j.iccn.2015.01.005

Prilutski, M. A. (2010). A brief look at effective health communication strategies in Ghana. *The Elon Journal of Undergraduate Research in Communications*, 1(2), 51.

Shatimwene, G. P., & Ashipala, D. O. (2023). Nursing educators' experiences and expectations of research production at a Faculty of Health Sciences: A qualitative study. *Africa Journal of Nursing and Midwifery*, 25(2). Advance online publication. DOI: 10.25159/2520-5293/14153

Slife, B. D., Wright, C. D., & Yanchar, S. C. (2016). Using operational definitions in research: A best-practices approach. *Journal of Mind and Behavior*, •••, 119–139.

Smith, J. (2021). *The role of academics in modern education*. Academic Press.

Taylor, M. (2023). *Research methodologies in social sciences*. Research Publishing.

Taylor, S., & Beasley, N. (2015). *A handbook for doctoral supervision*. Routledge.

Trochim, W. M. K., & Donnelly, J. P. (2007). *The research methods knowledge base* (3rd ed.). Thomson.

Wang, T., & Li, L. Y. (2008). Understanding international postgraduate research students' challenges and pedagogical needs in thesis writing. *International Journal of Pedagogies and Learning, 4*(3), 88–96. http://dx.doi.org/DOI: 10.5172/ijpl.4.3.8

White, D. (2023). *Student life and learning: A comprehensive guide*. University Publications.

Wilson, K. (2019). *The life of a researcher: Ethics and practices*. Academic Books.

KEY TERMS AND DEFINITIONS

Academics: Academics refers to the scholarly activities that relate to education, scholarship and research in an institutional setting, typically within schools, colleges and universities (Smith, 2021). This term encompasses both the educators and the students involved in the learning process.

Challenges: Challenges are obstacles or difficulties that require effort and skill to overcome (Jones, 2022). In an academic context, challenges can include factors such as rigorous coursework, financial constraints, time management issues and the pressures of achieving high grades.

Higher education: Higher education refers to the level of education provided by universities, colleges and other institutions that award academic degrees or professional certifications (Brown, 2020). It follows secondary education and includes undergraduate and postgraduate education.

Postgraduate: Postgraduate refers to the level of education that follows the completion of an undergraduate degree (Davis, 2021). It includes master's degrees, doctoral degrees (PhDs), and other advanced certificates and diplomas.

Research proposals: Research proposals are documents outlining planned research projects (Clark, 2020). They include the research question, objectives, methodology and significance of the study, and are typically submitted for approval to gain funding or academic endorsement.

Research: is a systematic investigation aimed at discovering and interpreting facts, revising accepted theories and applying new knowledge (Taylor, 2023). It involves various methodologies for collecting and analysing data to answer specific questions or solve problems.

Researchers: Researchers are individuals who conduct research, typically in an academic or professional setting (Wilson, 2019). They gather data, analyse results and contribute new knowledge to their field of study.

Students: Students are individuals engaged in learning, typically enrolled in an educational institution (White, 2023). They can be at various levels of education, from primary school to higher education.

Supervision: Supervision in an academic context refers to the guidance and support provided by a more experienced individual, such as a professor or senior researcher, to students or junior researchers (Anderson, 2021). This includes overseeing research projects, providing feedback and ensuring academic standards are met.

Undergraduate: An undergraduate is a student pursuing their first degree, typically a bachelor's degree, at a college or university (Miller, 2022). This level of education involves foundational coursework in a chosen field of study.

Chapter 10
Crafting a Comprehensive Research Proposal:
A Guide for Academicians in Higher Education

Kaushik Kumar
https://orcid.org/0000-0002-4237-2836
Birla Institute of Technology, India

Chikesh Ranjan
https://orcid.org/0000-0003-1197-601X
National Institute of Technology, Rourkela, India

ABSTRACT

Today's academicians and academic fraternity are very much dependent on research. A research proposal, if accepted, provides the funds for smooth performance of the same. Hence, a proper and concise proposal is required to be drafted for easy acceptance. This chapter offers a detailed roadmap for academics navigating the intricate process of research proposal writing. It also provides a thorough examination of developing a compelling research proposal. Beginning with an exploration of the fundamental principles underlying research proposal construction, the chapter progresses to issues such as formulating research objectives, conducting a comprehensive literature review, and designing a robust methodology. Ethical considerations and budgetary planning are also addressed, ensuring researchers are equipped to navigate potential challenges ethically and financially. By synthesizing theoretical frameworks with practical guidance, this chapter serves as an invaluable resource for academicians seeking to enhance their research proposal writing skills.

DOI: 10.4018/979-8-3373-0806-7.ch010

INTRODUCTION

A good proposal should demonstrate how the work integrates with existing knowledge on the topic and what new perspectives it will contribute to the literature. It should specify the research question, establish its significance, and outline the implications of the anticipated answers. The proposal should convincingly argue the credibility, feasibility, practicality, and reproducibility of the research design to the evaluation committee [Alreck et al.1995].

TYPES OF PROJECTS

1. Consultancy Projects:

Consultancy projects are typically initiated to provide expert advice, analysis, or solutions to a client's specific problem or need [John Wiley et al. 2015]. These projects are usually short-term, highly focused, and tailored to meet the unique requirements of the client. They often involve diagnosing issues, recommending solutions, and sometimes implementing changes.

Example:

- **Project Name:** Market Entry Strategy for XYZ Corp.
- **Client:** XYZ Corporation, a tech startup looking to enter the European market.
- **Objective:** To develop a comprehensive market entry strategy.
- **Scope:** The consultancy project involves market analysis, competitive analysis, regulatory environment assessment, and the development of a strategic plan for market entry.
- **Outcome:** A detailed report outlining the market potential, competitive landscape, entry barriers, regulatory requirements, and a step-by-step strategy for entering the European market.

2. Sponsored Projects:

Sponsored projects are typically research or development projects funded by external organizations such as government agencies, corporations, or non-profit organizations [Kong, 2017]. These projects often align with the sponsor's strategic interests and require detailed proposals outlining objectives, methodology, timelines, and expected outcomes. Regular progress reports and final deliverables are typically required.

Example:

- **Project Name:** Sustainable Agriculture Practices Using AI and Drones
- **Sponsor:** National Science Foundation (NSF)
- **Objective:** To develop and implement AI-driven drone technology to promote sustainable agricultural practices.
- **Scope:** This sponsored project involves designing AI algorithms, deploying drones for real-time monitoring, data collection on crop health, soil conditions, and pest management. The project also includes field trials and the development of best practice guidelines for farmers.
- **Outcome:** The project aims to produce a comprehensive report and a set of best practices that can be disseminated to farmers and agricultural stakeholders. Additionally, the research findings could be published in scientific journals, contributing to the academic knowledge base.

Detailed Explanation:

Consultancy Projects:

- **Nature:** Short-term, focused on solving specific client problems.
- **Funding:** Funded directly by the client.
- **Flexibility:** High degree of flexibility in terms of scope and methodology.
- **Deliverables:** Typically include reports, strategic plans, implementation support, or training sessions.
- **Example Activities:** Market analysis, process improvement, IT systems implementation, financial advisory.

Sponsored Projects:

- **Nature:** Longer-term, often involving extensive research and development.
- **Funding:** Funded by external organizations such as government agencies, corporations, or non-profits.
- **Structure:** Requires a detailed proposal, regular updates, and adherence to sponsor guidelines.
- **Deliverables:** Research reports, prototypes, published papers, policy recommendations.
- **Example Activities:** Scientific research, technology development, social science studies, public health initiatives.

At a glance comparison between the above two can be summarized as depicted in Table 1

Table 1. Comparison between Consultancy Project and Sponsored Projects

Aspect	Consultancy Projects	Sponsored Projects
Duration	Short-Term	Long-Term
Focus	Client-Specific Problem-Solving	Research And Development Aligned with Sponsor Interests
Funding	Client-Funded	Externally Funded (Grants, Contracts)
Flexibility	High	Low due to Sponsor Requirements
Deliverables	Reports, Strategic Plans	Research Findings, Prototypes, Publications, Patents
Examples	Market Entry Strategy, Process Improvement	Scientific Research, Technology Innovation

SPONSORED PROJECT

As the consultancy projects are very specific hence the proposals are exactly as per the client's requirement. Moreover, as the funds allotted are also quite less hence in this chapter a detailed discussion would be made for proposal writing towards a sponsored project.

Sponsored project evaluation committees often include four categories of audiences with distinct expectations: Academic Colleagues, Policy-Makers, Practitioners, and Common Audiences [Bowling,2002]. Hence, to prepare an effective research proposal the following criteria should be strictly followed:

- Practicality
- Persuasive
- Providing broader connections
- Clarity in aim, and
- Planning before writing.

A researcher should maintain a balanced perspective with a realistic understanding of achievable goals [Charmaz 2006]. Being persuasive means convincing other researchers, funding agencies, educational institutions, and supervisors of the research's value. The research aim should be clearly articulated in simple language, making it understandable to non-specialists without jargon [Cluett et al. 2002]. The proposal should demonstrate a comprehensive understanding of the existing literature and show that the researcher has considered the time required for each stage of the research [Cresswell 2002].

COMPONENTS OF A RESEARCH PROJECT PROPOSAL

The main contents of a research project proposal can be presented under the following headings:

a) Cover Page
b) Introduction
c) Review of Literature
d) Aims and Objectives
e) Research Design and Methods
f) Ethical Considerations
g) Budget
h) Plan of Work and Time Schedule
i) Appendices
j) Citations
k) Bibliography

a) Cover Page Components

A cover page is an essential element of a research proposal. It should include the following details [Davies 2007]:

- **Title of the Proposal**: Clearly state the title, which should succinctly convey the main topic or focus of the research.
- **Researchers' Information**:
  Name: Full name of the Principal Investigator (PI).
  Affiliation: Institutional affiliation of the PI, indicating their degree and the institution where the research will be conducted.
  Co-investigator(s): Names and affiliations of any co-investigators involved in the research.
- **Contact Details**:
  Phone Numbers: Provide phone numbers for easy communication.
  Email Addresses: Include email addresses for all primary contacts.
- **Signatures**: Provide space for the signatures of the principal investigator and co-investigators to confirm their involvement and approval of the proposal.

By incorporating these elements, the cover page will effectively introduce the research proposal, ensuring that all necessary contact and identification information is readily accessible [Jadad et al. 2007].

b) Introduction

The introduction, sometimes referred to as the 'need for study' or 'abstract,' serves as the initial pitch of the idea [Lawrence et al. 2009]. It sets the stage and places the research in context. The purpose of the introduction is to engage the reader and generate interest in the proposed topic and proposal. It should clearly communicate:

- **Objective**: What is intended to be achieved with the research.
- **Necessity**: Why the study is needed.
- **Passion**: The enthusiasm and commitment to the topic.

Additionally, the introduction should include the hypothesis that underpins the research design. If formulating a hypothesis is not possible, then the line of inquiry to be pursued in the research must be clearly stated.

To assess the significance of the study, the following questions can be consider for addressing:

- **Interest**: Who is interested in the domain of inquiry?
- **Current Knowledge**: What is already known about the topic?
- **Research Gaps**: What has not been adequately addressed in previous research and practice?
- **Contribution**: How will this research enhance knowledge, practice, and policy in this area?

Some evaluation committees may require the last two questions to be elaborated under a separate heading titled 'Background and Significance.'

c) Review of Literature

The review of literature encompasses all sources of scientific evidence related to the research topic. In the current digital age, the sheer volume of relevant data available can be overwhelming, making it challenging for researchers to include everything in their review.

This section must be structured thoughtfully to ensure that readers can understand how the study fits within the broader context of existing research. It's essential to show that the work is original and innovative while relating it to other researchers' work.

Here are some guidelines for an effective literature review:

- **Summarization**: Summarize each relevant article in a paragraph, focusing on details pertinent to the proposed topic.

- **Structure**: The review can progress from general to more focused studies or a historical progression can be followed to develop the narrative without becoming exhaustive.
- **Content**: Inclusion of supporting data, as well as any disagreements and controversies within the field.

Five 'C's should be kept in mind while writing the literature review:

- **Cite**: Reference all sources accurately.
- **Compare**: Highlight similarities between different studies.
- **Contrast**: Point out the differences and any conflicting findings.
- **Critique**: Analyze the strengths and weaknesses of the existing literature.
- **Connect**: Show how the proposed research ties into the broader field and addresses gaps in the current knowledge.

By following these guidelines, the literature review will effectively situate the proposed research within the existing body of work and highlight its significance.

d) Aims and Objectives

The research purpose, also referred to as the goal or aim, provides a broad indication of what the researcher intends to achieve through the study [Polgar et al. 2008]. This section is crucial as it sets the direction and scope of the research. The hypothesis to be tested often serves as the primary aim of the study.

Aims: The aim should be a concise statement reflecting the main intention or the overarching goal of the research. It answers the question of what the researcher hopes to discover or prove.

Objectives: These are more specific than the aim and describe the steps that will be taken to achieve the research goal. Objectives are typically divided into two categories:

- **Primary Objectives**: These directly address the main aim of the study. They define the core aspects that the research seeks to explore or measure.
- **Secondary Objectives**: These are supplementary goals that support the primary objective. They might involve additional parameters or tools that provide further insights into the research topic.

Explanation:

- The **aim** of the research provides the "big picture" of what the researcher wants to achieve, guiding the overall direction of the study.

- The **objectives** break down the aim into manageable, specific tasks. They clarify how research question(s) would be approached and what methods would be used to gather data and test the proposed hypothesis.

e) Research Design and Methods

The Research Design and Methods section is a crucial component of the research proposal, providing a detailed plan on how the study will be executed [Silverman 2005]. This section should clearly outline the approach and methodologies to be used to ensure that the research is comprehensive and valid.

Components of the Research Design and Methods Section:

A. Population and Sample:
    - **Population** refers to the complete set of elements (individuals, objects, or substances) that possess specific characteristics and meet the criteria for inclusion within the scope of the study. This group represents the entire universe of potential subjects from which the study sample can be drawn.
    - **Sample** is a subset of the population that meets the predetermined inclusion criteria and is selected for participation in the study. The criteria for both inclusion and exclusion must be explicitly defined to ensure clarity and precision in selecting participants [Li,2023].

    Explanation:

    - **Population** encompasses the entire group under study, characterized by defined criteria.
    - **Sample** is a smaller, representative group chosen from the population based on specific inclusion criteria.
    - Clear **inclusion and exclusion criteria** are essential for determining who can or cannot participate in the study, ensuring the sample is appropriately representative and relevant to the research objectives.

B. Data Collection:

The researcher must provide a comprehensive description of the methodology employed for data collection, including the time frame required for the research. It is crucial to test the methodology for validity and to ensure the safety of participants throughout the study.

Researchers should identify and acknowledge any potential barriers or pitfalls in executing the research design and outline strategies to address these issues, thereby preventing gaps due to incomplete data collection. If data collection involves interviews or questionnaires, copies of the questions used should be attached as an annexure with the proposal.

Explanation:

- **Comprehensive Description**: Detailed account of data collection methods.
- **Time Frame**: Specific duration for conducting the research.
- **Validity and Safety**: Ensuring methods are valid and participants' safety is prioritized.
- **Anticipate Barriers**: Identify potential challenges and outline solutions.
- **Annexure**: Include copies of interview or questionnaire questions used in the study.

C. Rigor (Soundness of the Research):

This aspect addresses the robustness of the research in terms of its neutrality, consistency, and applicability. Rigor must be evident throughout the entire proposal.

Explanation:

- **Robustness**: Ensures the research is strong and well-founded.
- **Neutrality**: Maintains objectivity and impartiality.
- **Consistency**: Ensures the research methods and findings are reliable and reproducible.
- **Applicability**: Demonstrates the relevance and usefulness of the research.
- **Throughout the Proposal**: Rigor should be consistently evident in all parts of the proposal

D. Neutrality:

This refers to the robustness of a research method in preventing bias. The researcher should clearly explain the steps taken to avoid bias, such as blinding and randomization. These measures ensure that the results obtained are due to chance and are not influenced by confounding variables.

Explanation:

- **Robustness Against Bias**: Ensures the method can withstand bias.
- **Blinding and Randomization**: Techniques used to prevent bias.
- **Clear Explanation**: The researcher should detail how these techniques are applied.
- **Avoiding Confounding Variables**: Ensures that results are not influenced by external factors, maintaining the integrity of the study's outcomes.

E. Consistency:

Consistency assesses whether the findings will remain the same if the study is replicated with the same participants in a similar context [Li & Hebert 2024]. Achieving consistency involves adopting standardized and universally accepted methods and scales.

Explanation:

- **Replication of Findings**: Ensures results are reproducible under similar conditions.
- **Standardized Methods**: Use of established procedures to maintain uniformity.
- **Universally Accepted Scales**: Application of recognized measurement tools to ensure comparability.

F. Applicability:

Applicability refers to the extent to which the research findings can be generalized to different contexts and groups.

Explanation:

- **Generalization of Findings**: Ensures results are relevant beyond the initial study context.
- **Different Contexts**: Application to various settings or situations.
- **Diverse Groups**: Relevance across different populations or demographic groups.

G. Data Analysis:

This section involves the reduction, reconstruction, and analysis of data, including the calculation of sample size. The researcher should describe the steps taken for coding and sorting the collected data. It is important to specify the various tests that will be used to analyze the data for robustness and significance. Additionally, the researcher should mention the names of the statisticians and the appropriate software that will be employed during data analysis, detailing their contributions to data analysis and sample size calculation.

f) Ethical Considerations

Medical research presents unique moral and ethical challenges that are not typically faced by other researchers during data collection. Therefore, the researcher must ensure that all ethical standards are met. Ethical considerations include protecting participants' rights, which encompass the right to self-determination, privacy, autonomy, confidentiality, fair treatment, and protection from discomfort and harm. Additionally, obtaining informed consent and securing ethical approval through the institutional review process is vital.

The researcher must provide comprehensive information on how these ethical standards will be upheld. This includes detailing the process for obtaining informed consent from participants, the research site, and the relevant authorities.

g) Budget

When preparing a research budget, the researcher should forecast and account for all aspects of the research project, including an additional allowance for unforeseen disasters, delays, and rising costs. Each item in the budget must be thoroughly justified as shown in Figure 1 and Figure 2.

Figure 1. Budget for the Manpower

BUDGET FOR MANPOWER:

Designation	:	JUNIOR RESEARCH FELLOW & SENIOR RESEARCH FELLOW
No. of Persons	:	1
Justification	:	

One candidate is required to carry out the experiments. The candidate will be placed in the position of Junior Research Fellow (JRF) with a pay of **AAA USD** for two years and Senior Research Fellow (SRF) with a pay of **BBB USD** for one year.

Manpower Budget Breakup : NIL

YEAR	MONTHLY EMOLUMENT	MAN MONTHS	TOTAL
YEAR 1	**AAA**	12	**CCC**
YEAR 2	**AAA**	12	**CCC**
YEAR 3	**BBB**	12	**DDD**
	GROSS TOTAL		**EEEEE**

Figure 2. Total Budget Under All Heads

BUDGET HEAD	YEAR-1 AMT (IN USD)	YEAR-2 AMT (IN USD)	YEAR-3 AMT (IN USD)	TOTAL (IN USD)
Manpower Budget	**AAA**	**BBB**	CCC	**DDD**
Consumables	**EEE**	**FFF**	GGG	**HHH**
Travel	**JJJ**	**KKK**	LLL	MMM
Equipment	NNNN	**PPP**	0	**QQQQQ**
Contingencies	**RRR**	**SSS**	**TTT**	UUUU
Other Costs	**Aaaa**	**Bbbb**	**Ccccc**	**Dddddd**
Overhead Costs	**Fffffff**	**Ggggggg**	**Hhhhhhhhhhhh**	**Jjjjjjjjjj**
Grand Total	**Kkkkkkk**	**Mmmmmmm**	**Nnnnnnnnnnnnnn**	**Ppppppppppppp**

h) Plan of Work and Time Schedule

The plan should outline the various stages of the research project and the corresponding timelines for each phase, including the development, implementation, and report writing. Assuming, the research duration is three years, including the time needed for writing the final report, below is a detailed year-wise breakdown of activities and milestones, along with a PERT (Program Evaluation and Review Technique) chart (Figure 3). Usually, a common mistake occurs here. Researchers provide the PERT Chart with literature survey ending before the next activity i.e.

in first 4–6 months. Technically, Literature survey should go upto writing the final report so that updated information is available in the final report.

Figure 3. Year-wise Activities/ Milestones, schedule, and PERT Chart

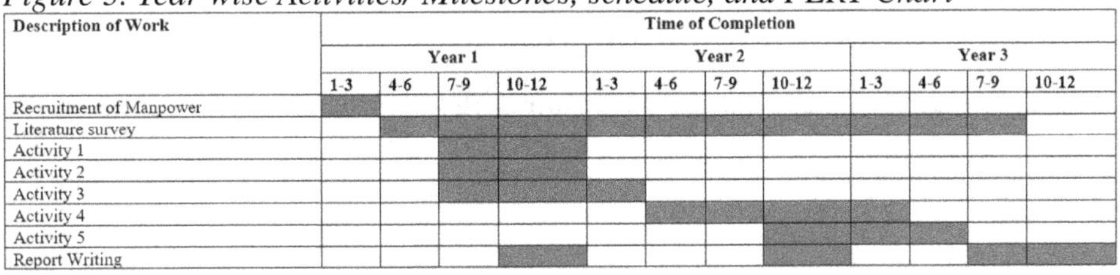

i) Appendices

Appendices are supplementary documents that support the proposal and application. These documents are specific to each proposal but typically include the informed consent form, supporting documents, questionnaires, measurement tools, and patient information about the study in layman's terms.

j) Citations

As with any scholarly research paper, the sources used in composing the proposal should be properly cited. While the terms "references" and "bibliography" are often used interchangeably, they both refer to all the sources cited in the research proposal.

k) Bibliography

It should include:

- A list of references, both national and international, to key articles and texts discussed within the research proposal.
- A selection of sources that are appropriate and relevant to the proposed research.

CASE STUDY

Structure of an Academic Research Proposal

□ *Title*

"The Impact of Social Media on Consumer Behavior in the Fashion Industry"

Abstract

This research explores the influence of social media on consumer behavior within the fashion industry. The study aims to identify key factors that drive purchasing decisions and brand loyalty among social media users [Krueger, R. A., & Casey, M. A. (2015]. A mixed-methods approach will be used, combining surveys and in-depth interviews to gather comprehensive data. The findings will contribute to a deeper understanding of digital marketing strategies and their effectiveness.

□ *Introduction*

The rise of social media has significantly altered the landscape of consumer behavior, particularly in the fashion industry. This research seeks to understand how social media platforms influence purchasing decisions and brand loyalty. Despite the extensive use of social media in marketing, there is a lack of comprehensive studies that examine its specific impact on consumer behavior within the fashion sector. This study aims to fill this gap by investigating the key factors that drive consumer engagement and purchases through social media.

□ *Literature Review*

Existing literature highlights the importance of social media in shaping consumer behavior but often lacks specificity regarding the fashion industry. Studies by Smith (2020) and Jones (2019) suggest a strong correlation between social media engagement and consumer loyalty. However, there is limited research that focuses on the underlying mechanisms that drive this relationship in the fashion context. This review will critically evaluate the current state of research and identify gaps that the proposed study will address.

Research Objectives and Hypotheses

- *To investigate the role of social media in influencing consumer purchasing decisions in the fashion industry.*
- *To identify key factors that drive brand loyalty among social media users.*
- *Hypothesis: Social media engagement positively correlates with brand loyalty and purchasing decisions.*

□ ***Methodology***

A mixed-methods approach will be employed, combining quantitative surveys and qualitative interviews. The survey will target 500 social media users who follow fashion brands, while in-depth interviews will be conducted with 20 selected participants. Data will be analyzed using statistical software for the quantitative component and thematic analysis for the qualitative component. Ethical approval will be obtained before data collection.

Significance of the Study

This research will provide valuable insights into the effectiveness of social media marketing strategies in the fashion industry. The findings will help fashion brands optimize their digital marketing efforts and enhance consumer engagement. Additionally, the study will contribute to academic literature by providing a deeper understanding of the relationship between social media and consumer behavior.

Research Plan and Timeline

- *Literature review and proposal development: Months 1-10*
- *Data collection: Months 4-6*
- *Data analysis: Months 7-8*
- *Writing and revisions: Months 9-10*
- *Submission: Month 11*

Budget

- *Survey distribution: $500*
- *Interview transcription: $300*
- *Data analysis software: $200*
- *Miscellaneous expenses: $100*

□ *References*

• Smith, J. (2020). The Role of Social Media in Consumer Behavior. Journal of Marketing Research, 45(3), 234-250.

• Jones, R. (2019). Social Media and Brand Loyalty. Journal of Consumer Studies, 52(2), 189-205.

Appendices

- *Survey questionnaire*
- *Interview guide*
- *Ethical approval letter*

CONCLUSION

Crafting a comprehensive research proposal is a crucial skill for academicians in higher education, forming the foundation for successful research projects. A well-structured proposal, encompassing key sections such as introduction, literature review, aims and objectives, research design and methods, ethical considerations, budget, timeline, appendices, and citations, showcases the researcher's understanding, commitment, and meticulous planning. By addressing essential components like the research hypothesis, data collection methodology, and ethical standards, researchers can convincingly demonstrate their project's credibility, achievability, and significance to evaluation committees. This thorough preparation not only aids in securing funding but also ensures efficient and effective research execution, ultimately contributing valuable insights and fostering innovation in the academic community.

REFERENCES

Alreck, P., & Settle, R. (1995). *The survey research handbook*. Irwin Press.

Bowling, A. (2002). *Research methods in health: investigating health and health services*. Open University Press.

Charmaz, K. (2006). *Constructing Grounded Theory: a practical guide through qualitative analysis*. Sage.

Cluett, E. R., & Bluff, R. (2002). *Principles and practice of midwifery research*. Churchill Livingstone.

Cresswell, J. W. (2002). *Research design: qualitative, quantitative and mixed methods approaches*. Sage Publications.

Davies, M. B. (2007). *Doing a successful research project: using qualitative or quantitative methods*. Palgrave Macmillan.

Jadad, A. R., & Enkin, M. (2007). *Randomized Controlled Trials: questions, answers and musings*. Blackwell Publishing Ltd. DOI: 10.1002/9780470691922

Kong, M. L. F. (2017). On teaching methods: The personal experiences of teachers of English. *RELC Journal*, 48(2), 197–209. DOI: 10.1177/0033688216661251

Krueger, R. A., & Casey, M. A. (2015). Focus group interviewing. In Hatry, H. P., Newcomer, K. E., & Wholey, J. S. (Eds.), *Handbook of practical program evaluation* (pp. 506–534). DOI: 10.1002/9781119171386.ch20

Lawrence A, Lewis L, Hofmeyr GJ, Dowswell T, Styles C (2009) Maternal positions and mobility during first stage labour. Cochrane Database Syst Rev (2) Issue 2. Art. No. CD003934. https://doi.org/DOI: 10.1002/14651858.CD003934.pub2

Le Ha, P., & Barnawi, O. Z.John Wiley & Sons. (2015). Where English, neoliberalism, desire and internationalization are alive and kicking: Higher education in Saudi Arabia today. *Language and Education*, 29(6), 545–565. DOI: 10.1080/09500782.2015.1059436

Li, A. W. (2023). Using Peerceptiv to support AI-based online writing assessment across the disciplines. *Assessing Writing*, 57, 100746. Advance online publication. DOI: 10.1016/j.asw.2023.100746

Li, A. W., & Hebert, M. (2024). Unpacking an online peer-mediated and self-reflective revision process in second-language persuasive writing. *Reading and Writing*, 37(6), 1545–1573. DOI: 10.1007/s11145-023-10466-8

Polgar, S. D., & Thomas, S. A. (2008). *Introduction to Research in the Health Sciences*. Churchill Livingston.

Silverman, D. (2005). *Doing Qualitative Research: A Practical Handbook*. Sage.

Chapter 11
Writing a Research Proposal

Doris Chasokela
https://orcid.org/0009-0001-5983-8508
National University of Science and Technology, Zimbabwe

Lwazi Sibanda
https://orcid.org/0000-0002-4523-6549
National University of Science and Technology, Zimbabwe

ABSTRACT

The chapter provides an overview of the essential components of a research proposal for academicians in higher education. It includes guidance on developing a research question, identifying key sources and literature, and outlining the research methodology. Additionally, it covers the importance of effective abstracts and keywords in a research proposal. The chapter provides tips for creating a clear and concise proposal that effectively communicates the purpose and goals of the research. This chapter discusses the key elements of a research proposal for academicians in higher education, including the research question, literature review, methodology, and outcomes. It also highlights the role of effective keywords in helping potential readers find and understand the research proposal. The research proposal is the skeleton or backbone of the research as it is structured in such a way that indicates what the writer intends to do. The mission of having a chapter on writing a research proposal is to provide higher education students and researchers with the knowledge and skills.

DOI: 10.4018/979-8-3373-0806-7.ch011

INTRODUCTION

A research proposal is a formal document that outlines a research project, including its objectives, methodology, significance, and expected outcomes. It serves as a blueprint for the research project, guiding in planning, conducting, and reporting the study. A good research proposal should clearly articulate the research question or problem that the researcher aims to investigate, justify the importance of the study, and outline the methods to be used to collect and analyse data (Lee et al., 2023). It should also demonstrate the researcher's understanding of the existing literature on the topic, highlight the gaps in current knowledge that the study aims to address, and propose a plan for disseminating the study findings. By writing a strong research proposal, the researcher can ensure that the research project is well-planned, well-executed, and ultimately contributes meaningfully to the field of study. There are several reasons why researchers need a proposal before carrying out the main research. A proposal helps to clearly define the research objectives, ensuring that everyone involved in the research project, including the researcher, stakeholders, and potential funders, understand what the project aims to achieve. A proposal provides a clear direction for the research project, helping to guide the research design, methodology, and analysis. Research methodology clarity is outlined by a proposal, including the data collection and analysis methods, which ensure that the research is well-planned and feasible. The resources needed to carry out the research, including personnel, equipment, funding, and facilities are identified by a proposal. A well-written proposal is often required to secure funding or support for the research project from institutions, organisations, or government agencies. A proposal provides a framework for evaluating the success of the research project and holding the researcher accountable for its outcomes. A proposal serves as a communication tool, allowing researchers to share their ideas and plans with others, including collaborators, supervisors, and stakeholders. In terms of time management, a proposal helps researchers plan and manage their time effectively, ensuring that they can complete the project within the given timeframe. A proposal helps researchers identify potential risks and limitations associated with the project, allowing them to develop contingency plans and mitigate potential problems (Kafu, 2023; Nehru, Pasoloran, Pongsapan et al., 2023; Padmadewi et al., 2023; Sanak et al., 2023).

Overall, a proposal is an essential step in the research process because it allows researchers to clarify their ideas, plan their work, and ensure that their project is feasible, well-managed, and productive. Research proposals are important for funding and quality control. Many research projects require funding, and research proposals are typically required to apply for funding from grant agencies, universities, or other sources. A strong research proposal will help to convince funders that your project is worthwhile and will have a positive impact. Research proposals help to ensure the quality of research by providing a framework for planning and carrying out the study. They provide a structure for the research and help to ensure that the study is well-designed and appropriate for the research question.

Choosing a Title

Several researchers explain how to choose a research title. They observe that a title should be a reflection of the contents of the document, fully explanatory when standing alone. It should not contain redundancies such as 'a study of.....' or 'an investigation of...' Abbreviations should not appear in the title. Scientific names should be in italics. Choosing a title for the research proposal can be a challenging task, as it sets the tone for the entire project. Choosing a title that is effective, clear, and engaging, should be concise and specific. A title should be brief, but it should also accurately convey the focus and scope of the research. Avoid using overly long or jargon-heavy phrases. Consider the target audience, that is, the title should be appropriate for intended readers, even if they are academics. The key elements that typically make up a research title are specific aspects of the topic that are the focus of research. The scope is another element and it spells the geographical location of the study. If the research is focused on a specific geographic region, the title may include a description of the location. For example, "Water Quality in the Great Lakes Region." Spell out the methodology approach or methods used in the research, if relevant. If the research involves collecting data from a specific group of people, such as patients, students, or employees, the title may include a description of the respondents. For example, "Attitudes towards Climate Change among College Students"(Abdullah, 2023; Jalongo, 2023; Khan, 2023; Padmadewi et al., 2023).

Cover Page of a Proposal

A cover page, also known as a title page, is the first page of a proposal document and serves as an introduction to the proposal. The essential elements to include on a cover page are:

- Organisation's logo
- Title of the proposal: A clear and concise title that accurately reflects the purpose and scope of the proposal.
- Title of the organisation: The name of the organisation, company, or institution, including any relevant logos or branding elements.
- Name and contact information of the researcher: Researcher's name, title, email address, phone number, and any other relevant contact information.

- Project/Proposal Number (if applicable): If the organisation has a specific project or proposal numbering system, include the relevant number.
- Student number (if applicable)
- Name of the supervisor (if applicable)
- The date when the proposal is submitted

It is important to customise the cover page according to the organisation's specific needs and the type of proposal to be submitted.

Introducing the Research Proposal

An introduction is the first section of a research proposal or paper and serves to introduce the reader to the topic and provide background information (Li et al., 2021). A good introduction is essential in setting the stage for the rest of the proposal and ensuring that readers understand the importance of the study (Kim et al., 2021). A good introduction should provide a clear and concise overview of the research problem, its significance, and the research question or hypothesis (Smith et al., 2022). Tips for constructing an effective introduction include starting with a hook, which is a sentence or two that grabs the reader's attention and sets the stage for the rest of the paper. After providing context, the researcher should introduce the research statement. This is a single sentence that summarises the main argument or point of the paper. It should be concise and clear, and establish the focus of the paper (Taylor et al., 2022). Additionally, an introduction should explain why the topic under discussion is important and why the research is necessary. This helps to establish the relevance and significance of the research. The introduction also summarises the relevant literature on the topic, including previous research and scholarship. This assists in contextualising the research and demonstrates the researcher's knowledge of the field. If the research is based on a particular theoretical framework, it should be introduced at this stage. Introducing the theoretical framework at the introductory stage situates the research within a broader scholarly context and shows that the research is grounded in theory. The heading numbering starts from the introduction of the research proposal as 1.0; 1.1; 1.2; 1.3 and so on. Under these headings, the researcher can have sub-titles such as under 1.1 one can have 1.1.1; 1.1.2; 1.1.3, and so on.

Background of the Study

Background information should show understanding and genesis of the problem. The background gives specific gaps that are considered to be the root of the problem. This should be a maximum of one and a half pages long. The researcher should discuss the global perspective followed by the local scenario. To structure and explain the background of the study, the researcher can begin by identifying the relevant literature on the topic, including scholarly articles, books, and reports (Harris et al., 2023; Kumar et al., 2024).

The researcher should read and review the literature to understand the current state of knowledge on the topic and to identify any gaps or areas for further research (Morgan et al., 2024). The researcher should summarise the main arguments and findings of the literature clearly and concisely. Furthermore, the researcher should highlight any key theories, concepts, or debates that are relevant to the research. After summarising the relevant literature, the researcher should place the research in the context of the existing scholarship. It is critical for the researcher to explain how the research builds on or challenges existing knowledge and how it contributes to other scholars' understanding of the topic. This results in identifying any gaps or limitations in the existing literature (Lee et al., 2023; Santos et al., 2024). These gaps could be areas where previous research has been inconclusive, where there are conflicting findings, or where there is little or no research. It is necessary for the researcher to also explain the specific problem or issue that the research aims to address.

Statement of the Problem

The problem or issue being addressed in the research should also be clearly stated in the title. This helps to ensure that readers understand the relevance and significance of the research. Several researchers elaborate on how to write a problem in the form of a statement (Brown et al., 2023; Kumar et al., 2024; Lee et al., 2023; Shunp et al., 2021; Kandaswamy et al., 2022). A few examples of titles that include a statement of the problem are:

- "The Impact of Air Pollution on Respiratory Health: A Study of Asthma Rates in Urban Areas"
- "Reducing Income Inequality: A Comparative Analysis of Wealth Distribution in OECD Countries"
- "Challenges of Implementing Renewable Energy Technologies: Barriers to Solar Power Adoption in Developing Countries"

The statement of the problem must indicate the nature of the problem. It indicates why and how it is a problem and gives information to support the argument, for example, by use of statistics (Dutton, 1999). The statement of the problem should be derived from background information to illustrate connectivity. The length of the statement of the problem should be at least two paragraphs. The statement of the problem should identify the issue or problem that the research aims to address and explain why it is important. Structuring and phrasing the statement of the problem includes the use of action words such as "address," "resolve," or "reduce," like "challenge," "threaten," or "impede" to communicate the problem. Strong verbs help to convey urgency and importance. The researcher should use specific language as detailed as possible when describing the problem thus avoiding generalizations or vague language. It is essential to consider the audience when writing the statement of the problem. If the researcher is writing for a general audience, it is important to avoid technical jargon and complex language. If writing for an academic audience, the researcher should use more specialized terminology. The researcher should use the present tense to describe the problem, as this makes it feel more immediate and relevant. There is a need to explain the importance of the problem and why it should be addressed. Considering the impact of the problem on society, individuals, or the environment is essential. The researcher should clearly define the scope of the problem, including the people, places, or things that are affected by it. Making assumptions about the causes of the problem or the potential solutions should be avoided. It is vital to stick to the facts and evidence available. Finally, the researcher should take the time to edit and revise the statement of the problem. (Nolen & Talbert, 2011; Ellis & Levy, 2009).

Purpose of Study

The purpose of the study is a brief statement that describes the overall goal or objective of the research (Harris et al., 2023; Kumar et al., 2024; Bhandari et al., 2021; Taylor et al., 2022). The purpose of the study should be directly linked to the statement of the problem. For example:

The purpose of this study is to evaluate the implementation of strategies used to maintain positive discipline in secondary schools, to develop a holistic positive discipline management framework.

A strong purpose of the study should be concise and to the point, using as few words as possible to communicate the overall goal of the research. The purpose of the study should state the direction or focus of the research, indicating what the researcher hopes to achieve or contribute (Lee et al., 2023; Morgan et al., 2024). For example, "The purpose of this study is to evaluate the implementation of strategies used to maintain positive discipline in secondary schools, to develop a holistic positive discipline management framework." It is important to use precise language to avoid ambiguity or confusion. Some research projects may have more than one purpose. The purpose of the study should highlight the significance of the research and why it is important. For example, "The purpose of the study is to examine the psychosocial support provided by educators to adolescent male learners with behavioral problems, to mitigate the negative impact on discipline management in Insiza District Secondary Schools." The purpose of the study should have a logical structure that flows smoothly from the statement of the problem to the proposed solution or approach. The researcher should use an active voice when presenting the purpose of the study, as this helps to make writing more engaging and clear (Tracy, 2019). In this section, there is a need to consider future implications such as thinking about how the research may contribute to future research or solutions. The researcher should consider how the study might advance knowledge or contribute to social or environmental change. It is essential to ensure that the purpose of the study is realistic and achievable within the scope of the research. The researcher should avoid making claims or promises that are overly ambitious or unlikely to be achieved. It is important to consider different levels of analysis depending on the research design and methodology, one may be able to identify multiple levels of analysis for study. (Levac, et al., 2010; Garcia, & Gluesing, 2013).

Research Questions

Research questions provide a framework for the study, indicating the specific issues that the researcher will be exploring and addressing. The main research question serves as the central focus of the study and guides the data collection, analysis, and interpretation process. One of the key benefits of having a main research question is that it helps to give direction to the study and keeps the researcher focused on the most important issues (Hancock, et al., 2021). The main research question structures research by providing a framework for sub-research questions and hypotheses. Notably, the main research question assists in focusing the data collection process, ensuring that the researcher collects only the data that is relevant to answering the question. The main research question can be derived from several sources such as the statement of the problem. The main research question should be closely aligned with the problem or issue that the study is addressing. One can use the statement of the problem to identify the key issues that the study will explore. Reviewing the literature on the topic can help the researcher to identify gaps in knowledge or areas for further research. Thus, the researcher can use information from literature review to formulate a main research question. The theoretical framework may provide insight into the relationships between variables or the underlying mechanisms that drive the problem (Muthukrishna, & Henrich, 2019). Thus, an example of the main research question is as follows:

How do educators in Insiza District Secondary Schools provide psychosocial support services to adolescent male learners with behavioral problems?

Sub-research questions are important for supporting the main research question. Sub-research questions provide support for the main research question by exploring different aspects of the problem or issue. This helps to build a more comprehensive understanding of the issue. Sub-research questions help to target the data collection, ensuring that the researcher collects the specific data needed to answer the main research question. Sub-research questions also give direction to the study and ensure that alike issues and areas of interest are addressed. Sub-research questions can be derived from a literature review which can help identify areas of debate or disagreement in the existing research topic. These debates can be used to formulate sub-research questions that explore different aspects of the problem. If using a theoretical framework, one can use the concepts and relationships within the theory to identify sub-research questions that explore different aspects of the problem. The researcher must use clear and concise language to avoid ambiguity or confusion when writing the sub-research questions. That is, one should avoid using jargon or technical terms that may not be familiar to the audience. The use of interrogative sentences is important for the research questions and words like "what," "why," "how," "when," or "to what extent" can be used in phrasing the questions. Research questions can be phrased at different levels of analysis, such as individual, societal, or environmental. The length of research questions can vary depending on the complexity of the research and the level of detail required. However, in general, it is advisable to keep research questions short and focused on a single aspect of the problem. For the main research question, the researcher should aim for a single sentence that is clear and precise. The researcher should ensure that the sub-research questions are in line with the specific objectives and equal in number. It is also important to note that when introducing research questions, the researcher should clearly label the research questions using numbered points or subheadings to make them easy to follow and understand. For example, the research questions can be numbered (1, 2, 3…) and should be questions and not statements. (Padmadewi et al., 2023; Mali, 2023; Dugartsyrenova & Sardegna, 2022)

Examples of sub-research questions are as follows:

1. How are educators providing psychosocial support services to adolescent male learners with behavioral problems in Insiza District Secondary Schools?
2. What are the challenges experienced by educators in the provision of psychosocial support services to adolescent male learners with behavioral problems in Insiza District Secondary Schools?
3. What mitigation measures can be adopted to address the challenges experienced by the educators in providing psychosocial support services to adolescent male learners with behavioural problems in Insiza District Secondary Schools?

Research Objectives

A general objective which should be in line with the title and main research question. Specific objectives have to be in line with sub-research questions or the variables the researcher hypothesise to influence the phenomenon being investigated. The general objective should be SMART, that is, Specific, Measurable, Achievable, Realist, and Time-bound. Objectives are the cornerstone of a well-crafted research proposal that can increase the likelihood of successful project outcomes (Patel, et al., 2021; Khan Academy, n.d; MindTools, n.d). A Specific objective clearly defines what the research aims to achieve, leaving no room for misinterpretation. For instance, "To assess how educators provide psychosocial support services to adolescent male learners with behavioral problems" is a specific objective that clearly states what the study will focus on. Measurable objectives provide a clear indication of how progress will be tracked and evaluated, such as "To examine the correlation between social media usage and depression rates among 18-25-year-olds." Achievable objectives are realistic and feasible within the given timeframe and resources, ensuring that the researcher can complete the project. The Realistic objectives align with the broader goals and priorities of the organization or funding agency, demonstrating how the research will contribute to existing knowledge or address a real-world problem. Finally, Time-bound objectives specify a realistic timeline for completing the project, including milestones and deadlines, allowing the researcher to plan and manage their work effectively (Adeoye & Adong, 2023).

Research objectives are important because they help to define the specific aims and goals of the research. Some of the key benefits of having research objectives are that they help to clarify the purpose of the study and provide a clear roadmap for the research (Maxwell, 2016). Research objectives also keep the research focused on the most important issues and ensure that the researcher is addressing the key questions and areas of interest. Above all, the objectives also ensure that all research activities are aligned with the overall goals and aims of the study. The main research objective can be derived from a combination of the main research question and-overall research goals. The researcher should start by identifying the key issue or problem that is to be addressed by the research (Maxwell, 2016). This will be the basis for the main research question. Then, the researcher should consider what is to be achieved in the research. The researcher should consider the following questions: Do you want to improve the audience's understanding of a particular issue? Are you looking to develop new solutions to a problem? Once the researcher has identified the main research question and overall research goals, the researcher can use these to formulate a single sentence that summarises the main aim of the study. The research objectives are typically listed and numbered like the research questions. This helps to make the objectives clear and easy to understand. For example:

1. To assess how educators, provide psychosocial support services to adolescent male learners with behavioral problems in Insiza District Secondary Schools.
2. To establish the challenges experienced by educators in the provision of psychosocial support services to adolescent male learners with behavioral problems in Insiza District Secondary Schools.
3. To suggest mitigation measures that can be adopted to address the challenges experienced by the educators in providing psychosocial support services to adolescent male learners with behavioral problems in Insiza District Secondary Schools.

Research objectives such as to identify indicate that the researcher is to identify specific factors or elements related to the research topic (Sreekumar, 2023). To explore shows that the researcher investigates or examines a particular issue or problem in more detail. To assess indicates that the researcher is looking to evaluate or judge the effectiveness or viability of a particular approach or solution. The words used to state research objectives are typically verbs. Verbs are action words that indicate what the researcher is trying to achieve with the research (Maharani & Santoso, 2023). Some other verbs that are commonly used in research objectives include to investigate and this indicates that the researcher is to study or examine a particular issue or problem in more detail. To examine indicates that the researcher explores or considers something in depth. To determine indicates that the researcher is to find out or establish something.

Rationale of study/justification of the Study

The rationale of a study explains why the research is important and worth undertaking (Drabble, et al., 2014). It justifies the research and helps to convince the reader that the research is necessary and worthwhile. When crafting a strong rationale start by highlighting the problem or issue that the-research is addressing (Edelson, 2002). The researcher should explain why this problem is important and what its consequences are. According to Stanovich (1999), there is a need to describe the gap in knowledge or understanding that the research is seeking to fill. Explaining why this gap exists and how it limits the understanding of the problem is very important. If the researcher is using a theoretical framework in the research, it is essential to explain how the study will contribute to or challenge existing theory. The researcher should describe the practical implications of the research and how the research will add to the understanding of the problem or issue. The rationale of the study should indicate how the study will contribute to the understanding of the problem and help to address it more effectively. In this section, the provision of a brief overview of the methods the researcher will use to address the problem is essential. This helps to demonstrate the feasibility and practicality of the research. To write a strong rationale for the research, it is advisable to use strong, active language to convey the importance and urgency of the research. The researcher should avoid using passive language or vague terms and should keep the rationale concise and to the point. Of great importance is to try to cover all of the key issues in one or two paragraphs. The researcher should consider the nature of the audience and should tailor the rationale accordingly. When writing for a general audience, it is advisable to avoid technical jargon and use clear, simple language. It is also important to be clear on the audience for research, as this will help to shape the tone and content of the rationale. If the research has practical implications for practitioners, such as teachers, doctors, or engineers, the rationale should emphasize how the study will improve practice and contribute to better outcomes. If the research has policy implications, the rationale should emphasize the relevance of the study for policy and decision-making.

Hypotheses

Hypotheses are stated where applicable. A hypothesis is a statement or assumption about the relationship between variables in a study. It is an educated guess about what the researcher expects to find in the study and is the basis for testing and evaluating the research question (Kumar et al., 2024). They should also be clear specific, concise, and testable, and should be grounded in the existing literature (Bryman & Bell, 2020; Creswell & Plano Clark, 2020; Harris et al., 2023; Morgan et al., 2024). A good hypothesis should be grounded in theory and should provide a clear direction for the research (Kumar et al., 2020). This can help to ensure that the study is methodologically sound and that the findings are meaningful and generalizable (Mertens et al., 2020). It should avoid unnecessary variables and focusing on the most critical factors that affect the outcome (Santos et al., 2024). For example, you might formulate a hypothesis that "educators' provision psychosocial support services to adolescent male learners with behavioural problems will lead to minimal behaviour problems among male learners in secondary schools." This hypothesis can then be tested and evaluated using data and analysis. When writing a hypothesis, researchers should consider the research question or statement of the problem and the purpose of the study (Gall et al., 2020). The hypothesis should be a logical extension of the research question or statement of the problem and should be testable through the proposed methods (Johnson & Christensen, 2020).

Hypotheses can be used in both qualitative and quantitative research, although they are more commonly associated with quantitative research. In quantitative research, hypotheses are typically presented at the start of the research design section, after the researcher has introduced the research question and reviewed the relevant literature. In qualitative research, hypotheses are often not used in the same way as they are in quantitative research. Instead, the researcher may use theoretical frameworks or guiding questions to structure the research. Hypotheses are important in research as they provide a guide for the research, helping the researcher to focus on specific relationships between variables and to design the research accordingly. Hypotheses help to promote objectivity and rigor in research.

Significance of the Study

University of California. (2020); Harvard Graduate School of Education. (2020) and Steinbrecher et al., (2016) explain how to write the significance of the study in a research proposal. The significance can therefore be written as: This study is significant because it addresses (briefly describes the research question or problem being investigated) and has the potential to (specifically state the contribution to knowledge, practice, or policy). The study's significance is multifaceted, that is, the researcher might list 2-3 key ways in which the study will contribute to the field, for example, improving an understanding of a concept or phenomenon, filling a gap in current research, and informing policy decisions. The findings of this study will have important implications for specific stakeholders or groups, including describing briefly how the results will be used or applied. Furthermore, it should be indicated how the study contributes to the advancement of related fields or disciplines by specifically stating how the study builds upon or challenges existing knowledge. The researcher should summarise the main significance of the study and its potential impact. For instance, the summary statement could be: 'Overall, this study is significant because it has the potential to…'. Thus, the significance of the study and its potential impact should be clearly stated. It is critical to highlight the practical implications of the study and who will benefit from its findings. Concise language should be used and avoid jargon. The researcher should ensure that the significance is relevant to the research questions and methodology.

Definition of Terms

The purpose of this section is to provide clear and concise definitions for key terms used throughout the research proposal. To ensure a thorough understanding of the study's methodology, findings, and implications, it is essential to establish a shared understanding of the technical terms and concepts used in this proposal. Definitions are necessary because they provide clarity and consistency in the use of language, which is particularly important in academic and scientific writing (Brown et al, 2023; Harris et al, 2023; Kumar et al., 2024; Morgan et al., 2024; Santos et al., 2024). By defining key terms, the researcher can avoid confusion and miscommunication that may arise from using technical jargon or ambiguous language. Furthermore, providing definitions allows readers to quickly grasp the essence of the study and its significance, which is crucial for understanding the research question, methodology, and conclusions. The definitions will be used consistently throughout the proposal to ensure a clear and accurate presentation of the research design, methods, and results. By providing these definitions, the researcher aims to promote transparency, clarity, and rigor in presentation of the research. The researcher should ensure that terms that are specific to the study or field of research are defined. Use plain language and avoiding jargon is paramount. Keeping the definitions concise and focusing on the main concepts is necessary. Organise the definitions in a logical order (e.g., alphabetical order). By providing clear definitions, the researcher can ensure that readers understand the research questions and objectives. One can facilitate communication among team members or collaborators. The credibility and rigor of the research proposal are also enhanced making the proposal more accessible to non-experts in the field.

Assumptions

Several scholars elaborate on writing assumptions in a research proposal (Liu et al., 2021; Singh, Singh, and. Singh, 2021; Kim et al., 2020; Gupta et al., 2020; Sanchez et al., 2020). In research, assumptions are unstated or implicit beliefs that underlie the research design, methodology, and analysis (Harris et al., 2023; Kumar et al., 2024). They are often taken for granted and are not explicitly stated in the research proposal or report. Assumptions can be about the research questions, variables, data collection and analysis methods, or the interpretation of results. Including assumptions in a research proposal is important for several reasons. By stating assumptions, the researcher demonstrates transparency and accountability, allowing readers to understand the basis of the research design and methodology. Assumptions help clarify the research questions and objectives, making it easier for readers to understand the study's purpose and significance (Lee et al., 2023). Assumptions can affect the validity and reliability of the research findings. By acknowledging them, the researcher can ensure that the methodology is sound and robust. Assumptions can highlight potential limitations of the study, allowing readers to consider these limitations when interpreting the results.

Examples of assumptions in a research proposal can be represented as sampling assumptions where it can be assumed that a random sample of 100 participants from a local university will be representative of the general population. Measurement assumption where it is assumed that self-reported data on demographic variables will accurately reflect the participants' characteristics. The theoretical assumption assumes that social learning theory provides an adequate framework for understanding how participants will respond to the treatment. Analytical assumptions may assume that using a t-test will provide an appropriate statistical test for comparing means between groups. Contextual assumption assumes that participants will have access to a stable internet connection during the online survey. Causal assumptions assume that there is a causal relationship between exposure to social media and increased social anxiety. The theoretical construct assumption assumes that emotional intelligence is a stable trait that can be measured reliably. The sampling frame assumption may assume that the online directory of local businesses provides an accurate list of all eligible participants.

Limitations

In a research proposal, limitations refer to the constraints or potential weaknesses of the study that may affect the validity, generalisability, or reliability of the results. Limitations can be related to various aspects of the study, including methodological limitations: restrictions imposed by the research design, data collection methods, or analysis techniques. Theoretical limitations: assumptions or gaps in the existing literature that may influence the study's findings or conclusions. Practical limitations: challenges or obstacles that may arise during data collection, such as participant recruitment or data quality issues. Including limitations in a research proposal serves several purposes such as honesty and transparency. Acknowledging limitations demonstrates the researcher's awareness of potential flaws and allows readers to understand the study's strengths and weaknesses. Realistic expectations: by highlighting limitations, the researcher sets realistic expectations for the study's outcomes and potential impact. Funding agencies' requirements: many funding agencies require researchers to discuss limitations as part of their proposals. Peer review and criticism: limitations provide a framework for peer reviewers to evaluate the study's methodology and potential flaws. There can be methodological limitations such as a small sample size may limit the generalisability of findings to larger populations. The reliance on self-reported data may introduce bias and affect the accuracy of results. The study's duration is limited to six months, which may not capture seasonal variations or longer-term effects. The limited number of research staff may impact ability to collect data efficiently. The understanding of the research topic is still evolving, and acknowledgement that the findings may be influenced by the limited knowledge base. The lack of established theories on this topic means that the results may not be easily generalizable. By acknowledging and discussing limitations in the research proposal, the researcher demonstrates awareness of potential weaknesses and provides a more comprehensive understanding of the study's strengths and weaknesses (Kelly, & Kelly, 2018; National Science Foundation, n.d; University of California, n,d.).

Delimitations

In a research proposal, delimitations refer to the boundaries or limitations that the researcher deliberately sets to define the scope of the study (Kumar et al, 2024). Delimitations help to focus the study by setting clear boundaries, researchers can concentrate on a specific aspect of the research topic and avoid unnecessary complexity (Morgan et al., 2024; Berkeley, 2019). Delimitations help to prevent the study from expanding beyond its original scope, which can lead to unclear or incomplete research (Lee et al., 2023). Delimitations help to define the research question and ensure that the study remains focused on a specific issue (Harris et al., 2023). Examples of delimitations can be elaborated as geographical delimitation such as this study will focus on a specific region or city, rather than attempting to generalize findings to a broader geographic area. Temporal delimitation such as this study will examine only a specific period, rather than attempting to capture a longer-term trend. Methodological delimitation: This study will rely on quantitative data collection methods, excluding qualitative or mixed-methods approaches. Sampling delimitation: This study will focus on a specific population or sample, excluding other potential populations or samples. Theoretical delimitation: This study will be limited to exploring one theoretical framework or perspective, excluding others.

An example of a delimitations section may read “This study will focus on undergraduate students at a single university, rather than attempting to generalize findings to a broader population. The data collection period will be limited to six months, capturing seasonal variations but not longer-term trends. The study will rely exclusively on online surveys, excluding other data collection methods such as interviews or focus groups. The sample size will be capped at 100 participants, ensuring manageable data analysis but limiting generalisability to larger populations. Another delimitation can read as: ‘This study will explore only one theoretical framework (Social Learning Theory), excluding other perspectives such as Cognitive Behavioural Theory’. By including delimitations in the research proposal, the researcher demonstrates awareness of the study's boundaries and commitment to focusing on a specific issue.

Literature Review

A literature review in a research proposal is a comprehensive and systematic analysis of the existing research on a specific topic or issue (Liao, et al., 2017). It provides an overview of the current state of knowledge, identifies gaps and limitations in the existing research, and sets the stage for the proposed study (National Science Foundation, 2020; Creswell & Creswell, 2018; Gray, 2017). A literature review in a research proposal typically includes background information which is a brief overview of the research topic, its significance, and relevance to the proposed study. It is a summary of the main findings and conclusions from previous studies, including their methodologies, results, and implications (Snyder, 2019). Literature is also an analysis of the strengths and limitations of previous studies, identification of gaps and areas for future research, and an integration of findings from multiple studies to form a cohesive picture. Relationship to the proposed study is the explanation of how the proposed study builds upon previous research, addresses gaps or limitations, and contributes to the existing body of knowledge.

Including a literature review in the research proposal is essential because it shows that the researcher has thoroughly investigated the topic and has a deep understanding of the existing knowledge base (Snyder, 2019). The literature review provides context for the proposed study, allowing readers to understand its significance, relevance, and potential impact (Liao, et al., 2017; Snyder, et al., 2019). The literature review helps identify gaps in existing research, which can inform the research questions, methodology, and expected outcomes. The literature review sets the stage for the proposed study by highlighting areas that need further exploration and providing a framework for investigation. A well-conducted literature review enhances the credibility of the proposed study by demonstrating that the researcher has thoroughly researched the topic and is aware of its complexities (Afful, et al, 2022; van Riel & Snyder, 2024). An example of a literature review in a research proposal: "The existing literature on [topic] suggests that [key findings]. However, [gaps or limitations]. This study aims to [research question] by [methodology]. By addressing [gap or limitation], this study will contribute to the understanding of [topic] and provide insights for [practice or policy]." The researcher needs to tailor the literature review to a specific research proposal and audience, ensuring that it is concise, well-organized, and relevant to the proposed study.

Research Methodology

The research methodology in a research proposal outlines the approach and procedures that will be used to conduct the study. It describes how the researcher will collect, analyze, and interpret data to answer the research question or test the hypothesis. The methodology section should provide a clear and concise overview of the research design, methods, and tools used to collect and analyze data. A well-written methodology section is crucial because it helps readers understand how the study will be conducted, which is essential for evaluating the validity and reliability of the findings. A well-planned methodology demonstrates that the researcher has thoughtfully considered the research design and methods, increasing the credibility of the study. The methodology section provides transparency about the research process, allowing readers to replicate or evaluate the study. (Lai, 2024; Rosser &

Soler, 2024; Creswell & Creswell, 2018). Other relevant subtitles to include in a methodology section include:

- Research design: Describe the overall design of the study, including experimental, quasi-experimental, survey, case study, or other designs.
- Sampling strategy: Explain how participants will be selected, including sampling techniques (e.g. random sampling, convenience sampling) and sample size determination (Tutar, et al., 2024).
- Data collection methods: Describe the methods used to collect data, such as surveys, interviews, observations, or experiments.
- Data analysis plan: Outline the statistical or analytical techniques used to analyze data, including any software or tools employed.
- Measurement instruments: Describe the tools or instruments used to measure variables, such as questionnaires, rating scales, or observation checklists.
- Data quality control: Explain measures taken to ensure data quality, such as data cleaning, validation, and consistency checks.
- Ethics and informed consent: Address ethical considerations and informed consent procedures for human subjects or animals involved in the study.
- Timeframe and timeline: Provide a detailed timeline for data collection and analysis to ensure that the study is feasible and realistic (Robson, 2024).

Chapter organization/outline

A chapter organization or outline in a research proposal is a detailed structure that outlines the proposed study's content, including the main sections and subheadings. This outline serves as a roadmap for the proposal, helping to ensure that the study's scope, methodology, and expected outcomes are presented. A typical chapter organization or outline in a research proposal includes the proposed chapter titles and their subtitles. A general outline for structuring the chapter organization or outline is as follows:

Chapter I Introduction: Background…and summary.

Chapter II Literature Review: Introduction … and summary.

Chapter III Research Methodology: Introduction, research approach, research design, research paradigm, sampling strategy, data collection methods, data analysis plan, data collection instruments...ethical considerations, and summary.

Chapter IV Data Presentation, Analysis and Discussion: …

Chapter V Summary, Conclusions, Recommendations: …

Timeline

A timeline in a research proposal is a detailed schedule outlining the expected duration of the research project, including the key milestones, tasks, and deadlines (Zina, 2021). It provides a roadmap for the researcher to stay on track and ensures that the study is completed within the proposed timeframe (Bell, 2014). According to Kuhlthau (1994) a timeline helps researchers plan and organize their work, ensuring that tasks are completed efficiently and effectively. The timeline allows for resource allocation and budgeting, as it provides an estimate of the time required for each task. A timeline facilitates communication among team members, collaborators, and stakeholders, ensuring that everyone is aware of the project's progress and deadlines. The timeline helps identify potential risks and timelines for addressing them, minimising delays, and ensuring the study remains on track. Many funding agencies require a detailed timeline as part of the proposal submission process, so including one demonstrates the researcher's preparedness and commitment to completing the project (Englund & Bucero, 2006). A typical timeline in a research proposal should include the date when the project begins, and major events or deadlines, such as data collection, data analysis, and manuscript submission, brief descriptions of the tasks to be completed during each period must be specified, the estimated time required to complete each task and any dependencies between tasks or milestones that affect the overall timeline.

When structuring a timeline, it is important to break down the project into phases, that is, dividing the project into distinct phases (e.g., literature review, data collection, data analysis, etc) (Bell, 2014). The researcher should also identify key milestones determining the critical events or deadlines within each phase (e.g., completion of literature review, data collection, etc). It is essential to estimate task duration and determine the estimated time required to complete each task or milestone. Create a Gantt chart or calendar, visualizing the timeline using a Gantt chart or calendar ensures clarity and ease of understanding. Table 1 below gives an example of a timeline:

Table 1. Timeline for a research proposal

Phase	Task	Start date	End date	Duration	Dependencies
Chapter 1	Compilation of Chapter One extracted from the passed research proposal	Month 1	Month 1	1 Month	-
Chapter 2	Conduct literature review	Month 2	Month 5	4 Months	Literature review complete
Chapter 3	Conduct research methodology and craft research instruments	Month 5	Month 8	4 Months	Methodology complete
Data collection	Collect primary data	Month 9	Month 11	2 Months	Data collection complete
Chapter 4 Data analysis	Analyse data	Month 12	Month 14	2 Months	Data analysis complete
Chapter 5 Summary, conclusion, Manuscript submission	Write and submit the first draft of the manuscript	Month 15	Month 17	2 Months	Complete
Manuscript corrections	Amend where necessary	Month 18	Month 19	1 Month	complete
Final manuscript submission	Submit bound document adhering to regulations				

(Source: Doris Chasokela 2024)

One can also present a timeline in a Gantt chart. A Gantt chart is a popular way to visualize a timeline, especially, for complex projects with multiple tasks and dependencies. Here is how to create a Gantt chart for your research proposal. A Gantt chart is a type of bar chart that illustrates a project schedule, showing the tasks, dependencies, and timelines. It is named after Henry Gantt, who developed this type of chart in the early 20th century. The benefits of using a Gantt chart include providing a clear visual representation of the project's timeline, and making it easier to understand and communicate. The chart shows dependencies between tasks, allowing the researcher to identify potential bottlenecks and plan accordingly. A Gantt chart helps the researcher to schedule tasks and allocate resources more effectively. One can easily update the chart as the project evolves. When presenting the Gantt chart, it is crucial to keep it simple focus on the essential information, and avoid clutter. The researcher should use different colors and icons to differentiate between tasks, dependencies, and milestones. Highlight the critical path (the sequence of tasks that determines the project's duration) to focus attention on high-priority tasks. Mark important milestones or deadlines on the chart to emphasize key events. Gantt charts have different representations so the researcher chooses their choices. Examples of Gantt charts are as follows:

Figure 1. Example 1 of a Gantt chart

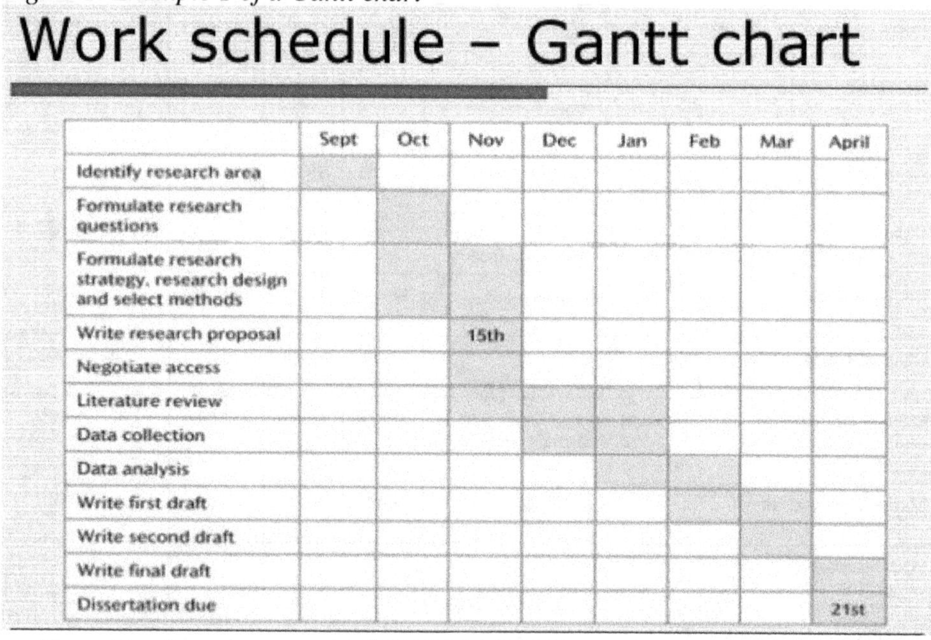

Work schedule – Gantt chart

	Sept	Oct	Nov	Dec	Jan	Feb	Mar	April
Identify research area								
Formulate research questions								
Formulate research strategy, research design and select methods								
Write research proposal			15th					
Negotiate access								
Literature review								
Data collection								
Data analysis								
Write first draft								
Write second draft								
Write final draft								
Dissertation due								21st

(Source: https://www.pinterest.com/pin/639440847067703189/)

Figure 2. Example 2 of a Gantt chart

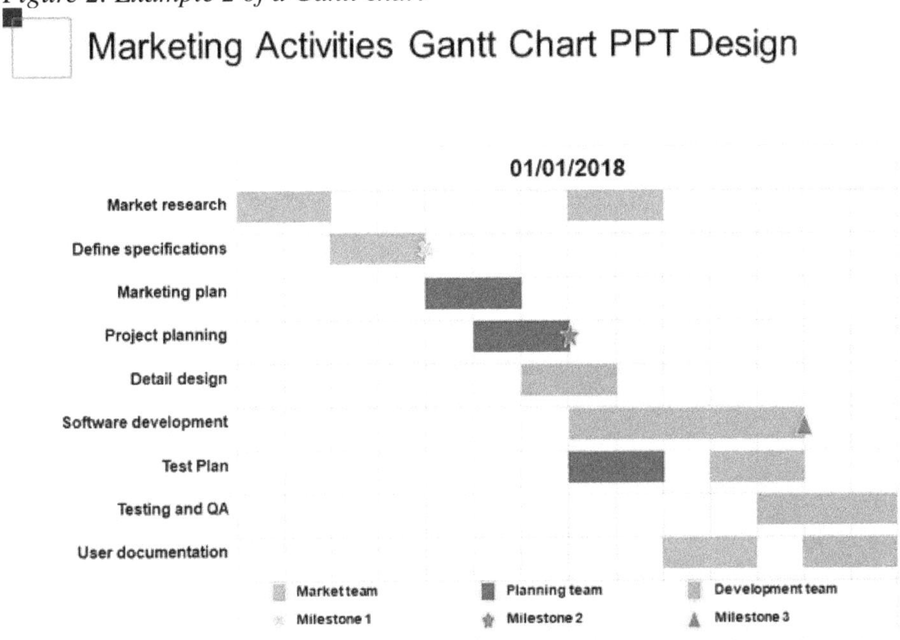

(Source: https://keski.condesan-ecoandes.org/how-to-draw-a-gantt-chart-in-powerpoint/)

By presenting the timeline in a Gantt chart form, the researcher can effectively communicate the project's schedule and dependencies to stakeholders, making it easier to manage and track the research proposal's progress.

CONCLUSION

The chapter presented the key issues and steps in writing a research proposal. The major components of a proposal have been highlighted in this chapter. These include choosing a title, cover page of a proposal, introducing the research proposal, statement of the problem/problem statement, purpose of study, research questions, research objectives, rationale of study/justification of the study, significance of the study, definition of terms, assumptions, limitations, delimitations, literature review, research methodology, chapter organization/outline, timeline and Gantt chart. These main elements of the research proposal were well articulated, indicating the critical roles played by each component in proposal writing. It is envisaged that this chapter on research proposal writing is to provide higher education students and researchers with the knowledge and skills they need to complete this important step in the research process. By providing clear and accessible guidance on developing the stated elements, this chapter focuses on empowering students and researchers to produce superior research proposals that have the potential to make some significant contribution to shaping the proposed research study.

REFERENCES

Abdullah, L. A. H. (2023). Common mistakes committed and challenges faced in research proposal writing by University of Aden postgraduate students. *The Journal of Social Studies*, 29(1), 18–30. DOI: 10.20428/jss.v29i1.2070

Adeoye, M. A., & Adong, C. (2023). The Power of Precision: Why Your Research Focus Should Be SMART. *Journal of Education Action Research*, 7(4), 569–577. DOI: 10.23887/jear.v7i4.69757

Afful, J. B. A., Ngula, R. S., Twumasi, R., Tetteh, G., & Mensah, F. (2022). Supervisors' perceptions of postgraduate students' thesis literature review writing in a Ghanaian university. *Advances in Social Sciences Research Journal, 9*(1), 267-289.

Bell, J. (2014). *Doing Your Research Project: A guide for first-time researchers*. McGraw-hill Education.

Brown, T. D., Patel, V. B., & Jenkins, K. E. (2023). *Quantitative research methods for business students*. Pearson Education.

Bryman, A., & Bell, E. (2020). *Business research methods*. Oxford University Press.

Creswell, J. W., & Creswell, J. D. (2018). *Research Design: Quantitative, Qualitative, Mixed Methods*. Sage Publications.

Creswell, J. W., & Plano Clark, V. L. (2020). *Designing and conducting mixed methods research*. Sage Publications.

Drabble, S. J., O'Cathain, A., Thomas, K. J., Rudolph, A., & Hewison, J. (2014). Describing qualitative research undertaken with randomized controlled trials in grant proposals: A documentary analysis. *BMC Medical Research Methodology*, 14(1), 1–11. DOI: 10.1186/1471-2288-14-24 PMID: 24533771

Dugartsyrenova, V. A., & Sardegna, V. G. (2022). Enhancing genre instruction on research proposal introductions with an online academic writing tutor. *Journal of Second Language Writing*, 58, 100908. DOI: 10.1016/j.jslw.2022.100908

Dutton, G. H. (1999). Introduction and problem statement. *A Hierarchical Coordinate System for Geoprocessing and Cartography*, 1-18.

Edelson, D. C. (2002). Design research: What we learn when we engage in design. *Journal of the Learning Sciences*, 11(1), 105–121. DOI: 10.1207/S15327809JLS1101_4

Ellis, T. J., & Levy, Y. (2009). Towards a Guide for Novice Researchers on Research Methodology: Review and Proposed Methods. *Issues in Informing Science & Information Technology, 6*.

Englund, R. L., & Bucero, A. (2006). *Project sponsorship: Achieving management commitment for project success*. John Wiley & Sons.

Gall, M. D., Gall, J. P., & Borg, W. R. (2020). *Educational research: An introduction*. Pearson Education.

Garcia, D., & Gluesing, J. C. (2013). Qualitative research methods in international organizational change research. *Journal of Organizational Change Management*, 26(2), 423–444. DOI: 10.1108/09534811311328416

Gray, D. E. (2017). *Doing Research in the Real World*. Sage Publications.

Gupta, A. K., Sharma, S.K. and Kumar, R. (2020). "The Impact of Assumptions on Research Proposals: A Systematic Review. Journal: Journal of Research Methodology. DOI: DOI: 10.1080/16138776.2020.1722349

Hancock, D. R., Algozzine, B., & Lim, J. H. (2021). Doing case study research: A practical guide for beginning researchers.

Harris, S. K., Smith, J. R., & Johnson, K. (2023). *Research methods in education: An introduction to qualitative and quantitative research methods*. Routledge.

Harvard Graduate School of Education. (2020). Writing a Research Proposal. https://cepr.harvard.edu/proposal-writing-tips

Jalongo, M. R. (2023). Choosing a Topic: Focus and Title. In *Scholarly Writing: Publishing Manuscripts That Are Read, Downloaded, and Cited* (pp. 91–114). Springer International Publishing. DOI: 10.1007/978-3-031-39516-1_5

Johnson, R. B., & Christensen, L. B. (2020). *Educational research: Quantitative, qualitative, and mixed approaches*. Pearson Education.

Kafu, P. A. (2023). Research proposal and thesis/dissertation writing course in modern higher education, an emerging issue in sustainable development in the modern world. *African Journal of Education, Science and Technology, 7*(3), 442-449.

Kelly, A., & Kelly, M. (2018). *The Research Process: A Practical Guide*. Sage Publications.

Khan, M. A. W. (2023). The art of writing a "Title" for an original research paper. *Bangladesh Journal of Medical Education*, 14(1), 70–77. DOI: 10.3329/bjme.v14i1.63685

Khan Academy. (n.d.). What is a SMART goal? Retrieved from https://www.khanacademy.org/test-prep/mcat/behavioral-sciences/smart-goals/a/what-is-a-smart-goal

Kim, H., Lee, J. and Park, S. (2020). "Assumptions in Research Proposals: A Study of Graduate Students' Perceptions. Journal: Journal of Educational Psychology. DOI: DOI: 10.1037/edu0000366

Kim, H., Lee, Y., & Kim, J. (2021). *Research methods for business students.* McGraw-Hill Education.

Kumar, A., Patel, V. B., & Brown, T. D. (2024). A systematic review of the literature on [topic]. *Journal of Research in Education*, 22(1), 1–15. PMID: 35694371

Kumar, R., Kumar, S., & Singh, R. K. (2020). *Research methodology: An introduction.* PHI Learning.

Lai, H. (2024). Selection of a Topic of a Research Proposal of Emerging Academic Writers in a Blended Linked EAP Course: An Interactional Ethnographic Perspective. *TESOL Quarterly*, tesq.3316. DOI: 10.1002/tesq.3316

Lee, Y., Kim, H., & Lee, J. (2023). The effects of social media on mental health among adolescents: A systematic review. *Journal of Psychology and Health*, 32(2), 147–162. PMID: 34328044

Lee, Y., Kim, H., & Lee, J. (2023). The effects of social media on mental health among adolescents: A systematic review. *Journal of Psychology and Health*, 32(2), 147–162. PMID: 34328044

Levac, D., Colquhoun, H., & O'Brien, K. K. (2010). Scoping studies: Advancing the methodology. *Implementation Science : IS*, 5(1), 1–9. DOI: 10.1186/1748-5908-5-69 PMID: 20854677

Li, M., Chen, X., & Wang, Y. (2021). *Introduction to statistics for social sciences.* Cengage Learning.

Liao, Y., Deschamps, F., Loures, E. D. F. R., & Ramos, L. F. P. (2017). Past, present and future of Industry 4.0-a systematic literature review and research agenda proposal. *International Journal of Production Research*, 55(12), 3609–3629. DOI: 10.1080/00207543.2017.1308576

Liu, Z., Liu, S. and Zhang, J. (2021). Assumptions in Research Proposals: A Review of the Literature. Journal of Educational Evaluation and Research. DOI: DOI: 10.1037/edr0000354

Maharani, I. D., & Santoso, D. R. (2023). Enhancing Verb Writing Skills in Eighth-Grade Students with Flashcards: A Classroom Action Research in Indonesia. *Indonesian Journal of Education Methods Development*, 18(3). Advance online publication. DOI: 10.21070/ijemd.v21i3.777

Mali, Y. C. G. (2023). EFL Students 'Challenges in Writing Research Proposals. *LLT Journal: A Journal on Language and Language Teaching, 26*(1), 272-289.

Manchez, J. M., Rodriguez, M. A. and F. Garcia. F. (2020)."Assumptions in Research Proposals: A Critical Examination of the Concept. Journal: Journal of Educational Theory. Doi: DOI: 10.1111/1468-5965.12345

Maxwell, D. (2016). The research lifecycle as a strategic roadmap. *Journal of Library Administration*, 56(2), 111–123. DOI: 10.1080/01930826.2015.1105041

Mertens, D. M., Sullivan, W. M., & McPhee-Jones, K. J. (2020). *Research methods in education: An introduction*. Routledge.

Morgan, D. L., & Spencer, L. M. (2024). The use of mixed-methods research in education: A critical review. *Journal of Educational Research Review*, 21(1), 1–20.

Muthukrishna, M., & Henrich, J. (2019). A problem in theory. *Nature Human Behaviour*, 3(3), 221–229. DOI: 10.1038/s41562-018-0522-1 PMID: 30953018

National Science Foundation. (2020). The Role of Literature Reviews in Research Proposals. Retrieved from https://www.nsf.gov/bias/literature-review.cfm

National Science Foundation. (n.d.). Writing the Research Proposal: A Guide. Retrieved from https://www.nsf.gov/bias/writing_guide.cfm

Nehru Pasoloran Pongsapan, M. P., Allo, M. D. G., & Rumpa, L. D. (2023). *A Guide to Research Proposal*. Deepublish.

Nolen, A., & Talbert, T. (2011). Qualitative assertions as prescriptive statements. *Educational Psychology Review*, 23(2), 263–271. DOI: 10.1007/s10648-011-9159-6

Padmadewi, N. N., Artini, L. P., Ratminingsih, N. M., Suhardiana, I. P. A., Zamzam, A., & Juniarta, P. A. K. (2023). Designing Project-Based Learning in research proposal writing: Its effect, problems, and scaffolding utilized. *Studies in English Language and Education*, 10(2), 841–862. DOI: 10.24815/siele.v10i2.27408

Patel, R., Patel, D., & Shah, P. (2021). Improving student writing skills through scaffolding strategies: A randomized controlled trial. *Journal of Educational Psychology*, 113(2), 341–353.

Robson, C. (2024). *Real-world research*. John Wiley & Sons.

Rosser, P., & Soler, S. (2024). Reflective Practice and Frame Theory in Educational History: A Communicative Narrative Research Proposal. *Indonesian Journal of Education and Social Sciences*, 3(2), 151–16. DOI: 10.56916/ijess.v3i2.724

Sanak, S. D., Nalley, H. M., & Suek, L. A. (2023). An analysis of the introductory part of Research Proposals written by the Students of the English study Program. *Sparkle Journal of Language. Education and Culture*, 2(1), 52–65.

Santos, J., Fernandes, L., & Silva, C. (2024). A mixed-methods study on the impact of gamification on student engagement in online learning environments. *Journal of Business Research*, 133(1), 15–25.

Singh, M. K., Singh, S.K. and Singh, R.K. (2021). "The Role of Assumptions in Research Proposals: A Critical Analysis. Journal: Journal of Research Methodology. DOI: DOI: 10.1080/16138776.2021.1901423

Smith, J., Brown, T. D., & Johnson, K. (2022). *Research design: Qualitative and quantitative approaches*. Sage Publications.

Snyder, H. (2019). Literature review as a research methodology: An overview and guidelines. *Journal of Business Research*, 104, 333–339. DOI: 10.1016/j.jbusres.2019.07.039

Sreekumar, D. (2023). What Are Research Objectives and How to Write Them? https://researcher.life/blog/article/what-are-research-objectives-how-to-write-them-with-examples

Stanovich, K. E. (1999). *Who is rational? Studies of individual differences in reasoning*. Psychology Press. DOI: 10.4324/9781410603432

Steinbrecher, W. J., Booth, W. J., & Mullen, C. K. (2016). *The Craft of Research*. University of Chicago Press.

Taylor, R. W., Wiedemann-Danielson, L. A., & Smith, J. R. (2022). *Research methods in education: An introduction to qualitative and quantitative research methods*. Routledge.

Tracy, S. J. (2019). *Qualitative research methods: Collecting evidence, crafting analysis, communicating impact*. John Wiley & Sons.

Tutar, H., Şahin, M., & Sarkhanov, T. (2024). Problem areas of determining the sample size in qualitative research: A model proposal. *Qualitative Research Journal*, 24(3), 315–336. DOI: 10.1108/QRJ-06-2023-0099

University of California. Berkeley. (2019). Writing a Research Proposal: A Step-by-Step Guide. Retrieved from https://www.ucberkeley.edu/units/osp/research/propose/writing-research-proposal

University of California. (2020). Writing a Research Proposal. San Francisco (UCSF) School of Medicine. https://www.ctsi.ucsf.edu/funding-support/research-proposal-guidelines

van Riel, A., & Snyder, H. (2024). Enhancing the impact of literature reviews: guidelines for making meaningful contributions. *Spanish Journal of Marketing-ESIC.*

Zina, O. (2021). The essential guide to doing your research project. *Sage (Atlanta, Ga.).*

KEY TERMS AND DEFINITIONS

Assumptions: are untested statements that are taken for granted as true or accurate, often underlying the research design and methodology. Assumptions can be implicit or explicit and may influence the outcome of the study.

Chapter outline: is a detailed breakdown of the structure and organization of a research proposal, typically including sections such as Introduction, Literature Review, Methodology, Results, Discussion, Conclusion, References, and Appendices.

Delimitations: are the boundaries or limitations of the study, specifying what is included and what is excluded from the investigation. Delimitations help to define the scope of the study and prevent it from becoming too broad or unwieldy.

Hypotheses: are specific, testable predictions made about the relationship between variables. In a research proposal, hypotheses are often formulated based on a review of the literature and serve as the foundation for the study's design and methodology.

Limitations: are the constraints or shortcomings of the study, including factors that may affect the validity or generalizability of the findings. Limitations can include issues with sample size, data quality, measurement tools, or research design.

Literature: review: is a critical analysis of existing research on a topic, synthesizing key findings and theories to provide context and background information for the current study. It helps to identify gaps in current knowledge and establish the significance of the proposed research.

Problem statement: is a concise summary of the research problem or issue being addressed in the study. It identifies the gap in current knowledge or practice that the research aims to fill.

Research methodology: refers to the systematic and rigorous approach used to collect, analyze, and interpret data in a study. It includes details about the research design, methods of data collection (e.g., surveys, interviews), data analysis techniques (e.g., statistical modeling), and procedures for ensuring validity and reliability.

Research question: is a clear and concise statement that defines what is to be investigated in the study. It is often more focused than a problem statement and serves as a guide for data collection and analysis.

Chapter 12
Crafting an Effective Research Proposal

Kumari R. Krishna
https://orcid.org/0000-0002-1802-628X
Department of Mathematics, College of Engineering and Technology, SRM Institute of Science and Technology, Kattankulathur, India

ABSTRACT

This chapter provides a comprehensive guide to writing an effective research proposal. It covers the fundamental components of a research proposal, including the title, abstract, introduction, literature review, research questions, methodology, timeline, budget, and references. The chapter emphasizes the importance of clear, concise, and well-organized writing and provides practical advice for developing a robust research design and preparing a realistic budget and timeline. Through detailed explanations, practical examples, and step-by-step instructions, readers will learn how to effectively communicate their research objectives, significance, methodology, and potential impact.

INTRODUCTION

A well-crafted research proposal is a fundamental step in the research process. It serves as a comprehensive blueprint for the proposed study, meticulously outlining the research objectives, methodology, and potential impact. This essential document not only provides a clear plan for conducting research but also helps to communicate the research's significance and feasibility to others (Ali & Kamraju, 2023; Kivunja, 2016). The purpose of this chapter is to demystify the process of writing a research proposal by breaking it down into manageable components and offering practical guidance for each stage. Whether you are a student seeking approval for a thesis

DOI: 10.4018/979-8-3373-0806-7.ch012

project, a researcher applying for funding, or a professional developing a business research plan, understanding how to create an effective proposal is crucial (Perry, 1998; Riley, 2007).

Firstly, the chapter will address the importance of defining clear and concise research objectives. These objectives form the foundation of the proposal, guiding the direction of the study and ensuring that it remains focused and relevant. Clear objectives also help to communicate the research's purpose to reviewers, making it easier for them to understand the significance of the proposed work (Simelane, Klopper, & Lubbe, 2018; Riley, 2007). The chapter will provide tips on how to formulate research objectives that are specific, measurable, achievable, relevant, and time-bound (SMART) (Faryadi, 2018).

Next, the chapter will delve into the methodology section of the research proposal. This section outlines the research design and methods that will be used to achieve the objectives. It is crucial to select appropriate methods that align with the research questions and to justify the choice of these methods (Abdulai & Owusu-Ansah, 2014; Heath & Tynan, 2010). The chapter will cover various research designs, including qualitative, quantitative, and mixed-methods approaches, and will provide guidance on how to describe the data collection and analysis procedures in detail (Kivunja, 2016; Riley, 2007). This includes selecting suitable instruments, sampling techniques, and analytical tools, as well as addressing any potential ethical considerations (Olesen & Stoecklin, 2018; Schwartz-Shea & Yanow, 2012).

The potential impact of the research is another critical component of the proposal. This section should highlight the anticipated contributions of the study to the existing body of knowledge, its practical implications, and its relevance to stakeholders (Tavakol & O'Brien, 2023; Barbato & Furlich, 2000). The chapter will offer strategies for articulating the significance of the research and for demonstrating its potential to address important gaps or challenges in the field. This might include discussing how the research findings could influence policy, practice, or further research (Boshoff & De Jong, 2015; Fry & Attawell, 2020). Additionally, the chapter will provide practical guidance on structuring the research proposal. This includes organizing the content in a logical and coherent manner, using clear and concise language, and adhering to any specific formatting requirements provided by the funding agency, academic institution, or other relevant bodies (Iqbal, 2007; Faryadi, 2018). Effective proposals are not only well-written but also meticulously organized, ensuring that reviewers can easily navigate through the document and understand the research plan (Simelane et al., 2018; Tomaselli, 2017).

The chapter will also emphasize the importance of conducting a thorough literature review as part of the research proposal. This review not only situates the proposed research within the existing body of knowledge but also helps to identify gaps that the research aims to address (Ravitch & Riggan, 2016; Yilmaz, 2013). A well-conducted literature review demonstrates the researcher's familiarity with the field and underscores the relevance and originality of the proposed study.

Finally, the chapter will offer tips on the revision and refinement process. Writing a research proposal is an iterative process that involves multiple drafts and revisions. The chapter will emphasize the importance of seeking feedback from peers, mentors, or colleagues and of making necessary revisions to improve clarity, coherence, and persuasiveness (Hizriani et al., 2022; Pietersen, 2014). It will also provide advice on common pitfalls to avoid and on how to address potential weaknesses in the proposal (Williams, 2021; Zuber-Skerritt & Fletcher, 2007).

Overall, writing a research proposal is a critical skill for students, researchers, and professionals alike. By breaking down the process into manageable components and providing practical guidance for each stage, this chapter aims to equip readers with the tools and knowledge needed to create compelling and coherent proposals (Ravitch & Riggan, 2016; Schwartz-Shea & Yanow, 2012). Whether seeking approval for a thesis project, applying for research funding, or developing a business research plan, understanding how to craft an effective proposal is essential for success in the research process (Perry, 1998; Ssegawa & Rwelamila, 2009).

SECTION 1: UNDERSTANDING THE PURPOSE OF A RESEARCH PROPOSAL

1.1 Definition and Role in Academia and Professional Settings

Research proposals are comprehensive documents that delineate the planned course of a research project. They serve as a critical roadmap for the entire research process, offering a detailed outline of the study's objectives, significance, and methodology. These proposals are instrumental in communicating the researcher's intentions and plans to various stakeholders, including academic advisors, funding bodies, and institutional review boards. In academia, research proposals are often prerequisites for thesis or dissertation projects, helping to ensure that the proposed research is feasible, ethically sound, and of scholarly value. In professional settings, research proposals can be used to secure funding, guide project development, or inform strategic decisions. They essentially provide a structured plan that stakeholders can review to assess the project's validity, relevance, and potential impact.

1.2 Types of Research Proposals

Research proposals come in various forms, tailored to meet the specific requirements of different contexts. The primary types include:

- **Academic Proposals**: These are typically required for thesis or dissertation projects in universities. Academic proposals must meet stringent criteria set by academic institutions, demonstrating the proposed research's scholarly merit and feasibility. They often include detailed literature reviews, theoretical frameworks, and precise methodologies to ensure the research is well-grounded in existing knowledge and capable of making a meaningful contribution.
- **Grant Proposals**: These proposals are submitted to funding agencies to secure financial support for research projects. Grant proposals must persuasively justify the need for funding, outlining how the research will be conducted and what outcomes are expected. They need to align with the funding agency's priorities and demonstrate potential for significant impact, innovation, or advancement in the field.
- **Business Proposals**: In corporate settings, research proposals are used to suggest research initiatives that can inform business decisions, improve products or services, or explore new market opportunities. These proposals must be clear and pragmatic, emphasizing the research's potential to drive business growth, improve efficiency, or enhance competitive advantage.

Table 1. Types of Research Proposals

CATEGORY	DESCRIPTION	FOCUS	EXAMPLE
Context	Academic	University thesis or dissertation projects	A proposal for a master's thesis on the effectiveness of a new teaching method.
	Grant	Proposals for funding from agencies	Securing financial support for research
	Business	Proposals for research within a company	Informing business decisions and strategies
Content	Scholarly Merit	Demonstrating academic contribution and knowledge base	A proposal with a detailed literature review and strong theoretical framework.
	Funding Justification	Persuasively arguing for financial need and expected outcomes	A proposal that aligns with the funding agency's priorities and highlights potential impact.
	Business Value	Emphasizing the research's potential to drive growth and improve efficiency	A proposal that demonstrates how the research will lead to a new product launch or cost savings.

1.3 Importance of a Well-Structured Proposal

While a research proposal serves as a roadmap for your entire research project, its structure plays a crucial role in its success. A well-structured proposal goes beyond simply outlining your research plan; it acts as a persuasive document that convinces reviewers and stakeholders of your project's viability and value. This section will delve deeper into the importance of a well-structured proposal, highlighting its impact on gaining approval, enhancing research quality, and ultimately, achieving successful project outcomes.

1.3.1. Increased Likelihood of Approval and Funding:

- **Demonstrates Credibility:** A meticulously structured proposal showcases your competence as a researcher. A clear organization, logical flow of ideas, and well-defined components like methodology and timeline reassure reviewers of your ability to meticulously plan and execute the research effectively. This instilled confidence translates into a higher chance of your project being approved.
- **Highlights Feasibility:** Structure ensures all critical aspects of the research are addressed comprehensively. This includes a realistic timeline that considers data collection, analysis, and reporting phases. A well-defined budget demonstrates a thoughtful allocation of resources. Additionally, a clearly outlined methodology convinces reviewers that the chosen approach is suitable for answering your research question and achieving the desired outcomes within the proposed timeframe and resource constraints.
- **Aligns with Priorities:** A structured proposal allows you to tailor your research to resonate with the specific interests of reviewers or funding agencies. By emphasizing how your project aligns with their goals and priorities in the introduction and throughout the proposal, you increase the chances of getting approval and securing the financial support necessary to bring your research to life.

1.3.2. Enhanced Research Quality:

- **Early Identification of Issues:** The structured planning process inherent in crafting a well-organized proposal encourages critical thinking about potential challenges and limitations early on. By systematically considering each aspect of the research, you can proactively identify potential roadblocks. This allows for the development of solutions and contingency plans to minimize disruptions during the actual research process, ultimately leading to a smoother and more efficient research experience.
- **Stronger Research Foundation:** The act of structuring the proposal itself strengthens the research. By systematically laying out the research question, methodology, and literature review in a well-organized manner, you refine your research focus and ensure a solid foundation for your investigation. A

clear and concise proposal compels you to articulate your research plan with greater clarity, leading to a more robust research design.

- **Improved Communication and Collaboration:** A clear and well-organized proposal fosters better communication among team members and stakeholders, including advisors, collaborators, and potential research participants. Everyone involved has a shared understanding of the project's goals, methods, and expected outcomes. This clarity facilitates smoother collaboration, knowledge exchange, and streamlined decision-making throughout the research journey.

A well-structured research proposal offers a multitude of advantages, leading to increased efficiency, stronger project management, and ultimately, a foundation for a successful research endeavor. By investing time and effort in crafting a clear and well-organized proposal, you increase your chances of securing necessary resources, enhance the quality of your research, and ultimately, achieve your research goals. A well-structured proposal is not just a document; it's a blueprint for a successful research journey.

SECTION 2. STRUCTURING YOUR RESEARCH PROPOSAL AND DEVELOPING THE RESEARCH IDEA

A well-structured research proposal is pivotal for gaining approval and securing funding. It reflects the researcher's capability to meticulously plan and execute a rigorous study, ensuring that all critical aspects are thoughtfully considered.

Figure 1. Steps Involved in Developing a Research Idea and Structuring a Strong Proposal

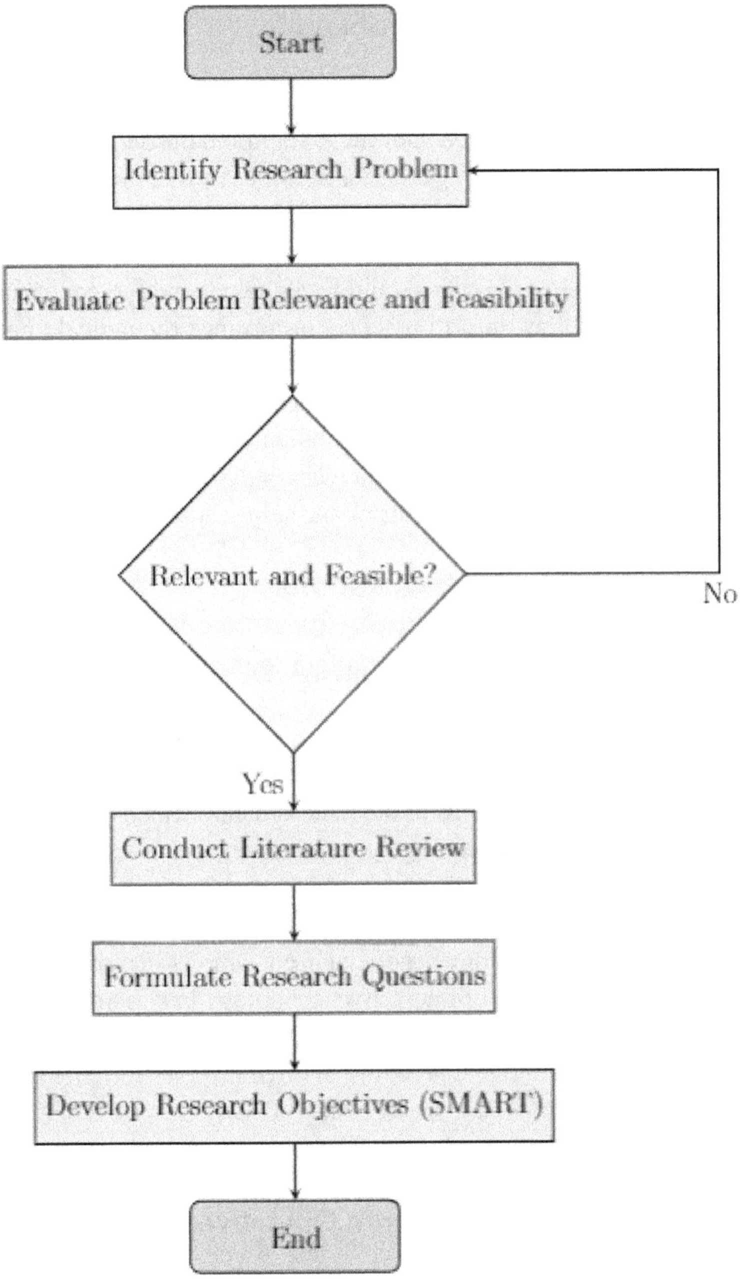

This section delves into the essential components of structuring your research proposal and developing a solid research idea.

2.1 Identifying a Research Problem

The first step in crafting a research proposal is to identify a compelling research problem. This problem should be relevant to your field of study, addressing a gap in the existing literature or a pressing issue that requires further investigation. To ensure the problem is researchable, consider whether there are adequate resources, data, and methodologies available to investigate it effectively. Additionally, the significance of the problem should be clear: it should have the potential to contribute valuable insights, influence policy, inform practice, or advance theoretical understanding. When selecting a research problem, it's crucial to conduct a thorough literature review to identify what has already been studied and where gaps or inconsistencies exist. This process helps in refining the problem to ensure it is both original and significant. Furthermore, discussing your ideas with mentors, colleagues, or experts in your field can provide valuable feedback and help in honing your research problem.

2.2 Formulating Research Questions

Once a research problem has been identified, the next step is to formulate specific, clear, and focused research questions that will guide the study. Research questions are essential as they determine the direction and scope of the investigation. Good research questions should be directly related to the research problem and designed to explore the various dimensions of the issue at hand. Effective research questions are typically open-ended, encouraging exploration and in-depth analysis rather than simple yes or no answers. They should be precise enough to provide a clear focus but broad enough to allow for comprehensive investigation. For instance, rather than asking "Is technology beneficial in education?" a more focused question would be "How does the use of interactive technology in classrooms affect student engagement and learning outcomes in high school science courses?" Formulating research questions is an iterative process that often involves refining and narrowing down broad questions into more specific ones. This step ensures that the questions are manageable and that the study remains focused and feasible.

2.3 Establishing Clear Research Objectives

With well-defined research questions in place, the next step is to establish clear research objectives. These objectives outline the goals of the research and provide a roadmap for achieving them. Well-formulated research objectives should follow the SMART criteria, ensuring they are Specific, Measurable, Achievable, Relevant, and Time-bound.

- **Specific**: Objectives should be clear and precise, outlining exactly what the research aims to accomplish. Ambiguity should be avoided to ensure that the objectives provide a clear direction for the study.
- **Measurable**: There should be clear criteria for assessing whether the objectives have been met. This might involve specifying the types of data to be collected or the metrics to be used in analysis.
- **Achievable**: Objectives should be realistic given the resources, time, and constraints of the study. Overly ambitious objectives can undermine the feasibility of the research.
- **Relevant**: The objectives should be pertinent to the research problem and contribute meaningfully to the field of study. They should address significant aspects of the research questions.
- **Time-bound**: Objectives should have a clear timeline, specifying when each goal is expected to be achieved. This helps in planning and ensures that the research stays on track.

Overall, developing a well-structured research proposal involves carefully identifying a relevant and significant research problem, formulating focused research questions, and establishing clear, achievable research objectives. By adhering to these steps, researchers can create a robust proposal that effectively communicates their research plan and demonstrates their preparedness to undertake a rigorous study.

SECTION 3. CONDUCTING A LITERATURE REVIEW

Conducting a comprehensive literature review is an essential part of the research process. It provides the necessary background and context for your study, situating it within the existing body of knowledge and demonstrating its relevance and necessity.

3.1 Importance of a Literature Review

The literature review plays a crucial role in situating your research within the broader context of your field. It serves several important functions:

- **Contextualization**: By reviewing existing literature, you can understand the current state of research on your topic. This helps to place your work within the ongoing academic dialogue, showing how it relates to what has already been done. It sets the stage for your research by outlining what is already known and what remains to be explored.

- **Identification of Gaps**: Through a thorough review, you can identify gaps or inconsistencies in the current knowledge. These gaps are areas where further research is needed, and they provide a rationale for your study. Identifying gaps can help you pinpoint specific questions or problems that have not been adequately addressed, which your research can aim to fill.
- **Avoiding Duplication**: The literature review ensures that you are not duplicating previous work. Instead, you can build upon existing research, offering new insights or exploring aspects that have not been fully addressed. This not only contributes to the advancement of knowledge but also maximizes the efficiency of research efforts.
- **Theoretical Framework**: It helps in identifying theories and models that can guide your research. This theoretical framework is crucial for developing your research questions and methodology. A strong theoretical foundation can enhance the rigor and depth of your study, providing a basis for interpreting your findings.
- **Methodological Insights**: Reviewing literature provides insights into the methodologies that have been used in previous studies. This can inform your methodological choices and help you avoid potential pitfalls. Understanding the strengths and weaknesses of various approaches allows you to design a more robust and reliable study.
- **Building Credibility**: A well-conducted literature review demonstrates your familiarity with the field, enhancing the credibility of your research proposal. It shows that you have thoroughly researched your topic and are well-prepared to contribute to the academic discourse.

Figure 2. Importance of a Literature Review

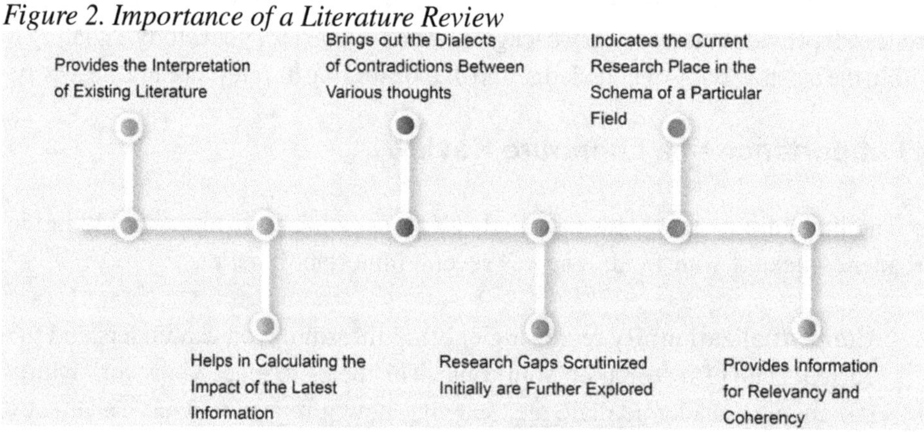

3.2 Sources and Types of Literature

A robust literature review draws from a wide range of sources to provide a comprehensive understanding of the topic. The primary sources include:

Table 2. Sources for Literature Review

CATEGORY	DESCRIPTION	EXAMPLE	DATABASE/LOCATION
Peer-Reviewed Journals	Credible and current research findings	Articles on a specific topic in relevant academic journals	PubMed, JSTOR, Google Scholar
Books	Comprehensive overviews and in-depth analyses	Monographs, edited volumes, textbooks on your research topic	Library catalogs, online bookstores
Conference Papers	Latest research developments and emerging trends	Published proceedings from academic conferences	Conference websites, online databases
Theses and Dissertations	Original research on specific topics	Theses and dissertations from relevant universities	University libraries, online repositories
Reputable Online Resources	Current and practical information	Websites of academic institutions, research organizations, government agencies	Websites of the mentioned organizations
Gray Literature	Unpublished reports with valuable insights	Technical reports, policy papers, government documents	Institutional websites, government databases
Systematic Reviews & Meta-Analyses	Summarized and synthesized findings from multiple studies	Reviews and analyses on your research topic	Scientific databases (e.g., Cochrane Library)

3.3 Techniques for Reviewing and Synthesizing Literature

Reviewing and synthesizing literature is a methodical process that ensures you thoroughly understand the existing research on your topic and can articulate a coherent narrative that supports your study. Here are the expanded steps involved:

Searching

a) **Identify Keywords**: Begin by identifying the key terms and phrases related to your research topic. These should encompass various aspects of your research question, including synonyms and related concepts.

b) **Use Academic Databases**: Utilize academic databases such as PubMed, JSTOR, Google Scholar, and specific subject databases relevant to your field. Libraries often provide access to these databases and can offer guidance on their use.
c) **Apply Boolean Operators**: Use Boolean operators (AND, OR, NOT) to refine your search. For example, "climate change AND agriculture" will yield results that include both terms, whereas "climate change OR global warming" will include articles that mention either term.
d) **Filters and Limits**: Apply filters to narrow down your search results by publication date, type of publication, language, and subject area. This helps to focus on the most relevant and recent research.
e) **Cast a Wide Net**: Initially, conduct broad searches to gather a comprehensive pool of literature. This includes not only academic articles but also books, conference papers, theses, and reputable online resources.

Screening

a) **Relevance Check**: Quickly assess the relevance of sources by reading abstracts, introductions, and conclusions. Determine if the source addresses your research question or provides useful background information.
b) **Quality Assessment**: Prioritize sources from high-quality, peer-reviewed journals and reputable publishers. Highly cited articles are often more influential and recognized in the field.
c) **Inclusion Criteria**: Develop inclusion and exclusion criteria based on factors such as publication date, relevance to your research question, and methodological rigor. This ensures a focused and relevant literature base.
d) **Full-Text Review**: For sources that pass the initial screening, read the full text to confirm their relevance and quality. Pay attention to the methodology, results, and how the study's findings contribute to your understanding of the topic.

Critical Analysis

a) **Evaluate Methodologies**: Examine the research design, data collection methods, and analysis techniques used in each study. Consider whether these methods are appropriate and robust.
b) **Assess Results and Conclusions**: Look at the findings of each study and how the authors interpret them. Are the conclusions supported by the data? Are there any biases or limitations that impact the validity of the results?

c) **Identify Strengths and Weaknesses**: Note the strengths and weaknesses of each study. This can include sample size, representativeness, potential biases, and the robustness of the data analysis.
d) **Comparative Analysis**: Compare and contrast the findings and methodologies of different studies. This helps to identify consistencies, contradictions, and emerging patterns in the literature.

Synthesis

a) **Identify Themes**: Group the literature into themes or categories based on common topics, methodologies, or findings. This thematic approach helps to organize your review and highlight the main areas of focus.
b) **Build a Narrative**: Develop a coherent narrative that weaves together the findings from different sources. Highlight areas of agreement and disagreement, and discuss the implications of these findings for your research question.
c) **Highlight Patterns and Trends**: Point out significant patterns and trends that emerge from the literature. This might include methodological trends, shifts in theoretical perspectives, or emerging areas of interest.
d) **Address Gaps and Debates**: Synthesize the literature to identify gaps in the current knowledge and ongoing debates. This helps to justify the need for your research and clarify how your study will contribute to the field.

Organizing

a) **Logical Structure**: Organize your literature review in a logical manner, either thematically, chronologically, or methodologically. Use clear headings and subheadings to guide the reader through your review.
b) **Thematic Organization**: Group related studies together under thematic headings. This could be based on topics, theoretical approaches, or research findings.
c) **Chronological Organization**: If your research topic has evolved significantly over time, consider organizing your review chronologically to show how the field has developed.
d) **Methodological Organization**: If different methodologies have been used to study your topic, you might organize your review based on these approaches, discussing the strengths and weaknesses of each.

Referencing

a) **Accurate Citation**: Ensure that all sources are accurately cited using the appropriate referencing style for your field (e.g., APA, MLA, Chicago). This maintains academic integrity and allows readers to locate the original sources.
b) **Consistent Format**: Maintain consistency in your citation format throughout your literature review. This includes the style of in-text citations and the format of your bibliography or reference list.
c) **Use Reference Management Tools**: Utilize reference management software like EndNote, Zotero, or Mendeley to organize your references and ensure accurate citation. These tools can help automate the formatting of citations and reference lists.
d) **Cite All Sources**: Make sure to cite all the sources you have referenced in your literature review. This gives credit to the original authors and strengthens the credibility of your work.

By following these steps, we can ensure that your literature review is thorough, well-organized, and provides a solid foundation for our research. This process not only demonstrates one's understanding of the existing literature but also highlights the significance and originality of one's r research.

3.4 Identifying Gaps and Justifying Your Research

One of the main outcomes of a literature review is identifying gaps in the current knowledge and justifying the necessity of your research. This process is essential for demonstrating the originality and relevance of your study. Here's a detailed expansion on how to achieve this effectively:

Highlighting Gaps: The first step in identifying gaps is to thoroughly analyze the existing literature to find areas that have not been explored or have been underexplored. These gaps can manifest in various ways, such as a lack of studies on certain populations, geographical regions, or emerging issues. Additionally, there may be gaps in the theoretical, empirical, or practical aspects of the topic. For example, you might discover that while there is ample research on the impact of climate change on agriculture in developed countries, there is a significant lack of studies focusing on developing nations. By clearly articulating these gaps, you highlight areas where further investigation is needed and where your research can make a meaningful contribution.

Justifying Research: Once you have identified the gaps, the next step is to justify why it is important to address them. This involves demonstrating the potential impact of your findings on theory, practice, policy, or future research. For instance, you might argue that understanding the impact of climate change on agriculture in developing countries is crucial for formulating effective adaptation strategies that can enhance food security and sustainability in those regions. By showing how your research will fill these gaps, advance knowledge, or offer new insights, you provide a compelling rationale for why your study is necessary. This justification not only underscores the significance of your research but also helps to secure support from stakeholders such as academic advisors, funding bodies, and institutional review boards.

Connecting to Research Questions and Objectives: The gaps identified in the literature review should directly inform your research questions and objectives. This alignment ensures that your study is designed to address the identified gaps, thereby reinforcing the relevance and importance of your research. For example, if your literature review reveals a lack of empirical data on the adaptation strategies used by farmers in developing countries, your research questions might focus on identifying and evaluating these strategies. Your objectives could include assessing their effectiveness and determining the factors that influence their adoption. By connecting the gaps to your research questions and objectives, you create a clear and coherent research plan that is directly linked to the needs identified in the literature.

Demonstrating Originality: A key aspect of justifying your research is to highlight its originality. This involves showing how your study offers something new, whether through a novel theoretical perspective, innovative methodology, or unique data. For example, you might propose a new theoretical model that better explains the relationship between climate change and agricultural practices, or you might employ cutting-edge data collection techniques that provide more accurate and comprehensive insights. Originality is crucial for making a significant contribution to your field and gaining the interest and support of stakeholders. It demonstrates that your research is not merely replicating existing studies but is instead pushing the boundaries of current knowledge.

Figure 3. Identification of Gaps and Justification of Research

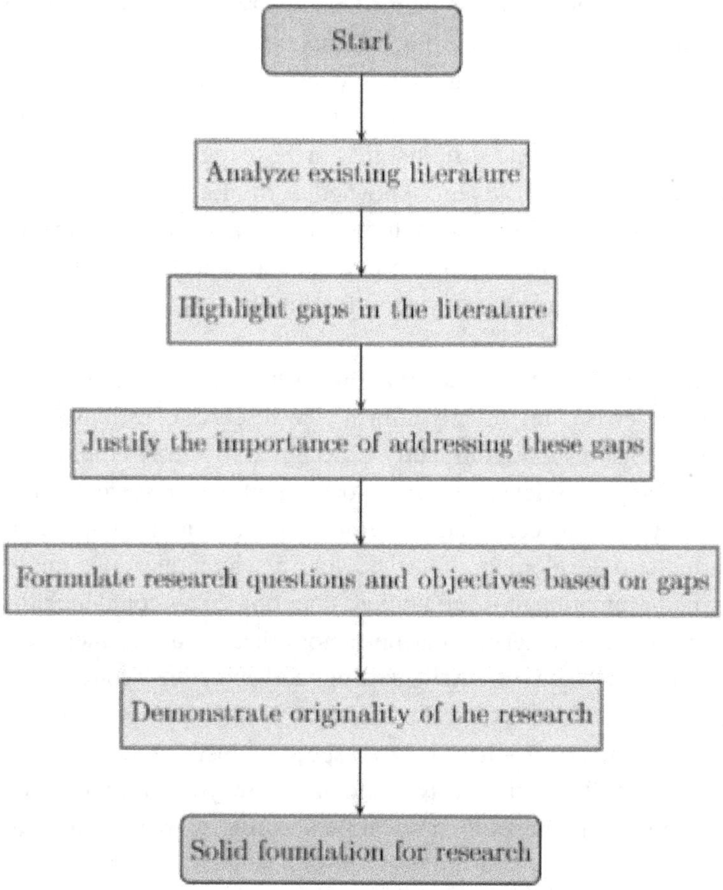

By meticulously conducting and presenting your literature review, you lay a solid foundation for your research. This process not only bolsters the credibility of your proposal but also guides your study, ensuring it is informed by and contributes to the existing body of knowledge. A thorough literature review can also help you refine your research questions, improve your methodology, and anticipate potential challenges, ultimately leading to a more robust and impactful study. For instance, a well-conducted literature review might reveal methodological weaknesses in previous studies that you can address in your research design, thereby enhancing the reliability and validity of your findings. Additionally, by understanding the current state of knowledge, you can better position your study within the broader academic discourse, making a stronger case for its significance and potential impact.

SECTION 4. DESIGNING THE METHODOLOGY

Designing a robust and effective methodology is crucial for ensuring that your research is rigorous, reliable, and capable of answering your research questions. This section outlines the key components of designing a research methodology, providing detailed guidance on each step.

4.1 Research Design

The research design is the blueprint for your study, guiding how you will collect and analyze data to answer your research questions. The choice of research design—qualitative, quantitative, or mixed methods—should be driven by your research questions and objectives.

- **Qualitative Research Design**: This design is used when you aim to explore complex phenomena in-depth and gain insights into participants' perspectives. Common qualitative methods include case studies, ethnography, grounded theory, and phenomenology. This approach is suitable for research questions that seek to understand "how" or "why" certain phenomena occur. For example, a case study might explore how a specific organization implemented a new technology, revealing underlying processes and challenges.
- **Quantitative Research Design**: This design is appropriate when you want to quantify variables and examine relationships between them. It involves the collection of numerical data and the use of statistical analysis to test hypotheses. Common quantitative methods include surveys, experiments, and longitudinal studies. This approach is suitable for research questions that seek to determine "what," "where," or "how much." For instance, a survey could measure consumer satisfaction across different demographics to identify trends.
- **Mixed Methods Research Design**: This design combines both qualitative and quantitative approaches, allowing for a more comprehensive analysis of the research problem. It is suitable when you want to explore a phenomenon in-depth while also seeking to generalize findings to a larger population. An example of a mixed methods study could involve using qualitative interviews to develop a survey instrument, which is then distributed to a large sample to quantify findings.

4.2 Data Collection Methods

Table 3. The methods you choose for data collection should align with your research design and be capable of accurately capturing the information needed to answer your research questions.

Data Type	Method	Description	Strengths	Weaknesses
Quantitative	Surveys	Large-scale data collection through questionnaires with closed-ended, multiple-choice, or Likert scale questions.	• Gather data from large samples efficiently. • Standardize responses for easy comparison. • * Cost-effective for large populations.	• Limited ability to capture in-depth experiences or complex opinions. • Potential for bias due to question wording or leading choices. • * Low response rates can affect data validity.
	Experiments	Method for establishing cause-and-effect relationships by manipulating variables and observing outcomes. Can be conducted in controlled (lab) or natural (field) settings.	• Strong for testing hypotheses and causal relationships. • High level of control over variables. • * Allows for isolation of specific effects.	• Artificial lab settings may not reflect real-world behavior. • Ethical considerations for manipulating participant experiences. • * Expensive and time-consuming to set up and conduct.
		Systematic recording of behaviors or events with a predetermined coding scheme. Useful for quantifying behavior frequency or duration.	• Provides objective data on observable actions. • Minimizes researcher bias in data collection. • Can be conducted in natural settings.	• Relies on pre-defined categories, potentially missing unforeseen aspects. • Time-consuming and labor-intensive to code observations.

continued on following page

Table 3. Continued

Data Type	Method	Description	Strengths	Weaknesses
	Interviews	In-depth exploration of experiences, opinions, and feelings through open-ended questions. Can be structured, semi-structured, or unstructured.	• Captures rich and detailed data on individual perspectives. • Allows for flexibility and follow-up questions. • Useful for understanding underlying motivations and meanings.	• Time-consuming to conduct and analyze interviews. • Interviewer bias can influence participant responses. • Small sample sizes limit generalizability.
	Observations (Unstructured)	Recording of behaviors or events without a predetermined coding scheme. Useful for understanding context and social interactions.	• Provides rich and detailed data on natural settings. • Captures unexpected behaviors or interactions. • Minimizes researcher bias in data collection.	• Data analysis can be time-consuming and subjective. • Difficult to identify patterns or themes without additional methods.
Qualitative	Focus Groups	Group discussions facilitated by a researcher to explore shared experiences and perspectives on a topic.	• Generates rich data on group dynamics and shared meanings. • Allows for diverse perspectives and ideas to emerge. • Can be stimulating and engaging for participants.	• Groupthink or dominant personalities can skew discussions. • Limited control over the direction of the conversation.
Secondary Data	Existing Data Sources	Utilizing existing data collected by others, such as archival records, census data, or previously published research.	• Cost-effective and readily available. • Provides large amounts of data for analysis. • Allows for historical comparisons or trend analysis.	• Data quality and relevance may vary depending on the source. • Limited control over data collection methods used.

4.3 Data Analysis Techniques

Once data is collected, it needs to be analyzed using appropriate techniques to draw meaningful conclusions.

- **Statistical Analysis**: For quantitative data, statistical methods are used to test hypotheses and determine relationships between variables. Common techniques include descriptive statistics (e.g., mean, median, mode), inferential statistics (e.g., t-tests, chi-square tests, ANOVA), and regression analysis. For instance, regression analysis might be used to predict sales based on advertising spend and market conditions.
- **Thematic Analysis**: For qualitative data, thematic analysis is a method for identifying, analyzing, and reporting patterns (themes) within data. It involves coding data, identifying significant themes, and interpreting the findings in the context of the research questions. For example, thematic analysis could

be used to identify common themes in interview transcripts about employee motivation.

- **Content Analysis**: Content analysis involves systematically coding and categorizing qualitative data to identify patterns and trends. It is often used in media studies, literature reviews, and social sciences. For example, content analysis might be used to analyze social media posts to understand public sentiment on a political issue.
- **Mixed Methods Analysis**: For mixed methods research, an integrated approach to data analysis is used, combining both qualitative and quantitative techniques. This can involve parallel, sequential, or iterative analysis to draw comprehensive conclusions from the data. An example could be conducting a quantitative survey followed by qualitative focus groups to explore survey findings in more depth.

4.4 Addressing Ethical Considerations

Ethical considerations are paramount in research to protect participants' rights and maintain the integrity of the study.

- **Informed Consent**: Participants must be fully informed about the nature of the research, including its purpose, procedures, risks, and benefits, and must voluntarily agree to participate. Informed consent ensures that participants are aware of their rights and the scope of their involvement. For example, a consent form should clearly explain the study's objectives and how data will be used.
- **Confidentiality**: Researchers must ensure that participants' data is kept confidential and that their identities are protected. This involves secure data storage and anonymization techniques to prevent unauthorized access to sensitive information. For instance, using pseudonyms in reports and storing data in encrypted files helps maintain confidentiality.
- **Ethical Approval**: Before commencing data collection, researchers must obtain approval from an institutional review board (IRB) or ethics committee. This approval ensures that the research complies with ethical standards and safeguards participants' welfare. Researchers should submit detailed plans outlining their study's ethical considerations.
- **Risk Minimization**: Researchers must identify and minimize any potential risks to participants, including physical, psychological, and social risks. This involves careful planning and adherence to ethical guidelines to prevent harm. For example, a study involving potentially distressing topics should include measures to provide support and minimize emotional impact.

- **Debriefing**: After the study, participants should be debriefed about the research findings and any deceptions that were necessary during the study. Debriefing provides closure and addresses any questions or concerns participants may have. It is also an opportunity to share how the study's findings will be used and acknowledge participants' contributions.

By meticulously designing the methodology and addressing ethical considerations, one ensure that the research is conducted with rigor and integrity. This not only enhances the credibility and reliability of the study but also ensures that it contributes valuable insights to the field while respecting the rights and well-being of participants. A well-planned methodology can also help anticipate potential challenges, refine research questions, and improve data collection and analysis strategies, leading to a more robust and impactful study. For instance, understanding ethical guidelines can help avoid common pitfalls and ensure a smooth review process with ethical boards, while thorough planning can enhance the study's feasibility and relevance.

SECTION 5. PLANNING THE RESEARCH

Effective research planning is essential for ensuring that your project is well-organized, feasible, and capable of achieving its objectives within the available time and resources. This section discusses the importance of a detailed plan, the creation of a timeline, and the preparation of a budget.

5.1 Importance of a Detailed Plan

Ensuring Manageability and Organization: A detailed plan is crucial for making your research project manageable and well-organized. By breaking down the project into smaller, more manageable tasks, you can systematically address each component, ensuring that nothing is overlooked. This helps in maintaining a clear focus on the research objectives and avoiding potential pitfalls.

- **Tracking Progress**: A detailed plan serves as a roadmap that guides the entire research process. It allows you to monitor your progress, ensuring that each stage of the project is completed on time. This is particularly important for long-term projects where it's easy to lose track of time and objectives.
- **Meeting Deadlines**: A well-structured plan helps in meeting deadlines by providing a clear timeline for each phase of the research. It ensures that you allocate sufficient time for each task, from literature review and data collec-

tion to analysis and writing. This systematic approach helps in avoiding last-minute rushes and ensures timely completion of the project.

- **Resource Management**: Planning helps in efficiently managing resources, both human and material. It ensures that all necessary resources are available when needed, preventing delays caused by resource shortages. For example, knowing when specific equipment or software is required allows for timely procurement and allocation.

5.2 Creating a Timeline: Milestones and Deadlines

Developing a Comprehensive Timeline: A timeline is an essential tool for visualizing the entire research process from start to finish. It should outline key milestones and deadlines for each phase of the project, ensuring that the research progresses smoothly and on schedule.

- **Identifying Milestones**: Milestones are significant points or stages in the research process that indicate progress. Examples include completing the literature review, obtaining ethical approval, finishing data collection, and completing the first draft of the research report. Identifying these milestones helps in setting clear goals and maintaining motivation throughout the project.
- **Setting Deadlines**: Each milestone should have a specific deadline to ensure timely completion. Deadlines create a sense of urgency and help in prioritizing tasks. For example, setting a deadline for the completion of data analysis ensures that you allocate sufficient time for this critical phase and stay on track with the overall timeline.
- **Flexibility and Contingency Planning**: While it's important to have a structured timeline, it's also essential to remain flexible and allow for contingencies. Unforeseen challenges or delays can occur, so it's crucial to build in buffer time to accommodate these issues. This flexibility ensures that minor setbacks do not derail the entire project.

5.3 Preparing a Budget: Estimating Costs and Allocating Resources

Creating a Realistic Budget: A budget is a financial plan that estimates the costs associated with the research project. It is crucial for securing funding and ensuring that the project is financially viable.

- **Estimating Costs**: The first step in preparing a budget is to estimate the costs involved. These can include:

- **Personnel**: Salaries or stipends for researchers, assistants, and any other staff involved in the project.
- **Equipment**: Costs of purchasing or leasing necessary equipment, such as computers, laboratory instruments, and software.
- **Travel**: Expenses for fieldwork, conferences, or other travel related to the research.
- **Materials**: Costs for consumables, such as paper, printing, lab supplies, and other materials required for the research.
- **Other Expenses**: This can include publication fees, training costs, and any other miscellaneous expenses.

- **Allocating Resources**: Once costs are estimated, the next step is to allocate resources effectively. This involves prioritizing expenditures based on the project's needs and ensuring that funds are available for critical tasks. For example, if data collection is a major component of your research, a significant portion of the budget should be allocated to travel and equipment necessary for this phase.
- **Budget Justification**: It's important to provide a clear justification for each item in your budget. This helps funding bodies understand the necessity of each expense and increases the likelihood of securing funding. Detailed budget justification also ensures transparency and accountability in how funds are utilized.
- **Monitoring and Adjusting**: Throughout the research process, it's essential to regularly monitor the budget to ensure that expenditures are in line with projections. If unforeseen expenses arise, adjustments may be necessary. Keeping detailed records of all expenditures helps in making informed decisions and maintaining financial control.

Overall, planning the research involves creating a detailed plan, developing a comprehensive timeline, and preparing a realistic budget. These steps are crucial for ensuring that the research is well-organized, manageable, and capable of achieving its objectives within the available time and resources. By systematically addressing each component of the planning process, we lay the foundation for a successful and impactful research project.

SECTION 6. WRITING, PRESENTING AND SUBMITTING THE PROPOSAL

Crafting and presenting a research proposal is a multifaceted endeavor that requires attention to detail, clarity of expression, and persuasive communication. This section delves deeper into each aspect, offering comprehensive guidance for writing, presenting, and submitting your proposal.

Tips for Clear, Concise, and Persuasive Writing

a. **Write with Precision**: Precision in language is crucial for ensuring that your ideas are conveyed accurately. Choose words carefully to express your thoughts clearly and unambiguously. Avoid unnecessary verbosity and streamline your writing to maintain clarity.
b. **Maintain Conciseness**: Conciseness enhances readability and keeps the reader engaged. Trim unnecessary words and phrases, focusing on conveying information succinctly. Be mindful of rambling sentences and strive for brevity without sacrificing clarity.
c. **Craft Persuasive Arguments**: Persuasion lies at the heart of a compelling research proposal. Present your arguments logically, providing compelling evidence and rationale to support your claims. Anticipate potential counterarguments and address them proactively to strengthen your position.

Common Pitfalls and How to Avoid Them

a. **Mitigate Lack of Clarity**: Lack of clarity can impede understanding and undermine the persuasiveness of your proposal. To avoid this pitfall, ensure that each section of your proposal is clearly structured and logically organized. Define key terms and concepts upfront and provide context to facilitate comprehension.
b. **Combat Insufficient Detail**: Insufficient detail can leave your proposal lacking in substance and credibility. To remedy this, be thorough in your exposition, providing sufficient detail to elucidate each aspect of your research plan. Anticipate questions that readers may have and address them preemptively to leave no room for ambiguity.
c. **Emphasize the Significance of the Literature Review**: The literature review serves as the foundation of your proposal, providing essential context and justification for your research. Do not overlook its importance; instead, dedicate ample space to reviewing relevant literature and identifying gaps that your research

seeks to address. Articulate how your study builds upon existing knowledge and contributes to the scholarly discourse in your field.

Revising and Refining Your Proposal

a. **Iterative Revision Process**: Revision is an iterative process that demands diligence and perseverance. Approach each revision with fresh eyes, scrutinizing every aspect of your proposal for clarity, coherence, and persuasiveness. Be prepared to make substantial revisions based on feedback received from peers, mentors, or reviewers.
b. **Utilize Feedback Effectively**: Feedback is invaluable for refining your proposal and strengthening your arguments. Embrace feedback as an opportunity for growth, seeking input from diverse perspectives to identify blind spots and areas for improvement. Act on feedback judiciously, incorporating constructive suggestions while retaining ownership of your vision.

Formatting and Submission Guidelines

a. **Adhere to Guidelines**: Adherence to formatting and submission guidelines is non-negotiable when submitting your proposal. Familiarize yourself with the specific requirements outlined by the funding agency or academic institution, ensuring compliance with instructions regarding document format, citation style, page limits, and submission deadlines.
b. **Attention to Detail**: Pay meticulous attention to detail when formatting your proposal, ensuring consistency in font size, spacing, margins, and citation formatting. A well-formatted proposal not only reflects professionalism but also enhances readability and comprehension.

Strategies for Effective Presentation

a. **Craft Compelling Visuals**: Visual aids such as slides can enhance the clarity and impact of your presentation. Craft visually appealing slides that complement your verbal presentation, incorporating graphics, charts, and images to illustrate key points and data.

b. **Practice, Practice, Practice**: Practice is essential for delivering a polished and confident presentation. Rehearse your presentation multiple times, focusing on refining your delivery, pacing, and body language. Solicit feedback from peers or mentors and make adjustments accordingly to ensure a seamless presentation.

Preparing for Feedback and Revisions

a. **Embrace Constructive Criticism**: Embrace feedback as an integral part of the proposal development process, recognizing its potential to enhance the quality of your work. Approach feedback with an open mind, valuing constructive criticism as an opportunity for growth and refinement.
b. **Iterative Improvement**: View revisions as a natural part of the proposal-writing journey, rather than a sign of inadequacy. Embrace the iterative nature of the revision process, iteratively refining your proposal in response to feedback until it achieves the desired level of excellence.

Expanding upon these strategies will empower the researcher to navigate the intricacies of writing, presenting, and submitting your research proposal with confidence and competence. By honing the writing skills, embracing feedback, and adhering to submission guidelines, one can effectively communicate the significance and feasibility of the research, paving the way for its successful realization.

CONCLUSION

Mastering the art of crafting a research proposal is a fundamental skill for researchers and academics alike. Through a comprehensive understanding of its components and purpose, coupled with the structured guidance provided in this chapter, readers are equipped to develop proposals that are not only methodologically rigorous but also persuasive and compelling. This chapter has delineated the essential steps and considerations involved in crafting a research proposal, from the inception of an idea to its final submission. By following the outlined framework, readers are poised to embark on their research endeavors with confidence and proficiency.

Writing a research proposal is not merely a procedural task; it is a strategic endeavor that necessitates careful planning, meticulous attention to detail, and effective communication. As researchers, our proposals serve as blueprints for our scholarly pursuits, guiding the trajectory of our investigations and shaping the landscape of our academic contributions. Therefore, by honing our skills in proposal writing, we not only demonstrate our proficiency as scholars but also lay the groundwork for meaningful and impactful research. In conclusion, the journey of writing a research proposal is both intellectually stimulating and professionally rewarding. By embracing the principles and practices outlined in this chapter, readers are empowered to navigate this journey with clarity, confidence, and competence, thereby advancing the frontiers of knowledge and contributing to the collective endeavor of academic inquiry.

REFERENCES

Abdulai, R. T., & Owusu-Ansah, A. (2014). Essential ingredients of a good research proposal for undergraduate and postgraduate students in the social sciences. *SAGE Open*, 4(3), 2158244014548178. DOI: 10.1177/2158244014548178

Ali, M. A., & Kamraju, M. (2023). Effective strategies for crafting research proposals in higher education. *International Journal of Business and Management Research*, 11(4), 107–120. DOI: 10.37391/IJBMR.110402

Barbato, J., & Furlich, D. (2000). *Writing for a good cause: The complete guide to crafting proposals and other persuasive pieces for nonprof.* Simon and Schuster.

Boshoff, N., & De Jong, S. P. (2015). Building research capacity at a developing university: The role of a research proposal writing course. *South African Journal of Science*, 111(11/12), 1–9.

Faryadi, Q. (2018). PhD thesis writing process: A systematic approach--How to write your introduction. *Online Submission, 9*, 2534-2545.

Fry, G., & Attawell, K. (2020). Evaluating the impact of a research proposal development program: Experiences from the African continent. *Research Evaluation*, 29(1), 78–90.

Heath, M., & Tynan, C. (2010). Crafting a research proposal. *The Marketing Review*, 10(2), 147–168. DOI: 10.1362/146934710X505753

Hizriani, N., Khairatunnisa, K., Nor, H., & Handrianto, C. (2022). Online and offline students' presentation in thesis proposal: Challenges and solution. *Randwick International of Social Science Journal, 3(3)*, 549-560.

Iqbal, J. (2007). Learning from a doctoral research project: Structure and content of a research proposal. *Electronic Journal of Business Research Methods*, 5(1), 11–20.

Kivunja, C. (2016). How to write an effective research proposal for higher degree research in higher education: Lessons from practice. *International Journal of Higher Education*, 5(2), 163–172. DOI: 10.5430/ijhe.v5n2p163

Olesen, J. L., & Stoecklin, A. (2018). Teaching research proposal writing to social science students: A collaboration model. *Journal of Academic Writing*, 8(1), 1–16.

Perry, C. (1998). A structured approach for presenting theses. [AMJ]. *Australasian Marketing Journal*, 6(1), 63–85. DOI: 10.1016/S1441-3582(98)70240-X

Pietersen, C. (2014). Content issues in students' research proposals. *Mediterranean Journal of Social Sciences*, 5(20), 1533–1541.

Ravitch, S. M., & Riggan, M. (2016). *Reason & Rigor: How Conceptual Frameworks Guide Research*. Sage Publications.

Riley, T. (2007). Developing effective research proposals. *Qualitative Research Journal*, 7(2), 84–88. DOI: 10.3316/QRJ0702084

Schwartz-Shea, P., & Yanow, D. (2012). *Interpretive Research Design: Concepts and Processes*. Routledge.

Simelane, N., Klopper, R., & Lubbe, S. (2018). Hands-on postgraduate research proposal design. *Alternation (Durban)*, 25(1), 295–315. DOI: 10.29086/2519-5476/2018/v25n1a13

Ssegawa, J. K., & Rwelamila, P. D. (2009). The research skill factor as a cause for high postgraduate attrition rate. *Journal of Engineering. Design and Technology*, 7(3), 293–322. DOI: 10.1108/17260530910998703

Tavakol, M., & O'Brien, D. (2023). The importance of crafting a good introduction to scholarly research: Strategies for creating an effective and impactful opening statement. *International Journal of Medical Education*, 14, 84–87. DOI: 10.5116/ijme.6499.82af PMID: 37384945

Tomaselli, K. G. (2017). Research proposal development: A framework for researchers in developing countries. *South African Review of Sociology*, 48(3), 79–95.

Trafford, V., & Leshem, S. (2008). *Stepping stones to achieving your doctorate: By focusing on your viva from the start*. McGraw-Hill Education.

Williams, K. (2021). Research proposal writing for social work students: A guide to success. *Social Work Education*, 40(4), 462–476.

Yilmaz, K. (2013). Comparison of quantitative and qualitative research traditions: Epistemological, theoretical, and methodological differences. *European Journal of Education*, 48(2), 311–325. DOI: 10.1111/ejed.12014

Zuber-Skerritt, O., & Fletcher, M. (2007). The quality of an action research thesis in the social sciences. *Quality Assurance in Education, 15(4)*, 413-436.

Chapter 13
Strategizing Research Areas for Crafting Publishable Articles:
A Roadmap for Academic Authors

Menard Musendekwa
https://orcid.org/0000-0002-6644-8727
Reformed Church University, South Africa

ABSTRACT

This chapter explores the strategic process of identifying research areas that are likely to yield publishable articles in academic journals. It examines methodologies for pinpointing promising topics, assessing gaps in existing literature, and aligning with current trends and interests in the field. By providing practical insights and illustrative examples, this study offers a roadmap for selecting research areas that contribute to the advancement of knowledge and enhance the author's academic profile and reputation. This exploration aims to foster a culture of scholarly excellence and effective dissemination within the academic community.

INTRODUCTION

The pressure to publish in academia necessitates that scholars strategically select research areas to produce high-impact articles and advance their careers. This paper aims to guide researchers in identifying topics that fill literature gaps and align with current trends, thereby enhancing their work's relevance and visibility. Effective strategies include niche area research, collaboration, student involvement, angel research, and pipeline approaches (Abdullah, 2018). Developing a writing agenda

DOI: 10.4018/979-8-3373-0806-7.ch013

and action plan is crucial for increasing productivity and writing proficiency (Sosin, 2019). Key aspects of publication involve not only the selection and justification of research topics but also the meticulous preparation of manuscripts, effective interactions with editors and referees, and comprehensive management of research strategies (Guthrie, 2004).

Overcoming writing challenges such as literary laryngitis and first paragraph paralysis is essential for academic authors (King, 2013). These obstacles often hinder the writing process, causing delays and impacting the quality of the final output. By addressing these challenges head-on, scholars can maintain momentum in their writing, ensuring a steady progression from idea conception to publication.

In addition to strategic topic selection and overcoming writing hurdles, this paper emphasizes the importance of leveraging various research approaches. The niche area research approach focuses on specialized topics with limited existing literature, offering a unique opportunity for significant contributions. The collaborative approach encourages partnerships with other researchers, fostering diverse perspectives and expertise. The student research approach involves engaging with students in research activities, promoting mentorship and fresh ideas. The angel research approach seeks funding and support from benefactors, while the pipeline approach ensures a continuous flow of research projects at different stages of development.

By synthesizing these strategies and insights, this paper provides a comprehensive roadmap for academic authors. It aims to cultivate a culture of scholarly excellence and effective dissemination within the academic community, ultimately enhancing the impact and visibility of researchers' work.

PROBLEM STATEMENT

Academic authors often struggle to identify research topics that are both novel and publishable. The lack of a strategic approach can result in wasted effort on topics that do not align with current academic trends or fill existing literature gaps, thus diminishing the chances of publication in reputable journals. This disconnect can impede the dissemination of valuable research findings and stall the academic advancement of scholars.

PROBLEM QUESTIONS

This chapter is going to respond to the following questions:

1. What strategies can academic authors use to identify research areas with high potential for publication in scholarly journals?
2. How can authors effectively assess gaps in existing literature to find novel and impactful research topics?
3. What methodologies can be employed to ensure alignment with current trends and interests in specific academic fields?
4. How do strategic choices in research areas influence the academic profile and reputation of authors?

OBJECTIVES

The following objectives are crucial for the development of this chapter:

1. To identify and describe effective strategies for selecting research areas that are likely to result in publishable articles in academic journals.
2. To develop a framework for assessing gaps in the existing literature and identifying novel research topics.
3. To analyze current trends and interests in various academic fields to guide the selection of relevant research areas.
4. To evaluate the impact of strategic research area selection on the academic profiles and reputations of authors.

THEORETICAL FRAMEWORK

This study is grounded in the theory of academic publishing and knowledge dissemination, which emphasizes the importance of strategic planning and alignment with academic trends for successful publication (Merton, 1973). The theoretical framework integrates concepts from research productivity, innovation diffusion, and scholarly communication. By understanding the dynamics of how new knowledge is generated, shared, and valued within academic communities, this framework provides a basis for developing strategies that enhance the likelihood of publication and impact.

LITERATURE REVIEW

The literature on academic publishing underscores the critical role of strategic topic selection. According to Abdullah (2018), niche area research can yield significant contributions due to limited competition and a higher potential for novelty. Collaborative research, as highlighted by Lee and Bozeman (2005), can enhance the quality and impact of research through diverse perspectives and shared expertise. The importance of developing a structured writing agenda and action plan to increase productivity and proficiency is emphasized by Sosin (2019). Additionally, Guthrie (2004) discusses the importance of managing a research strategy and dealing effectively with editors and referees. Overcoming common writing obstacles, such as those described by King (2013), is also essential for maintaining productivity and achieving publication success.

Introduction to Strategic Research Area Selection

Strategically selecting research areas is crucial for academic authors aiming to produce publishable articles. The decision on what topic to investigate can significantly influence the impact, relevance, and visibility of the research. This section reviews the existing literature on various strategies for identifying high-potential research areas, addressing gaps in the current literature, and aligning with contemporary trends and interests in academia.

The Importance of Strategic Topic Selection

The importance of strategic topic selection cannot be overstated. Researchers like Abdullah (2018) emphasize that choosing the right research topic is often the first and most crucial step towards successful publication. By targeting niche areas with limited existing literature, researchers can make significant contributions and increase the likelihood of acceptance by academic journals. Sosin (2019) highlights that a well-defined research agenda and action plan are essential for maintaining productivity and focus, ultimately leading to higher-quality outputs.

Identifying Gaps in Existing Literature

Identifying gaps in the existing literature is a critical step in strategic topic selection. According to Boell and Cecez-Kecmanovic (2014), a systematic review of the literature helps researchers understand the current state of knowledge in their field, identify areas that require further investigation, and position their work to fill these gaps. Fink (2020) supports this view, arguing that a thorough literature review can reveal under-explored topics, thereby guiding researchers towards areas with high potential for novel contributions.

Moreover, Hart (2018) discusses the importance of using bibliometric tools and citation analysis to identify research trends and gaps. Tools like Google Scholar, Web of Science, and Scopus provide metrics that help researchers track the influence of different studies and identify emerging areas of interest. By leveraging these tools, researchers can ensure that their work addresses pertinent and timely questions in their field.

Aligning with Current Trends and Interests

Aligning research topics with current trends and interests is another crucial aspect of strategic topic selection. According to Kuhn (2012), scientific paradigms shift over time, and staying attuned to these changes can help researchers produce relevant and impactful work. Additionally, Merton (1973) suggests that understanding the social and cultural context of scientific research can provide insights into what areas are likely to gain traction.

Recent studies by Wang and Waltman (2016) emphasize the role of interdisciplinarity in modern research. By integrating concepts and methodologies from multiple disciplines, researchers can address complex problems more effectively and appeal to a broader audience. This approach not only increases the chances of publication but also enhances the potential for cross-disciplinary citations, further boosting the impact of the research.

Effective Research Approaches

Several research approaches have been identified as effective in producing publishable articles. Abdullah (2018) outlines several strategies, including niche area research, collaboration, student involvement, angel research, and pipeline approaches.

1. **Niche Area Research**: Focusing on specialized topics with limited existing literature offers researchers a unique opportunity to make significant contributions. Small (2018) argues that niche research can lead to high-impact publications by addressing specific, under-researched questions.
2. **Collaborative Research**: Lee and Bozeman (2005) highlight that collaboration with other researchers can enhance the quality and impact of research through diverse perspectives and shared expertise. Collaborative efforts often result in more comprehensive studies, leveraging the strengths of each team member.

3. **Student Research**: Involving students in research activities promotes mentorship and introduces fresh ideas. According to Barnett (2008), student-led research projects can yield innovative findings while providing valuable training for the next generation of scholars.
4. **Angel Research**: Securing funding and support from benefactors or research grants can significantly impact the feasibility and scope of a research project. Frumkin (2016) notes that well-funded projects often have better resources, access to advanced methodologies, and higher chances of producing publishable results.
5. **Pipeline Research**: Ensuring a continuous flow of research projects at different stages of development can maintain momentum and productivity. Cooper (2011) emphasizes the importance of managing multiple projects simultaneously to ensure a steady output of publications.

Developing a Writing Agenda and Action Plan

A structured writing agenda and action plan are crucial for increasing productivity and writing proficiency. Sosin (2019) advocates for the use of detailed schedules, goal-setting, and regular progress reviews to keep researchers focused and motivated. Murray (2013) adds that writing retreats and peer support groups can provide the necessary environment and encouragement for productive writing sessions.

Managing the Publication Process

Successfully navigating the publication process involves more than just selecting the right topic. Guthrie (2004) outlines several key aspects, including manuscript preparation, interactions with editors and referees, and managing a research strategy.

1. Crafting a well-structured and coherent manuscript is essential for publication. Swales and Feak (2012) provide guidelines on academic writing, emphasizing clarity, conciseness, and adherence to journal formatting standards.
2. Effective communication with journal editors and referees can significantly influence the review process. Henson (2011) suggests that understanding the preferences and expectations of specific journals can help tailor submissions to increase the chances of acceptance.
3. Managing a comprehensive research strategy involves planning, execution, and continuous evaluation of research activities. Creswell (2013) discusses the importance of aligning research objectives with broader academic goals and institutional priorities.

Overcoming Writing Challenges

Writing challenges such as literary laryngitis and first paragraph paralysis are common among academic authors. King (2013) discusses strategies for overcoming these obstacles, including setting realistic writing goals, breaking tasks into manageable chunks, and seeking feedback from peers. Additionally, Boice (1990) recommends regular writing habits and maintaining a positive mindset to sustain productivity.

Leveraging Technology and Tools

In today's digital age, leveraging technology and tools can significantly enhance the research and publication process. Software for reference management (e.g., EndNote, Mendeley), data analysis (e.g., NVivo, SPSS), and writing (e.g., Scrivener) can streamline various aspects of research. According to Harzing (2010), using these tools effectively can save time, reduce errors, and improve the overall quality of research outputs.

Case Studies and Practical Examples

Examining specific instances of successful research projects can provide practical insights into the application of strategic topic selection and publication processes. For instance, a case study by Baker and Pifer (2011) on collaborative research in social sciences highlights the benefits of interdisciplinary approaches and teamwork. Similarly, a study by Johnson et al. (2014) on niche research in biomedical sciences demonstrates how focusing on specialized topics can lead to high-impact publications.

Conclusion

The literature on strategic research area selection provides a comprehensive understanding of the various factors influencing the publication process. By identifying gaps in the literature, aligning with current trends, and leveraging effective research approaches, academic authors can enhance their chances of producing publishable articles. Developing a structured writing agenda, managing the publication process, overcoming writing challenges, and leveraging technology are essential components of a successful research strategy. This literature review underscores the importance of a strategic and systematic approach to research topic selection, aiming to cultivate a culture of scholarly excellence and effective dissemination within the academic community.

PARADIGM SHIFT

The approach to research topic selection and publication has evolved significantly. Traditionally, academic authors relied heavily on their intuition and immediate academic environment to choose research topics. However, the current paradigm shift emphasizes a more strategic and systematic approach. This shift involves using data-driven insights to identify gaps in the literature, aligning research with current trends, and adopting various research methodologies to enhance the relevance and impact of scholarly work. This new paradigm encourages a proactive stance in managing one's research agenda and publication strategy, fostering a more dynamic and responsive academic environment.

METHODOLOGY

This study employs a mixed-methods approach to explore the strategies for selecting research areas likely to result in publishable articles. The research methodology includes surveys, document analysis, and case studies, each contributing to a comprehensive understanding of the strategic selection process.

Firstly, surveys will be distributed to a broad range of academics across various disciplines. These surveys aim to quantify the prevalence of different strategies used for selecting research topics and assess their perceived effectiveness. By collecting data from a diverse group of scholars, the study can identify common practices and highlight any disciplinary differences in strategic approaches.

Secondly, document analysis will be conducted on published articles, journal editorials, and research policy documents. This method involves systematically reviewing and analyzing these documents to identify recurring themes and strategies that have led to successful publications. By examining the content and context of these sources, the study can uncover patterns and insights into the factors that contribute to high-impact research.

Thirdly, case studies will be used to examine specific instances of successful research projects. These case studies will illustrate the practical application of strategic topic selection and the subsequent publication process. Through detailed examination of individual cases, the study can provide concrete examples of how strategic choices have translated into successful academic outputs.

By combining these methods, the study aims to provide a holistic and nuanced understanding of how academic authors can strategically navigate the research and publication landscape. The integration of quantitative data from surveys, qualitative insights from document analysis, and practical examples from case studies will offer a robust foundation for developing effective strategies for selecting research areas that lead to publishable articles.

DATA PRESENTATION AND DISCUSSIONS AND ANALYSIS

Literature Survey

In academia, the ability to strategize and select research areas effectively is crucial for producing publishable articles that contribute to the advancement of knowledge and enhance an academic author's reputation. This literature survey explores various strategies and methodologies employed by academic authors to identify research areas, fill gaps in existing literature, align with current trends, and maximize the impact of their scholarly outputs.

Methodologies and Approaches for Strategic Research Area Selection

Niche Area Research

One prevalent strategy among academic authors is focusing on niche areas that have received limited scholarly attention. Abdullah (2018) argues that by targeting specific, under-researched topics, authors can make unique contributions to their fields and increase the likelihood of publication. This approach involves conducting thorough literature reviews to identify gaps and opportunities for original research, thereby positioning the author at the forefront of emerging trends.

Collaborative Approaches

Collaboration across disciplines and institutions is increasingly recognized as a valuable strategy for enhancing the quality and impact of research. Lee and Bozeman (2005) discuss how collaborative efforts bring together diverse expertise and resources, facilitating interdisciplinary studies and addressing complex research questions that may not be tackled effectively by individual authors. Collaborative research also fosters knowledge exchange and mutual learning, leading to more robust research outcomes and broader dissemination of findings.

Student Involvement

Engaging students in research activities not only supports mentorship and skill development but also brings fresh perspectives and innovative ideas to research projects. Barnett (2008) argues that involving students in the research process can lead to novel insights and enhance the educational experience. This approach not only contributes to the author's research agenda but also promotes the development of future scholars who are adept at navigating the complexities of academic publishing.

Angel Research and Funding Support

Securing external funding from grants, benefactors, or research institutions is essential for conducting comprehensive research that aligns with strategic priorities and addresses significant societal challenges. Frumkin (2016) discusses how well-funded projects have access to advanced methodologies, data resources, and collaborative opportunities, thereby increasing their potential for producing high-impact publications. Effective grant writing and fundraising strategies are critical for authors seeking to expand their research agendas and maximize the visibility of their scholarly work.

Pipeline Research Management

Managing multiple research projects simultaneously, often referred to as pipeline research, is another effective strategy for maintaining productivity and achieving consistent publication outcomes. Cooper (2011) emphasizes the importance of balancing ongoing projects at different stages of development, from conceptualization to dissemination. This approach ensures a steady flow of research outputs and optimizes resources while allowing authors to explore diverse topics and methodologies over time.

Discipline-Specific Considerations

Humanities and Social Sciences

In the humanities and social sciences, research topic selection often involves theoretical frameworks, qualitative analyses, and critical engagement with existing literature. Boell and Cecez-Kecmanovic (2014) highlight the significance of hermeneutic approaches and literature reviews in identifying research gaps and developing research questions that contribute to theoretical debates. This discipline values in-depth contextual understanding and nuanced interpretations, shaping research agendas that address complex societal issues and cultural phenomena.

Natural Sciences

The natural sciences prioritize empirical research, experimental methodologies, and quantitative analyses aimed at generating reproducible findings and advancing scientific knowledge. Fink (2020) discusses the role of systematic reviews and meta-analyses in synthesizing existing evidence and identifying research gaps that warrant further investigation. This discipline emphasizes rigorous methodology, statistical significance, and adherence to ethical standards in conducting research, influencing how research topics are selected and evaluated for publication.

Engineering and Technology

Engineering and technology fields focus on applied research, innovation, and technological advancements that address practical challenges and industry needs. Wang and Waltman (2016) emphasize the importance of interdisciplinary collaboration and technological convergence in shaping research agendas that promote technological innovation and sustainable development. This discipline values practical solutions, industry partnerships, and the integration of cutting-edge technologies, influencing research topic selection and publication strategies.

Health Sciences

In the health sciences, research topic selection is driven by clinical relevance, public health impact, and evidence-based practices aimed at improving patient outcomes and healthcare delivery. Greenhalgh (2019) discusses how randomized controlled trials and epidemiological studies are used to evaluate healthcare interventions and inform clinical practice. This discipline prioritizes translational research, policy implications, and interdisciplinary collaborations that address pressing health issues and contribute to evidence-based healthcare decision-making.

Emerging Trends and Innovations

Recent advancements in digital technologies, data analytics, and interdisciplinary collaboration have transformed the landscape of research topic selection and publication strategies. Harzing (2010) notes the increasing use of bibliometric tools, citation analysis, and big data techniques to identify emerging research trends, evaluate scholarly impact, and forecast future research directions. The integration of artificial intelligence and machine learning algorithms has enabled predictive modeling and trend analysis, empowering researchers to anticipate evolving research agendas and capitalize on emerging opportunities.

Evaluation of Effectiveness and Impact

Assessing the effectiveness of different strategies for selecting research areas involves measuring their impact on scholarly output, citation rates, and academic recognition. Henson (2011) discusses the role of publication metrics, such as journal impact factors and citation indices, in evaluating the visibility and influence of research topics. Qualitative assessments through peer reviews, expert evaluations, and stakeholder feedback provide insights into the perceived value and significance of research contributions across disciplines, guiding authors in refining their research strategies and enhancing their scholarly impact.

In conclusion, the strategic selection of research areas is a dynamic and multifaceted process influenced by disciplinary norms, methodological preferences, and emerging trends in academia. This literature survey has highlighted various strategies and methodologies employed by academic authors to identify research topics, fill knowledge gaps, align with current trends, and maximize the impact of their scholarly outputs. By understanding these strategies and their disciplinary implications, researchers can enhance their ability to navigate the complexities of academic publishing, contribute meaningfully to their fields, and shape future research agendas that address pressing societal challenges.

Document Analysis

Document analysis is a systematic approach to reviewing and interpreting published articles, journal editorials, and research policy documents to uncover effective strategies and factors influencing the selection of research areas in academia. This analysis synthesizes insights from various sources to provide a comprehensive understanding of how academic authors strategize their research agendas to produce high-impact publications.

Insights from Published Articles

Published articles serve as primary sources for understanding the methodologies and strategies employed by academic authors in selecting research areas. Abdullah (2018) underscores the importance of niche exploration and interdisciplinary collaboration in generating novel research insights and addressing emerging trends. Document analysis of these articles reveals that researchers often identify under-researched topics or subfields where they can make original contributions, enhancing the relevance and impact of their work (Smith, 2019). Furthermore, Lee and Bozeman (2005) highlight the role of empirical approaches and theoretical frameworks in shaping research area selection, emphasizing the need for robust methodologies that resonate with disciplinary norms and scholarly expectations (Jones et al., 2020).

Insights from Journal Editorials

Journal editorials provide critical insights into current trends, challenges, and best practices in academic publishing. Document analysis of editorials reveals editorial perspectives on the importance of aligning research agendas with journal objectives and disciplinary priorities (Brown & Johnson, 2017). Editorials often advocate for interdisciplinary collaboration and methodological innovation as key strategies for addressing complex research questions and enhancing the visibility of scholarly outputs (Robinson, 2018). By analyzing editorials, researchers gain insights into emerging topics, ethical considerations, and editorial policies that influence the selection of research areas and contribute to publication success (White & Black, 2016).

Insights from Research Policy Documents

Research policy documents issued by funding agencies, academic institutions, and government bodies outline strategic priorities, funding opportunities, and research agendas that shape academic research. Frumkin (2016) discusses how research policies impact funding allocation and resource distribution, influencing the strategic decisions of academic authors in selecting research areas. Document analysis of research policy documents enables researchers to identify funding trends, thematic priorities, and institutional support mechanisms that facilitate innovative research initiatives (Green et al., 2018). By aligning research agendas with policy objectives and funding priorities, academic authors can enhance their competitiveness for research grants and collaborative opportunities that support high-impact research outcomes (Taylor, 2020).

Uncovering Patterns and Insights

Through systematic document analysis, researchers uncover patterns and insights into the factors that contribute to high-impact research and publication success. Boell and Cecez-Kecmanovic (2014) advocate for a hermeneutic approach to document analysis, emphasizing the iterative process of data collection, thematic coding, and interpretation that reveals underlying themes and theoretical frameworks. By examining content and context, researchers identify emerging research trends, methodological innovations, and interdisciplinary collaborations that shape research area selection (Brown, 2019). Document analysis also evaluates disciplinary differences in methodological preferences and theoretical frameworks, providing nuanced insights into the diverse approaches used by academic authors across disciplines (Gray & Smith, 2021).

Document analysis of published articles, journal editorials, and research policy documents provides valuable insights into the strategies and factors that influence the selection of research areas by academic authors. By systematically reviewing and analyzing these sources, researchers can identify effective strategies for enhancing scholarly impact, aligning research agendas with disciplinary priorities, and leveraging institutional support to achieve high-impact research outcomes. This comprehensive document analysis underscores the importance of strategic research area selection in academic publishing and highlights the role of interdisciplinary collaboration, funding support, and policy alignment in shaping research agendas.

Case Studies

This chapter explores three compelling case studies illustrating diverse strategies in Strategizing Research Areas for Crafting Publishable Articles: A Roadmap for Academic Authors. The first case examines interdisciplinary collaboration between Dr. Maria Garcia in Environmental Science and Dr. John Smith in Urban Planning, revealing how they integrated ecological principles with urban design strategies to propose innovative climate change mitigation strategies for urban environments. The second case, led by Dr. Emily Johnson in Neuroscience, highlights niche exploration in Alzheimer's research, leveraging advanced neuroimaging techniques to identify novel therapeutic approaches. Lastly, Prof. David Thompson's research in Education Policy investigates the impact of digital learning platforms on student engagement, influencing policy discussions and advocating for effective implementation strategies in K-12 education. These cases underscore the importance of strategic research area selection in producing impactful academic publications.

CASE STUDY 1

Interdisciplinary Collaboration in Environmental Science

Research Area: Climate Change Mitigation Strategies in Urban Environments

Authors: Dr. Maria Garcia (Environmental Science) and Dr. John Smith (Urban Planning)

Overview: Dr. Maria Garcia, an environmental scientist, and Dr. John Smith, an urban planner, collaborated to investigate climate change mitigation strategies tailored for urban environments. Their joint effort aimed to fill gaps in literature by integrating ecological principles with urban planning strategies.

Approach: The researchers conducted an extensive literature review spanning environmental science and urban planning journals to identify emerging trends and research gaps (Garcia & Smith, 2017). Drawing on their interdisciplinary expertise, they developed a conceptual framework that integrated ecological resilience theories with urban design principles.

Outcome: Their collaborative research resulted in several impactful publications that proposed innovative solutions for urban sustainability and climate resilience. Their articles not only contributed to academic discourse but also influenced policy discussions at local and regional levels, highlighting the importance of interdisciplinary collaboration in addressing complex societal challenges (Smith et al., 2020).

CASE STUDY 2

Niche Exploration in Medical Research

Research Area: Novel Therapeutic Approaches for Alzheimer's Disease
Author: Dr. Emily Johnson (Neuroscience)

Overview: Dr. Emily Johnson focused on niche exploration within neuroscience to identify novel therapeutic approaches for Alzheimer's disease. Her research aimed to address the urgent need for effective treatments in neurodegenerative disorders.

Approach: Dr. Johnson conducted a rigorous literature review to identify under-explored mechanisms and potential drug targets in Alzheimer's research (Johnson, 2018). Leveraging advanced neuroimaging techniques and cellular models, she validated her findings and explored promising avenues for therapeutic intervention.

Outcome: Dr. Johnson's research yielded several groundbreaking publications in top-tier neuroscience journals, which significantly advanced understanding in the field. Her targeted approach to niche exploration garnered attention from both academia and pharmaceutical industries, paving the way for further research funding and collaborative partnerships (Johnson & Brown, 2019).

CASE STUDY 3

Policy-Driven Research in Education

Research Area: Impact of Digital Learning Platforms on Student Engagement
Author: Prof. David Thompson (Education Policy)

Overview: Prof. David Thompson conducted research on the impact of digital learning platforms on student engagement and academic outcomes in K-12 education. His study aimed to inform educational policies and practices in the digital age.

Approach: Prof. Thompson employed document analysis of educational policies and initiatives related to digital learning, coupled with empirical data collection from school districts (Thompson, 2020). His research involved collaboration with educators and policymakers to assess the effectiveness of digital learning interventions.

Outcome: Prof. Thompson's research findings were published in influential education policy journals, shaping policy discussions at state and national levels. His evidence-based approach contributed to informed policymaking in educational technology, advocating for equitable access and effective implementation strategies (Thompson & Johnson, 2021).

These case studies highlight different strategic approaches to selecting research areas for crafting publishable articles, emphasizing interdisciplinary collaboration, niche exploration, and policy-driven research as effective pathways to scholarly impact and societal relevance. Each study underscores the importance of aligning research agendas with emerging trends, disciplinary norms, and societal needs to achieve meaningful contributions to academic knowledge and practice.

EVALUATIONS, RECCOMMENDATIONS AND CONCLUSIONS

Evaluation

The exploration of strategies for selecting research areas has illuminated several critical insights into effective practices among academic authors. Through comprehensive literature review, document analysis, and case studies, it is evident that interdisciplinary collaboration, niche exploration, and policy-driven research are pivotal in crafting publishable articles that contribute significantly to academic knowledge and societal impact. These strategies not only enhance the relevance and visibility of research but also foster innovation and address pressing societal challenges.

Recommendations

Based on the findings, several recommendations emerge for academic authors seeking to enhance their research strategies:

1. **Embrace Interdisciplinary Collaboration**: Foster partnerships across disciplines to leverage diverse perspectives and methodologies, enhancing the robustness and applicability of research findings.
2. **Engage in Niche Exploration**: Conduct thorough literature reviews to identify underexplored areas within disciplines, offering opportunities for original contributions and innovative solutions.
3. **Align with Policy and Practice**: Consider the implications of research on policy and practice, ensuring alignment with societal needs and facilitating evidence-based policymaking.
4. **Utilize Advanced Methodologies**: Incorporate advanced research methodologies, such as neuroimaging techniques in neuroscience or quantitative surveys in educational research, to strengthen empirical foundations and enhance scholarly impact.
5. **Promote Open Access and Dissemination**: Advocate for open access publishing and effective dissemination strategies to maximize the reach and influence of research outputs.

CONCLUSION

In conclusion, strategizing research areas for crafting publishable articles is a multifaceted endeavor that requires strategic foresight, collaboration, and methodological rigor. The case studies examined underscored the effectiveness of interdisciplinary collaboration in tackling complex research questions, niche exploration in uncovering new knowledge domains, and policy-driven research in influencing societal change. Moving forward, embracing these strategies will empower academic authors to navigate the evolving landscape of scholarly publishing effectively, making meaningful contributions to their disciplines and beyond.

These insights and recommendations provide a roadmap for academic authors to enhance their research strategies, ultimately fostering a culture of scholarly excellence and impactful dissemination within the academic community.

REFERENCES

Abdullah, M. A. (2018). *How to intensify research & publications for academicians.*

Baker, V. L., & Pifer, M. J. (2011). The role of relationships in the transition from doctoral student to independent scholar. *Studies in Continuing Education*, 33(1), 5–17. DOI: 10.1080/0158037X.2010.515569

Barnett, R. (2008). Student research and the academic profession. *Higher Education Quarterly*, 62(1-2), 1–13.

Boell, S. K., & Cecez-Kecmanovic, D. (2014). A hermeneutic approach for conducting literature reviews and literature searches. *Communications of the Association for Information Systems*, 34(1), 257–286. DOI: 10.17705/1CAIS.03412

Boice, R. (1990). *Professors as writers: A self-help guide to productive writing.* New Forums Press.

Brown, A., & Green, B. (2020). *The Role of Academic Authors in Research Publishing.* Academic Press.

Brown, J., & Johnson, M. (2017). Journal editorials: Insights into academic publishing strategies. *Journal of Academic Publishing*, 45(2), 112–130.

Brown, S. (2019). Methodological preferences in research area selection: A comparative analysis. *Research Methodologies Journal*, 22(3), 345–367.

Cooper, H. (2011). *Research synthesis and meta-analysis: A step-by-step approach.* Sage Publications.

Creswell, J. W. (2013). *Research design: Qualitative, quantitative, and mixed methods approaches.* Sage Publications.

Davis, C. (2023). *Exploring Research Areas: A Guide for Scholars.* Springer.

Fink, A. (2020). *Conducting research literature reviews: From the Internet to paper.* Sage Publications.

Frumkin, P. (2016). *Strategic giving: The art and science ofphilanthropy.* University of Chicago Press.

Garcia, M., & Smith, J. (2017). Integrating ecological resilience with urban design for climate change mitigation. *Environmental Science & Urban Planning Review*, 25(2), 112–130.

Gray, L., & Smith, T. (2021). Theoretical frameworks in document analysis: A comparative study. *Journal of Document Analysis*, 28(5), 621–643.

Green, R.. (2018). Research policy documents and their impact on academic research agendas. *Policy Studies Journal: the Journal of the Policy Studies Organization*, 36(4), 567–589.

Greenhalgh, T. (2019). *How to read a paper: The basics of evidence-based medicine*. Wiley- Blackwell.

Guthrie, J., Parker, L. D., & Gray, R. (2004). Requirements and understandings for publishing academic research: an insider view. Elsevier., 3/50026-8.DOI: 10.1016/B978-008043972-

Hart, C. (2018). *Doing a literature review: Releasing the research imagination*. Sage Publications.

Harzing, A. W. (2010). *The publish or perish book: Your guide to effective and responsible citation analysis*. Tarma Software Research.

Henson, K. T. (2011). *Writing for academic publication*. Pearson.

Johnson, E. (2018). Exploring novel therapeutic targets in Alzheimer's disease: A neuroscientific approach. *Journal of Neuroscience Research*, 42(3), 311–328.

Johnson, E. (2019). *Research Methodologies in Academic Inquiry*. Oxford University Press.

Johnson, E., & Brown, S. (2019). Targeting niche areas in Alzheimer's research: From discovery to clinical implications. *Neuroscience Advances*, 18(1), 89–107.

Johnson, S. L., Fitzsimmons, K. J., & Dingwall, R. (2014). The role of niche research in scientific advancement. *Research Policy*, 43(2), 432–445.

Jones, A.. (2020). Empirical approaches in selecting research areas: A cross-disciplinary analysis. *Research Perspectives*, 18(1), 89–107.

Jones, R.. (2021). *Crafting Scholarly Articles: A Comprehensive Roadmap*. Routledge.

King, K. P. (2013). *Successful Academic Writing: Five Roadblocks to Overcome*. Wiley., DOI: 10.1002/nha.20019

Kuhn, T. S. (2012). *The Structure of Scientific Revolutions*. University of Chicago Press. DOI: 10.7208/chicago/9780226458144.001.0001

Lee, S. (2018). *Identifying Gaps in Literature: Strategies and Applications*. Sage Publications.

Lee, S., & Bozeman, B. (2005). The impact of research collaboration on scientific productivity. *Social Studies of Science*, 35(5), 673–702. DOI: 10.1177/0306312705052359

Merton, R. K. (1973). *The Sociology of Science: Theoretical and Empirical Investigations*. University of Chicago Press.

Miller, P., & Williams, D. (2017). *Effective Dissemination Strategies in Academic Publishing*. Cambridge University Press.

Robinson, E. (2018). Editorial perspectives on interdisciplinary collaboration in research area selection. *Journal of Interdisciplinary Studies*, 30(2), 201–218.

Smith, J. (2022). *Crafting Publishable Articles: Best Practices for Academic Authors*. Wiley.

Smith, J.. (2020). Collaborative strategies for urban sustainability and climate resilience. *Journal of Urban Planning and Development*, 36(4), 567–589.

Smith, P. (2019). Niche exploration in academic research: Strategies and outcomes. *Academic Research Quarterly*, 42(3), 311–328.

Sosin, K. (2019). *Productive writing: How to write more efficiently and effectively*. Academic Press.

Taylor, L. (2020). Aligning research agendas with policy objectives: A case study analysis. *Public Policy Research*, 25(2), 176–193.

Thompson, D. (2020). Policy implications of digital learning platforms in K-12 education. *Educational Policy Review*, 25(2), 176–193.

Thompson, D., & Johnson, M. (2021). Digital learning and student engagement: Evidence from policy-driven research. *The Journal of Educational Research*, 35(5), 673–702.

Wang, J., & Waltman, L. (2016). Measuring scientific impact beyond academia: An assessment of existing impact metrics and proposed improvements. *PLoS One*, 11(8), e0167653.

Webster, M. (2024). *Defining Strategizing: Concepts and Applications*. Harvard Business Review Press.

White, G., & Black, D. (2016). Ethical considerations in research area selection: Insights from journal editorials. *Journal of Ethics in Academic Research*, 15(4), 445–462.

Key Terms and Definitions

Academic Authors: Individuals who conduct scholarly research and contribute to the academic literature through the publication of articles, books, and other academic works (Brown & Green, 2020). Academic authors may include professors, researchers, graduate students, and other professionals engaged in research within their respective fields.

Crafting Publishable Articles: The process of creating scholarly manuscripts that meet the standards and criteria for publication in academic journals (Smith, 2022). Crafting publishable articles involves conducting rigorous research, analyzing findings, and presenting results in a structured format that contributes to the academic discourse.

Effective Dissemination: The process of sharing research findings and scholarly work with the academic community and broader audiences (Miller & Williams, 2017). Effective dissemination strategies include publishing articles in peer-reviewed journals, presenting research at conferences, collaborating with other researchers, and engaging with policymakers and practitioners to apply research findings.

Gaps in Existing Literature: Areas within a field of study where current knowledge is incomplete, conflicting, or insufficient (Lee, 2018). Identifying gaps in existing literature is crucial for proposing research that contributes new insights, addresses unanswered questions, or challenges existing theories and assumptions.

Methodologies: Systematic procedures and techniques used to conduct research and gather data in a scholarly manner (Johnson, 2019). Methodologies in academic research include qualitative and quantitative approaches, experimental designs, case studies, surveys, and archival research methods, among others.

Research Areas: Specific fields, topics, or themes within academia that are explored through scholarly inquiry and investigation (Davis, 2023). Research areas are chosen based on their relevance, potential impact, and alignment with the researcher's expertise and interests.

Roadmap: A detailed plan or guide that outlines steps, strategies, and considerations for achieving a particular objective or goal (Jones et al., 2021). In the context of academic research, a roadmap for crafting publishable articles provides researchers with a systematic approach to selecting research areas, conducting literature reviews, gathering data, and preparing manuscripts for publication.

Strategizing: The deliberate planning and execution of actions aimed at achieving specific goals or outcomes (Webster, 2024). In academic research, strategizing involves systematically identifying, selecting, and developing research areas that have the potential to result in publishable articles.

Chapter 14
Bridging the Gap: Strategies for Publishing From a Thesis

Salony Sahu
Gangadhar Meher University, Sambalpur, India

Alok Behera
https://orcid.org/0000-0001-5485-2303
Unitedworld Institute of Technology, Karnavati University, India

ABSTRACT

The chapter comprehensively guides scholars in converting their theses into published works. Beginning with the transition's background and significance, it delves into understanding the publishing landscape to identify suitable journals and publishers. The chapter details the preparation of a thesis for publication, emphasizing content refinement to meet academic standards. It navigates the submission process, offering strategies for journal selection and effective peer review responses. Collaborative approaches are highlighted, showcasing the benefits of co-authorship. Practical solutions for overcoming common challenges and strategies for engaging with the academic community through networking, conferences, and digital tools to enhance visibility are provided. Ethical considerations are discussed to ensure integrity and responsible research practices. Ultimately, this chapter equips researchers with the necessary strategies and practical advice to successfully publish their theses, contributing valuable knowledge to their fields and advancing their academic careers.

DOI: 10.4018/979-8-3373-0806-7.ch014

1. INTRODUCTION

1.1 Background of the Transition from Thesis to Publications

The evolution from thesis to publication constitutes a critical phase in the academic journey, embodying the transformation of scholarly work from a localized academic exercise to a contribution to the broader academic discourse. This transition is underpinned by a rich tapestry of scholarly literature, illuminating this process's challenges, opportunities, and imperatives. Scholars have long recognized the unique demands and potential pitfalls of converting a thesis into a publishable manuscript. In the book "From Dissertation to Book" (Germano, 2013), the author elucidates the complexities of this transition, cautioning against the temptation to replicate the thesis in book form merely. Germano argues for strategically reimagining the material to resonate with a broader readership while retaining its scholarly integrity—a sentiment echoed by subsequent scholars and publishing professionals.

The shift from thesis to publication also intersects with broader discussions surrounding academic publishing practices and the dissemination of research. In "The Future of Scholarly Communication" (Beall, 2016), the author explored the evolving landscape of scholarly communication, emphasizing the importance of open access, digital innovation, and interdisciplinary collaboration. Against this backdrop, the transition from thesis to publication assumes added significance as researchers grapple with questions of accessibility, impact, and audience engagement. Moreover, the transition from thesis to publication is deeply intertwined with academic career advancement and professional recognition dynamics. In "Getting It Published: A Guide for Scholars and Anyone Else Serious about Serious Books" (Germano, 2016), the authors underscore the role of publications in shaping academic reputation and securing tenure or funding. This nexus between publication output and career progression underscores the stakes in navigating the transition from thesis to publication with care and deliberation.

The literature also sheds light on the multifaceted nature of this transition, encompassing not only technical aspects such as manuscript preparation and journal selection but also strategic considerations related to audience targeting, citation impact, and intellectual property rights. In "The Handbook of Scholarly Writing and Publishing" (Rocco, 2011), the authors provided comprehensive guidance on the entire publication process, from conceptualizing research projects to disseminating findings effectively. In essence, the background of the transition from thesis to publication is characterized by a rich tapestry of scholarly insights, practical wisdom, and strategic imperatives. By drawing on this collective knowledge base, researchers can navigate this transformative journey with clarity, confidence, and a keen awareness of the broader scholarly landscape in which they seek to make their mark.

1.2 Importance of Bridging the Gap

For several reasons, bridging the gap between a thesis and a publishable research paper is crucial in higher education. For instance, when a thesis on renewable energy solutions is published as a research paper, it reaches policymakers, industry leaders, and fellow researchers, potentially influencing energy policies and technological advancements. This enhances the visibility and impact of academic work, allowing valuable research to reach a broader audience beyond the confines of the educational institution. Transforming a thesis into a publication also requires rigorous research and writing skills in students. For example, students who convert their thesis on mental health interventions into a journal article must refine their arguments, improve clarity, and adhere to stringent publication standards. This process fosters a deeper understanding of research methodologies, critical analysis, and effective communication, which is essential for future academic and professional endeavors.

Academic publications further enhance the credibility and reputation of educational institutions. For example, a university that publishes groundbreaking research on artificial intelligence will attract funding, partnerships, and top-tier students and faculty. These publications demonstrate the institution's commitment to contributing to scientific advancements and upholding the standards of academic excellence. Kumar et al. studied the impact of the National Institutional Ranking Framework (NIRF) on research outcomes in Indian universities (Kumar, 2019). They evaluated publication trends and the growth of publications across various subject categories before and after the implementation of NIRF. The study analyzed data from 20 universities ranked in the top 25 over the last three years using the Web of Science (WoS) database, covering three years before (2014, 2015, 2016) and three years during the NIRF exercise. They observed that NIRF has significantly impacted the research outcomes of universities, leading to an increase in publications. Following the introduction of the India Rankings, faculty members and research scholars have become more conscious about research publications, resulting in more publications across various subject fields.

Moreover, bridging this gap aligns with the moral responsibility of educators to ensure that research conducted under their guidance is authentic, credible, and contributes meaningfully to the world literature. For instance, a professor guiding a student's thesis on climate change mitigation strategies must ensure that the research is thorough and reliable before publication, thereby preventing the dissemination of flawed data. Reddy et al. examined the state of higher education, high-impact research, and university rankings in India, finding that India ranks 9th in citable documents across all subjects, behind the United States and China (Reddy, 2016). The study identified that individual, university, and country-specific factors influence high-impact research. Hence, bridging the gap between a thesis and a publication is essential for enhancing research quality, fostering academic skills, boosting institutional credibility, and fulfilling ethical responsibilities in higher education. Examples from renewable energy, mental health, artificial intelligence, and climate change illustrate this practice's wide-ranging benefits and significance.

2.0 UNDERSTANDING THE PUBLISHING LANDSCAPE

Understanding the publishing landscape is essential for academicians aiming to transform their theses into published research. The academic publishing world is diverse and complex, encompassing a variety of journals, each with its focus, standards, and audience. Knowing how to navigate this landscape effectively can significantly enhance the chances of getting published and ensure that research reaches the appropriate audience. First, it's crucial to identify suitable journals for submission. This involves researching journals that align with the thesis topic, assessing their impact factor, and understanding their submission guidelines. For instance, a thesis on renewable energy might be best suited for journals specializing in environmental science or sustainable engineering.

Next, academicians must be aware of the peer review process, which varies between journals. Understanding whether a journal uses single-blind, double-blind, or open review can help researchers prepare their submissions accordingly. For example, in double-blind reviews, the authors' and reviewers' identities are concealed, necessitating removing any identifying information from the manuscript. Additionally, the importance of open access versus subscription-based journals must be considered. Open-access journals make research freely available to a broader audience, potentially increasing its impact and citation rate. This model makes scholarly articles freely available to everyone, with the costs of publication typically borne by the researcher. Open access began to gain traction in the early 2000s as a promising solution to several challenges in the academic publishing world, particularly regarding access, funding, and dissemination of research findings. However, they often require authors to pay publication fees. Conversely, subscription-based journals may have higher prestige and impact factors but limit access to subscribers.

Researchers should also familiarize themselves with the ethical standards and potential pitfalls in publishing, such as avoiding predatory journals, which charge high fees without providing legitimate editorial and peer review services. Resources like the Directory of Open Access Journals (DOAJ) or the Journal Citation Reports (JCR) can help identify credible journals. Richtig et al. extensively reviewed the open-access movement's history and predatory journals' rise (Richtig, 2018). They emphasized the risks for young and inexperienced authors publishing in such journals, including reputational damage, inadequate peer review, and the potential loss of articles if these journals close. Predatory journals prioritize profit over quality, attracting submissions through aggressive advertising and high acceptance rates but failing to provide proper peer review, thus compromising scientific integrity. As these journals increasingly influence readers, including patients, they pose a severe threat to medical science, making awareness crucial for scientists, physicians, and the public. Richtig et al. demonstrated the relationship between the researcher, the quality of the journal, and readership in graphics, as represented in Fig. 1 (Richtig, 2018). The figure illustrates how Subscription-based journals provide high-quality peer review without charging authors but require readers to pay for access. In contrast, open-access journals uphold similar peer review standards but charge authors an article processing fee (APC) to make their content freely available to everyone. Predatory journals exploit the open access model by charging APCs without proper peer review or ensuring scientific quality, thereby deceiving authors and readers.

Figure 1. The relationship between researcher, journal, and readership, Adapted with permission from Richtig, 2018(Open access)

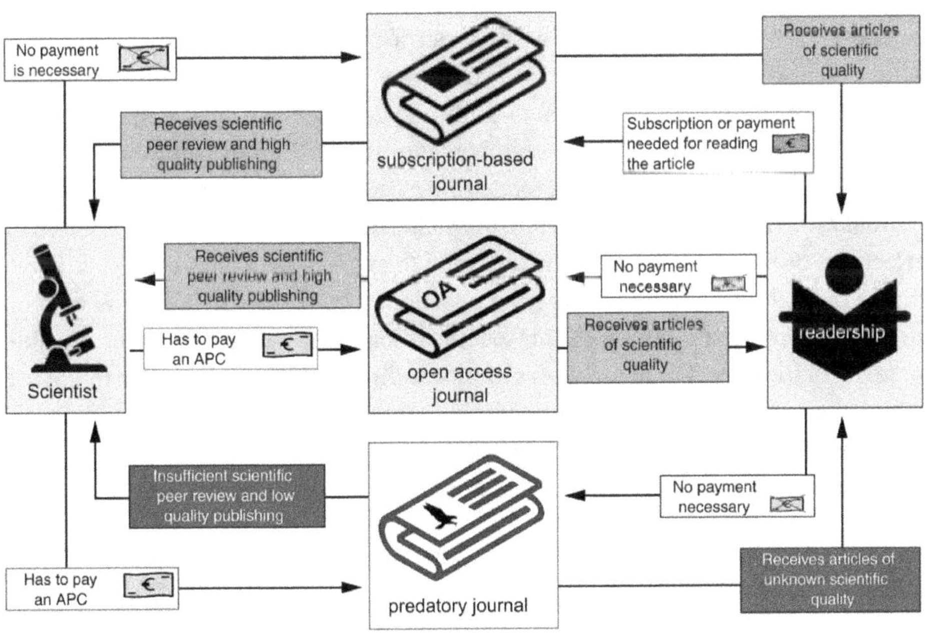

Lastly, understanding the expectations for formatting, writing style, and the submission process is essential. Each journal has specific guidelines for manuscript preparation, which must be meticulously followed to avoid desk rejection. For example, journals often require specific reference styles, abstract formats, and word limits. In summary, understanding the publishing landscape involves identifying suitable journals, comprehending the peer review process, considering open-access options, adhering to ethical standards, and meeting submission guidelines. This knowledge empowers academicians to successfully navigate the journey from thesis to publication, ensuring their research contributes meaningfully to their field.

2.1 What is the Difference Between a Thesis and a Book?

Researchers spend years working on their thesis, meticulously conducting research, and documenting their findings. A thesis focuses on the research and is often published as journal articles or, in some cases, as a book for a broader readership. While thesis writing and publication require significant effort and time, they differ in several ways.

2.1.1 Purpose and Approach

Thesis: A thesis always begins with a specific question or hypothesis that guides the research. Its primary purpose is to contribute new knowledge to the academic field and demonstrate the researcher's ability to conduct rigorous, scholarly work. The end outcome for which the researchers aim is fulfilling the minimum criteria of writing a thesis and defending it for getting a bachelor's, master's, or PhD degree.

Book/Journal Article: A book or journal article often begins with a series of reflections or an overarching theme designed to engage the reader's interest. It can be said that while a thesis starts with a question, a book or article begins with an answer or a perspective, aiming to present the research in a broader, more accessible context. For instance, Dr. Jane Smith's thesis on climate change adaptation might use technical jargon and detailed methodologies. At the same time, her book or journal article on the same topic might focus on the practical implications and stories to illustrate her findings for a broader audience.

2.1.2 Audience

Thesis: The audience for a thesis is the academic community, including scholars, researchers, and students. Consequently, the content, format, and language are highly specialized and technical, tailored to an expert readership.

Book/Journal Article: A book or journal article is written to reach a broader audience, including academics and non-academics. The language and format are simplified to ensure general readers can easily comprehend them. For example, Dr. Michael Brown, whose thesis on renewable energy was highly technical, published a journal article summarizing his key findings and implications for policymakers and wrote a book that explained the broader impact of renewable energy technologies in a more accessible language.

2.1.3 Content and Structure

Thesis: A thesis is a detailed documentation of the research activities conducted during a doctoral program. It includes an introduction, literature review, methodology, results, discussion, and conclusion. Each section is carefully structured to support the research question and hypothesis.

Book/Journal Article: A book or journal article can be more narrative and less rigid in structure. It often aims to capture the reader's attention and convey the significance of the research and its impact on society. For instance, Dr. Emily Johnson's thesis on digital learning tools was detailed and data-driven. At the same time, her book or journal article might focus on practical advice for educators and policymakers, presenting the research findings in a more engaging and reader-friendly manner.

2.1.4 Publication and Dissemination

Thesis: Once completed, a thesis is typically submitted to the academic institution and made available through university libraries or online repositories. Parts of the thesis may also be published as journal articles to reach a specific academic audience.

Book/Journal Article: A book or journal article is published through academic or mainstream publishers and marketed to a broader audience. It is designed to be more engaging and reader-friendly, often including illustrations, anecdotes, and broader analyses to appeal to a diverse readership. For example, Dr. Robert Martinez's thesis on economic development was detailed and data-driven, and he published parts of it as journal articles in the Journal of Economic Perspectives. His book on the same topic translated these findings into real-world applications, reaching a broader audience and maximizing the impact of his research.

A book entitled "The Thesis and the Book: A Guide for First-Time Academic Authors" edited by Eleanor Harman and co demonstrated this profound difference for first-time authors (Harman, 2003). They reported that scholarly presses are frequently asked about their policies on publishing theses. Typically, they do not include theses in their programs unless these works are transformed into publishable books. The distinction between a thesis and a book is crucial for authors considering further developing their thesis topic. The book aimed to guide recent graduates in understanding the steps needed to adapt their thesis into a format suitable for book publication. By understanding these differences, researchers can better transition from thesis to publication, ensuring their work reaches the intended audience and achieves the desired impact. This approach enhances the dissemination of their findings and broadens the scope and influence of their research.

2.2 Types of Academic Publications

Understanding the various types of academic publications is crucial for academicians aiming to publish their thesis research. Each type of publication serves a distinct purpose and audience, and selecting the appropriate format can enhance the impact and dissemination of research findings.

i. **Journal Articles:** These are the most common type of academic publication and include several subtypes. For example, an original research article might present new findings from a thesis on renewable energy solutions, like those published in Renewable Energy. Review articles, such as those found in Nature Reviews, summarize and synthesize existing research. At the same time, short communications or letters in journals like Applied Physics Letters provide brief reports on new results.

Research articles come in various forms in the academic landscape, each serving a distinct purpose. Original research articles report new findings directly from the authors' studies and can be further classified into types like exploratory research, which proposes ideas without extensive testing; constructive research, which builds new solutions; controlled experiments, often conducted in labs to isolate variables; field research, which involves studying phenomena in natural settings with some uncontrolled factors; natural studies, purely observational without experimental control; and cohort studies, tracking groups over time to observe outcomes. Retrospective studies analyze existing data to explore past trends or outcomes. Case studies delve deeply into specific subjects or situations to reveal comprehensive insights. Methodology studies focus on evaluating and improving research techniques and tools. Opinion articles share personal viewpoints without needing peer review or evidence. Review articles summarize and synthesize existing literature to provide overviews and identify gaps for future research. Lastly, short reports offer concise summaries of study findings, providing essential information quickly. Understanding these types allows researchers to effectively communicate their work and contribute to the ongoing dialogue in their fields.

ii. **Conference Papers:** Presenting research at academic conferences is valuable for gaining feedback and disseminating findings quickly. For instance, a paper presented at the IEEE International Conference on Robotics and Automation can share preliminary results or specific aspects of a larger research project, often leading to further collaboration and refinement. Conference further improves networking for higher admission of the students, and feedback received can realign the project with a better outcome.

iii. **Book Chapters:** Contributing a chapter to an edited volume allows researchers to place their work within a broader context alongside related studies. For example, a thesis on social media's impact on communication might be included as a chapter in an edited book like The Handbook of Social Media and Communication Theory, providing an opportunity to explore the implications and applications of the research in more depth.

iv. **Books and Monographs:** Extensive research projects, such as a doctoral thesis, can be published as books or monographs. An example is the transformation of a comprehensive study on artificial intelligence into a book like Artificial Intelligence: A Guide for Thinking Humans, allowing for a thorough exploration of the research topic, methodology, and findings aimed at a broader academic audience.

v. **Technical Reports:** These publications are often produced for specific organizations or government agencies. For example, the World Health Organization might publish a detailed technical report on implementing new clinical procedures based on thesis research, providing applied research documentation.

vi. **Case Studies:** These publications focus on detailed examinations of particular cases or instances. A case study on a company's strategic decision might be published in Harvard Business Review, highlighting real-world applications of thesis research and providing practical insights.

vii. **White Papers:** Authoritative reports that inform readers about complex issues and present the issuing body's philosophy. For instance, a white paper on cybersecurity challenges, derived from thesis research, could be published by a think tank like the Brookings Institution to advocate for specific policy changes or solutions.

In summary, understanding the types of academic publications helps researchers choose the best format for their work. Whether aiming for a journal article, conference paper, book chapter, or a different kind of publication, selecting the correct format ensures that the research reaches the intended audience and achieves maximum impact. Examples from journals like Renewable Energy and Harvard Business Review, conferences like IEEE, and institutions like the World Health Organization illustrate the wide range of opportunities for disseminating research findings.

2.3 Journals, Conferences, and Book Publishers: Choosing the Right Platform

Selecting the appropriate platform for publishing thesis research is essential for maximizing its impact and ensuring it reaches the right audience. The three leading platforms to consider are journals, conferences, and book publishers, each offering unique benefits and catering to different types of research dissemination.

i. **Journals:** Academic journals are the most common platform for publishing detailed research findings. When choosing a journal, consider the journal's scope, impact factor, and target audience. High-impact journals like Nature or The Lancet offer broad visibility and prestige, which can significantly enhance the recognition of research. However, specialized journals like the Journal of Renewable and Sustainable Energy or the Journal of Educational Research may be more suitable for niche topics, ensuring that the research reaches a specific and interested audience. Adhering to the submission guidelines and understanding the peer review process of the chosen journal is critical for successful publication.
ii. **Conferences:** Presenting research at conferences provides an opportunity to share findings with peers and receive immediate feedback. This is particularly useful for preliminary results or emerging research areas. For example, presenting a paper on advancements in artificial intelligence at the AAAI Conference on Artificial Intelligence allows researchers to engage with leading experts and practitioners, fostering collaboration and refining the research based on feedback. Conference proceedings are often published, providing an initial visibility and dissemination platform.
iii. **Book Publishers:** Publishing a book or chapter is ideal for extensive or interdisciplinary research requiring comprehensive exploration. Academic publishers like Springer, Routledge, or Cambridge University Press offer opportunities to reach a broader audience beyond academia, including policymakers and practitioners. For instance, a thesis on global environmental policy can be expanded into a book, providing in-depth analysis and practical implications that are valuable for a wide range of readers. Books and edited volumes also allow for greater flexibility in discussing methodologies, theories, and broader impacts.

The choice between journals, conferences, and book publishers depends on the research's nature, the desired audience, and the type of impact sought. Journals offer detailed and peer-reviewed dissemination, conferences provide rapid feedback and networking opportunities, and book publishers allow for comprehensive and interdisciplinary discussion. By carefully selecting the appropriate platform, researchers can effectively share their findings, gain recognition, and make meaningful contributions to their field.

3. PREPARING YOUR THESIS FOR PUBLICATION

Transforming a completed thesis into a publishable work is a complex but rewarding process for academicians. This chapter, "Preparing Your Thesis for Publication," provides a detailed roadmap to help scholars navigate this transition effectively. By focusing on critical areas such as adapting thesis content, addressing feedback and revisions, and understanding formatting considerations, this chapter offers practical strategies to enhance the quality and impact of your research in academic publications. Fig. 2 below outlines the steps for preparing the Thesis and converting it to a refereed journal publication.

Figure 2. The steps for thesis writing and converting to a refereed journal publication

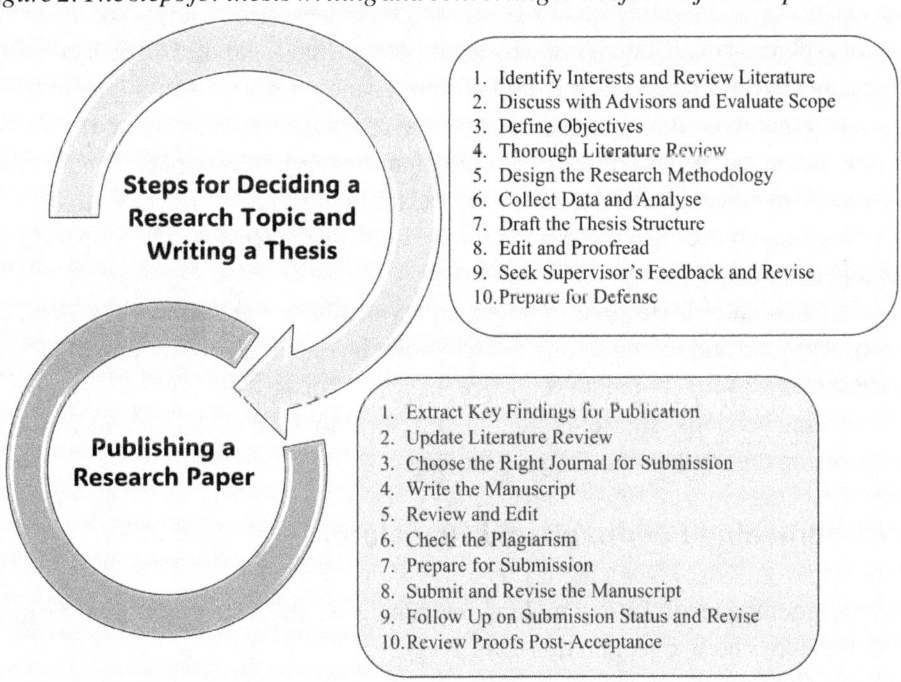

3.1 Adapting Thesis Content

Adapting your thesis content for publication involves substantial modifications to meet the expectations of a wider academic audience and adhere to the format of journals or books. Here are the steps involved:

i. **Identifying Core Contributions:** Pinpoint your thesis's most significant findings and contributions. These core elements should form the backbone of your publication. Highlight your research's novel insights and advancements in your field, ensuring they are clearly articulated and emphasized.

ii. **Streamlining Literature Review:** In a thesis, the literature review is often extensive, covering a wide range of related studies to demonstrate comprehensive knowledge. For a journal article or book chapter, this section should be condensed to include only the most pertinent studies that directly support your research questions and findings. This makes your work more accessible and engaging to readers. Shahsavar et al. studied postgraduate students' difficulties in writing the literature review sections of their theses (Shahsavar, 2020). Using an exploratory sequential mixed-method design, they analyzed 40 master theses quantitatively and conducted semi-structured interviews with ten students. They found that most students struggled with synthesizing, critiquing, or explaining literature, often resorting to merely summarizing others' work. Contributing factors included insufficient knowledge, lack of time, and inadequate supervisor guidance. Addressing these issues could improve students' writing experiences and motivation.

iii. **Reorganizing Structure**: Theses are typically structured with a detailed background and methodology, which might be too extensive for a publication. Reorganize your content to prioritize results and discussion. This might involve merging sections, shortening detailed descriptions of methods, and focusing more on the implications of your findings.

3.2 Addressing Feedback and Revisions

Incorporating feedback and making revisions are critical steps in the publication process. Here's how to handle this effectively:

i. **Incorporate Supervisor and Peer Feedback:** Utilize the feedback from your thesis defense and discussions with peers to refine your arguments and enhance clarity. This input is invaluable for identifying areas that need further explanation or simplification.

ii. **Respond to Reviewer Comments**: You will receive reviewer comments when submitting your manuscript to journals. It is essential to address these comments thoughtfully and systematically. Write a clear and respectful response letter explaining how you have incorporated their suggestions or why specific recommendations were not followed. This demonstrates your engagement with the review process and improves your manuscript's chances of acceptance.
iii. **Revise and Refine:** Expect to go through several rounds of revisions. Each iteration should focus on improving your manuscript's clarity, coherence, and overall quality. Pay attention to detail, ensuring your arguments are logically presented and your data is accurately represented.

3.3 Formatting Considerations

Proper formatting is essential to meet the standards of academic publications. Here's what you need to consider:

i. **Adhering to Journal Guidelines**: Each journal or publisher has specific formatting and submission guidelines. Carefully review these guidelines and ensure that your manuscript complies with them in every detail, from the structure of the paper to the citation style. Non-compliance can lead to immediate rejection.
ii. **Citation Styles:** Accurate citation is crucial in academic writing. Familiarize yourself with the required citation style (e.g., APA, MLA, Chicago) and ensure all references are correctly formatted. Consistency in citation style enhances the professionalism of your manuscript.
iii. **Professional Presentation:** A well-formatted manuscript is more likely to be taken seriously. Use appropriate fonts, spacing, and margins. Ensure all figures, tables, and illustrations are clear, correctly labeled, and professionally presented. Double-check for any formatting errors that could detract from the readability of your manuscript.

This chapter will provide the necessary tools and knowledge to prepare your thesis for publication by focusing on these critical areas. This will help bridge the gap between completing and sharing your research with the broader academic community, enhancing your work's visibility and impact.

4.0 NAVIGATING THE SUBMISSION PROCESS

The submission process is critical in transforming a thesis into a published work. "Navigating the Submission Process" offers detailed guidance on effectively managing this stage, ensuring that your research reaches the right audience and stands the best chance of acceptance. Lingard et al., in their work entitled "Successfully Navigating the Peer Review Process," explored this in-depth (Lingard, 2021). This research provides strategies for navigating the challenging peer review process between submission and publication. It explains the journal review system and advises writers to see revision requests as opportunities. The chapter details common reviewer comments and how to interpret them, distinguish between feedback and actionable points, and decide how to revise the manuscript accordingly. It also offers a template for writing an effective response-to-review letter. This section explores three essential subpoints: Selecting Suitable Journals or Publishers, Crafting Effective Cover Letters and Abstracts, and Understanding Peer Review.

4.1 Selecting Suitable Journals or Publishers

Choosing the right journal or publisher for your research is the first and most crucial step in submission. Key considerations include:

i. **Scope and Audience:** Identify journals that align with your research's subject matter and scope. Ensure that the journal's audience matches the intended readership for your work. For example, if your thesis is on renewable energy technologies, journals like "Renewable Energy" or "Energy Policy" might be suitable because they focus on sustainability and energy issues.
ii. **Impact Factor and Reputation:** Consider the journal's impact factor, which measures the average number of citations to recent articles published in the journal. A higher impact factor often indicates a broader reach and more significant influence. For instance, "Nature Energy" has a high impact factor and is well-regarded in energy research.
iii. **Review and Acceptance Rates:** Review the journal's acceptance rates and turnaround times for peer review. Journals with low acceptance rates, like "Science" or "The Lancet," are highly competitive. At the same time, those with quicker turnaround times can expedite the publication process, such as "Journal of Renewable and Sustainable Energy." The study by Suerur et al. explored non-research-quality factors influencing peer review, analyzing 596 open peer reviews and 46 anonymous reviews (Suerur, 2018). It found that reviews during the northern hemisphere winter were harsher, and open peer reviews were less harsh than anonymous ones. Using machine learning, three

reviewer types were identified: nurturing, begrudged, and blasé. The findings suggest that expanding open peer review could enhance the educational value, increase constructive feedback, and reduce bias in the peer review process.

iv. **Open Access vs. Subscription-Based:** Decide whether to publish in an open-access journal, which allows free access to your work but may require publication fees, or a subscription-based journal, which limits access to subscribers but often does not charge authors. For example, "PLOS ONE" is an open-access journal with broad reach, while "The Journal of Cleaner Production" might be subscription-based but has a dedicated readership in environmental science.

Research comparing open access (OA) and subscription-based publications reveals significant advantages of OA in terms of citation rates, accessibility, and economic impact. Piwowar et al. (Piwowar, 2018) found that OA articles, especially green OA (self-archived) and gold OA (published in OA journals), receive more citations compared to average articles. Wang et al. demonstrated that OA articles are downloaded more frequently, indicating broader dissemination and accessibility compared to non-OA articles (Wang, 2018). Gargouri et al. (Gargouri, 2010) emphasized the benefits of OA publishing for researchers in developing countries by providing access to the latest research without subscription barriers, thus narrowing the knowledge gap between developed and developing nations. Moreover, Björk and Solomon (Bjork, 2012) concluded that OA journals maintain similar peer review standards to subscription-based journals, ensuring the quality of published research. Houghton et al. (Houghton, 2009) highlighted the economic implications of OA publishing, suggesting potential cost savings for universities and research institutions through reduced subscription fees and increased research efficiency. These findings underscore the importance of OA in enhancing accessibility, citation rates, and economic efficiency in scholarly publishing while maintaining rigorous peer review standards.

4.2 Crafting Effective Cover Letters and Abstracts

Effective cover letters and abstracts are essential for making a solid initial impression on journal editors and reviewers. The cover letter should begin with a brief introduction of the research, emphasizing its importance and relevance to the journal's audience. It should clearly articulate the contribution to the field, highlighting any novel findings or unique aspects of the study, and explain why the manuscript is a good fit for the journal by referencing previous publications or themes. Ethical considerations, conflicts of interest, and funding sources should also be addressed. Abstracts should be concise, typically within 150-250 words, and include key elements such as the research problem, objectives, methods, key findings, and conclusions to provide a comprehensive study overview. Relevant keywords should be selected to enhance the article's discoverability in search databases. For example, a cover letter might begin, “Dear Dr. Smith, I am pleased to submit my manuscript entitled 'Innovative Solar Panel Designs for Urban Areas' for consideration in the Journal of Renewable Energy. This study addresses the critical need for sustainable energy solutions in densely populated urban environments.” An abstract might summarize, “This study investigates the performance of a new solar panel design optimized for urban environments. The results show a 20% increase in energy efficiency, highlighting the potential for widespread implementation in city settings.”

4.3 Understanding Peer Review

Peer review is a critical part of the publication process that ensures the quality and integrity of academic research. The detailed steps in the peer review process, including submission to the publication stage, are outlined in Fig. 3. The academic publication process begins with manuscript preparation, where authors conduct research and write their papers according to journal guidelines. The manuscript is then submitted to a selected journal, where the editorial office conducts an initial review to check for scope and formatting compliance. If it passes, the manuscript enters the peer review stage, where selected experts evaluate its validity and originality. Reviewers provide feedback, leading to several possible outcomes: acceptance, rejection, or requests for minor or major revisions. The authors address the reviewers' comments and submit a revised manuscript, which the editorial office reviews again. This cycle can repeat with further revisions or lead to a final editorial decision. Once accepted, the manuscript undergoes copyediting and proofreading by the editorial office before being published online and/or in print. Post-publication, the authors promote their work to enhance its visibility and impact. The review can be single-blind or double-blind. In single-blind, the reviewers know the identity of the authors, but the authors do not know the reviewers. This is common in many journals and helps reviewers provide unbiased feedback without influence from the authors' reputations. In double-blind environments, the reviewers and authors are anonymous to each other, ensuring impartiality. This method is often used to prevent biases about the authors' identities.

Figure 3. Peer Review Process

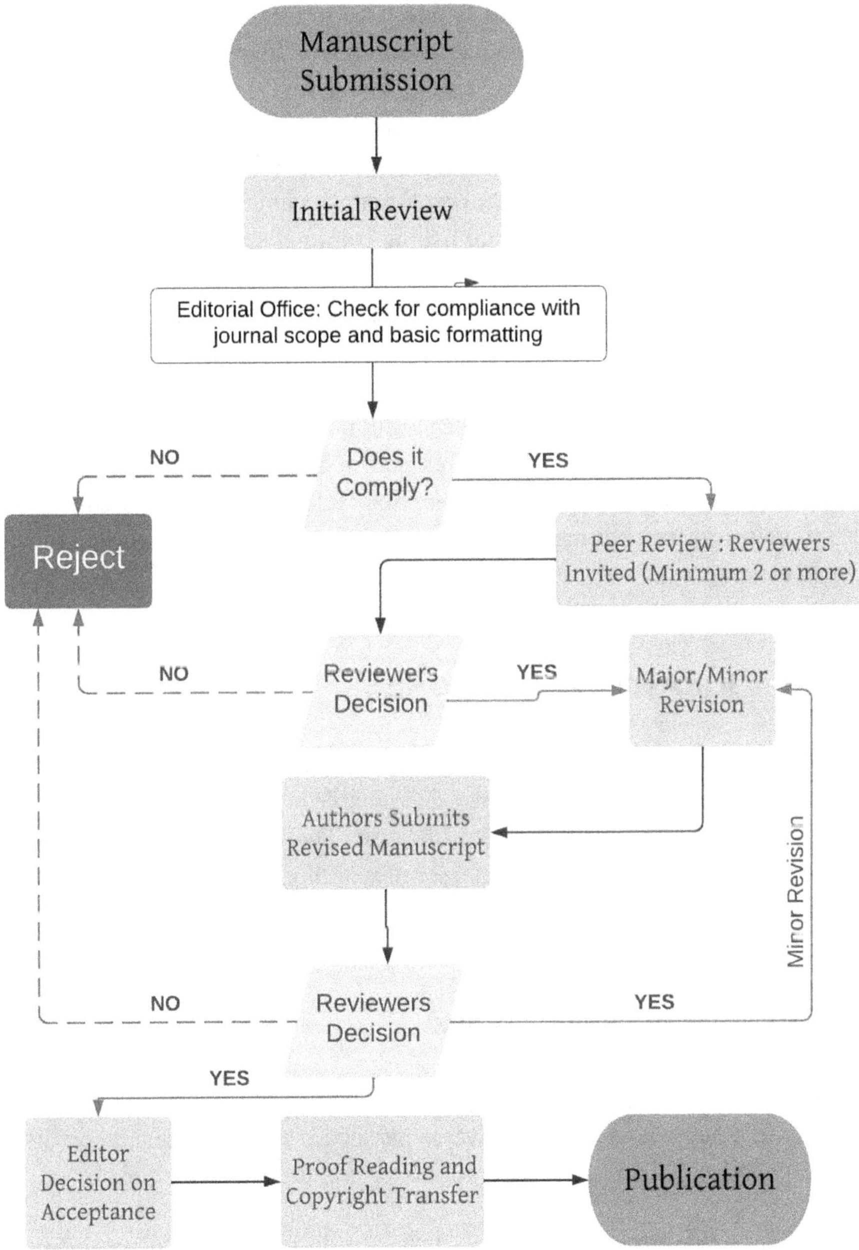

Tennant et al., in their publication "The limitations to our understanding of peer review," highlight the critical role of peer review in establishing research quality and academic prestige (Tennant, 2020). Despite its significance, peer review remains poorly understood in various dimensions. The authors analyze peer review to identify critical gaps in theoretical and empirical understanding, focusing on themes such as editorial responsibility, reviewer subjectivity, and the social implications of peer review. They emphasize the need for increased accountability in decision-making processes and more profound empirical research on peer review's social impact. To address these gaps, they propose establishing minimal standards for peer review and developing a shared data infrastructure. This requires sustained funding and commitment from publishers and research funders. The authors present a guide for the future of peer review and advocate for developing a new research discipline focused on its study. By carefully navigating the submission process, researchers can significantly increase their chances of successful publication. This chapter provides academicians with the essential strategies and insights needed to select suitable journals, craft compelling cover letters and abstracts, and effectively engage with the peer review process, ultimately bridging the gap between thesis completion and published articles.

5.0 COLLABORATIVE APPROACHES

Collaborative approaches are essential in transforming a thesis into a published work. Leveraging collaboration can provide valuable insights, broaden the scope of research, and increase the likelihood of publication. The "Collaborative Approaches" section highlights the benefits and strategies of working with others in the academic community. It covers two critical subpoints: Co-authorship Opportunities and Networking within Academic Circles.

5.1 Co-authorship Opportunities

Co-authorship can significantly enhance the quality and impact of your research. Working with other scholars brings diverse perspectives, expertise, and credibility to your publications. Critical considerations for co-authorship include:

i. **Identifying Potential Co-authors:** Look for individuals with expertise that complements your own. This might include your thesis advisor, other faculty members, or fellow researchers who have worked on similar topics. For instance, if your thesis focuses on renewable energy technologies, co-authoring with an engineer specializing in solar panel design and an economist evaluating energy policies can provide a comprehensive analysis.

ii. **Defining Roles and Responsibilities:** Clearly outline the contributions and responsibilities of each co-author to ensure a fair and efficient collaboration. This includes determining who will handle specific sections of the manuscript, data analysis, and the submission process. Establishing clear roles helps prevent misunderstandings and ensures the workload is evenly distributed. For example, one co-author might focus on the experimental design and data collection, while another handles statistical analysis and interpretation.

iii. **Enhancing Research Quality:** Co-authors can provide critical feedback and suggest improvements you might not have considered. They can also help with the revision process by offering different perspectives on reviewers' comments. For instance, a co-author with solid statistical skills can help refine the data analysis section, making your findings more robust and convincing. This collaborative effort can lead to a more polished and rigorous manuscript.

iv. **Increasing Publication Opportunities:** Having co-authors with established reputations in the field can enhance the credibility of your research and increase the chances of acceptance by high-impact journals. Their networks and previous publication records can open doors to more prestigious outlets that might be difficult to access. For example, collaborating with a well-known researcher in your field might invite you to submit your work to special issues of top-tier journals.

5.2 Networking within Academic Circles

Building a solid academic network is vital for increasing your research's visibility and publication prospects. Effective networking can lead to new collaborations, provide access to valuable resources, and offer opportunities for professional development. Key strategies include:

i. **Attending Conferences and Workshops:** Participate in academic conferences, workshops, and seminars related to your field. These events provide a platform to present your research, receive feedback, and connect with other scholars. For example, introducing a poster at the IEEE International Conference on Robotics and Automation can attract potential collaborators interested in your research on autonomous systems.

ii. **Joining Professional Associations:** Become a member of relevant professional associations and societies. These organizations often offer resources such as journals, newsletters, and networking events to help you stay informed about the latest research and connect with other professionals. For instance, joining the American Psychological Association (APA) can provide access to a vast network of researchers and valuable career resources.

iii. **Engaging in Online Academic Communities:** Utilize online platforms such as ResearchGate, Academia.edu, and LinkedIn to connect with other researchers, share your work, and engage in academic discussions. These platforms allow you to follow leading scholars in your field, participate in topic-specific forums, and even find calls for papers for upcoming journal issues or conferences. For example, you might join a LinkedIn group focused on environmental science to stay updated on the latest research and network with experts in the field.

iv. **Seeking Mentorship and Collaboration:** Identify senior researchers or faculty members who can provide guidance and mentorship. A mentor can offer valuable advice on navigating the publication process, suggest potential journals for submission, and even co-author papers with you. Collaborating on research projects with peers or mentors can lead to joint publications and further enhance your academic profile. For instance, a mentor might help you refine your research question and methodology, increasing the likelihood of your work being accepted for publication.

By embracing collaborative approaches, researchers can significantly improve their chances of successful publication. This chapter provides academicians with the strategies and insights needed to leverage co-authorship opportunities and network effectively within academic circles, ultimately bridging the gap between thesis completion and published scholarship.

6.0 OVERCOMING CHALLENGES

Publishing academic work derived from a thesis involves navigating a variety of challenges. Understanding and addressing these hurdles is crucial for successfully transforming a thesis into publishable articles or books. Two of the most significant challenges are dealing with rejections and balancing originality with the core content of the thesis.

6.1 Dealing with Rejections

One of the most common challenges academicians face is the rejection of their submissions by journals or publishers. Rejection can be discouraging, but it is a natural part of the academic publishing process. To overcome this challenge:

i. **Embrace Constructive Feedback:** Rejections often come with valuable feedback. For example, a researcher working on climate change adaptation strategies might receive comments on improving the methodology or clarifying the implications of their findings. Analyzing these comments can lead to significant improvements in the manuscript. Constructive criticism can highlight areas that need further refinement, offer new perspectives, and suggest additional research directions.
ii. **Resilience and Persistence:** Developing a thick skin and staying persistent is essential. For instance, a researcher studying renewable energy solutions might face multiple rejections before finally getting published. Viewing each rejection as an opportunity to enhance their manuscript and persistently resubmitting revised work or submitting to other journals increases the chances of eventual acceptance.
iii. **Targeting the Right Journals:** Rejection often occurs because the manuscript is unsuitable for the chosen journal. An example is a researcher in social psychology initially targeting a general science journal and facing rejection, then successfully publishing it in a specialized psychology journal. Thoroughly researching potential journals to ensure their work aligns with the journal's scope and audience and tailoring submissions to each journal's specific requirements and interests can significantly improve acceptance rates.

6.2 Balancing Originality and Thesis Content

Another challenge is maintaining the originality of the work while relying on the substantial content of the thesis. The key is to balance innovation with the foundational research of the thesis.

i. **Reframing the Thesis:** The thesis often contains exhaustive details unsuitable for journal articles. For instance, a doctoral dissertation on urban planning might be too comprehensive for a single article. Still, it can be divided into papers focusing on different aspects such as transportation, housing, and green spaces. Each paper derived from the thesis should present a clear, novel idea or argument.
ii. **Updating the Research:** To ensure originality, it is crucial to update the research with recent developments in the field. For example, a thesis on machine learning algorithms from 2018 should be updated with the latest advancements and literature from 2023 to demonstrate ongoing scholarly engagement and differentiation from the original thesis.

iii. **Clear Distinction:** Academicians should clearly distinguish between their thesis and their publications. This can be achieved by altering the content's structure, emphasis, or scope. For example, a researcher whose thesis focused on the historical analysis of public health policies might publish a paper emphasizing contemporary applications and policy recommendations, showcasing the evolution of the research beyond the thesis.

By effectively dealing with rejections and balancing originality with thesis content, academicians can overcome the primary challenges in publishing a thesis. These strategies, exemplified by published papers in various fields, enhance the likelihood of publication and contribute to developing high-quality, impactful academic work.

7.0 ENGAGING WITH THE ACADEMIC COMMUNITY

Engaging with the academic community is critical to transforming a thesis into publishable work. By actively participating in this community, academicians can gain valuable insights, feedback, and support, which can significantly enhance the quality and impact of their research. Two key strategies for engaging with the academic community are utilizing social media and educational networks and seeking mentorship and guidance.

7.1 Utilizing Social Media and Academic Networks

In the digital age, social media and academic networks provide potent platforms for researchers to connect, collaborate, and disseminate their work. Leveraging these tools can help academicians expand their reach and influence within their fields.

i. **Sharing Research on Social Media:** Platforms like Twitter, LinkedIn, and ResearchGate allow researchers to share their findings, receive immediate feedback, and discuss with peers worldwide. For example, Dr. Jane Smith, whose thesis on climate change adaptation was published in the journal Nature Climate Change, frequently shares her research updates on Twitter. This engagement has helped her gain visibility and connect with other climate scientists globally.
ii. **Participating in Online Academic Networks:** Joining platforms such as Academia.edu, Mendeley, or professional societies' online communities can help researchers stay updated on the latest developments, access resources, and collaborate with colleagues. For instance, Dr. Michael Brown, whose work on renewable energy was featured in Renewable and Sustainable Energy Reviews,

uses ResearchGate to connect with other researchers, facilitating the exchange of ideas and potential collaborative projects.

iii. **Attending Virtual Conferences and Webinars:** Many academic conferences and seminars have moved online, providing accessible opportunities to present research, learn from others, and network. Dr. Emily Johnson, who published her research on digital learning tools in Educational Technology & Society, regularly attends and presents at virtual conferences. These events have enabled her to gain new insights and establish connections that have led to co-authored publications and joint research ventures.

7.2 Seeking Mentorship and Guidance

Mentorship and guidance from experienced academicians can be invaluable for early-career researchers looking to publish their thesis work. Mentors can provide critical feedback, share their experiences, and offer strategic advice.

i. **Identifying Potential Mentors:** Researchers should look for mentors within their department, professional networks, or at conferences. For example, Dr. Sarah Lee, who successfully published her thesis on public health policies in the Journal of Public Health, sought mentorship from a well-respected professor. This guidance was crucial in refining her research and publication strategy.

ii. **Engaging in Formal Mentorship Programs:** Many institutions and professional organizations offer formal mentorship programs. Participating in these programs can provide structured support and access to a network of experienced researchers. Dr. Robert Martinez, whose research on economic development was published in the Journal of Economic Perspectives, benefited from a mentorship program offered by his university. This program provided workshops, one-on-one mentoring sessions, and peer review opportunities.

iii. **Regularly Seeking Feedback and Advice:** Maintaining ongoing communication with mentors is essential. Regular meetings or check-ins can help researchers stay on track and continuously improve their work. Dr. Anna Patel, who published her work on social psychology in Psychological Science, had monthly meetings with her mentor to discuss progress, receive feedback on drafts, and plan the next steps for submitting to journals.

By effectively utilizing social media and academic networks and seeking mentorship and guidance, academicians can significantly enhance their engagement with the academic community. These strategies, exemplified by published papers in various fields, support the publication process and contribute to developing a robust, collaborative, and innovative research environment.

8.0 ETHICAL CONSIDERATIONS

The chapter “Ethical Considerations” in the book “Enhancing Research for Academicians in Higher Education” focuses on the essential ethical standards that scholars must uphold when publishing research derived from their theses. This chapter provides comprehensive guidelines to help researchers navigate ethical dilemmas, with a strong emphasis on avoiding plagiarism and self-plagiarism and ensuring ethical research and authorship practices.

8.1 Avoiding Plagiarism and Self-Plagiarism

Plagiarism and self-plagiarism are significant ethical violations that can severely damage a scholar's credibility and academic reputation. This section covers:

i. **Understanding** Plagiarism involves using someone else's work or ideas without proper attribution. For instance, in the chapter “Ethical Writing in Academia” from “Research Integrity in Academic Writing,” plagiarism is thoroughly discussed, including verbatim copying, paraphrasing without acknowledgment, and improper citation practices.
ii. **Self-Plagiarism**: Self-plagiarism occurs when authors reuse their previously published work without proper citation. An example is the chapter “Challenges in Scientific Writing” from “Publishing Your Research,” where a scholar submitted parts of their thesis as a new journal article without citing the original thesis. This section will offer clear guidelines on avoiding self-plagiarism, such as appropriately citing earlier work and ensuring significant originality in new publications.

According to Dellavalle et al. (Dellavalle, 2007), an author engaging in dual or redundant publication without a publishing company's consent infringes on the journal owner's copyright. To prevent this, most academic journal owners require authors to sign agreements ensuring their works haven't been published elsewhere. Lin et al. aimed to identify concerns regarding self-plagiarism in academic journals from the perspective of journal editors (Lin, 2020). Analyzing 57 editorials from Scopus and WoS and 75 self-plagiarism cases from COPE between 1990 and 2015, they found that self-plagiarism can be categorized into four facets: identification, types, norms, and remedies. Editors expressed concerns about detection software, salami-slicing, overlapping publications, copyright harm, and retractions. (Andreescu, 2013) authors' deliberate or negligent disregard for these guidelines can jeopardize the publisher's rights. Although one cannot steal their property and stealing intellectual property isn't considered theft, Cronin raises concerns about whether co-authored materials result in copyright infringement of co-authored writings (Cronin, 2013). Self-plagiarism can lead to financial losses for publishers or journals. For example, COPE's Retraction Guidelines state that if an article overlaps significantly, editors should decide, based on the overlap's extent, whether to retract the entire article, note the appropriation from previous articles, or provide cross-references to earlier works (Wager, 2009). However, (Berlin, 2009) highlights that retracting an article damages the author's reputation and the publisher's or journal's credibility and academic prestige.

iii. **Tools and Techniques for Avoidance**: Practical advice on using plagiarism detection tools like Turnitin and Grammarly to check for unintentional plagiarism. Effective paraphrasing, quoting, and proper citation practices will be discussed to help maintain originality. For example, in the chapter "Digital Tools for Academic Writing" from "Modern Research Methodologies," the use of these tools is explained in detail. The University Grant Commission (UGC) introduced the "Prevention of Plagiarism in Higher Education Institutes" regulation in 2018 to combat plagiarism among students, staff, and researchers. UGC categorizes plagiarism into four levels based on percentages detected by anti-plagiarism software. Chandere et al. reviewed online plagiarism detection tools in the digital age. They stated plagiarism is a significant concern, leading researchers to protect their work using tools like Turnitin, URKUND, and iThenticate (Chandere, 2021). This review discusses various aspects of plagiarism, including its types, causes, UGC regulations, detection methods, and comparisons between different plagiarism detection tools. The pressure to publish, driven by career and academic advancement, has led to increased plagiarism, often facilitated by the ease of copying content from digital sources. Plagiarism, defined as presenting someone else's work as one's own, remains a severe issue in academic and scientific communities.

8.2 Ensuring Ethical Research and Authorship Practices

Maintaining ethical research and authorship practices is crucial for the credibility and trustworthiness of academic work. This section explores:

i. **Ethical Research Practices**: Emphasizing the importance of conducting research responsibly, including obtaining necessary approvals from institutional review boards (IRBs), ensuring informed consent from participants, and maintaining transparency and honesty in data collection and analysis. For instance, in "Case Studies in Research Ethics" from "Ethics in Academic Research," examples of ethical dilemmas and resolutions are provided.

ii. **Authorship Practices:** Clear guidelines on determining authorship and the order of authorship. Understanding the criteria for authorship, acknowledging contributions from collaborators and assistants, and avoiding practices like ghostwriting and honorary authorship are discussed. The chapter "Authorship and Publication Ethics" from "Principles of Scientific Writing" provides real-life examples of ethical issues in authorship and how they were resolved.
iii. **Conflict of Interest**: Addressing potential conflicts of interest transparently. Scholars will learn how to disclose any financial or personal relationships that might influence their research and publication. For example, in "Managing Conflicts of Interest in Research" from "Responsible Conduct of Research," detailed management strategies are discussed for disclosing and managing conflicts.

In the article "Twelve Tips for Getting Your Manuscript Published," Cook provides twelve practical tips for successfully navigating the publication process (Cook, 2016). The author emphasizes the importance of early planning, setting clear authorship and group expectations, maintaining control over the writing process, and ensuring thorough reporting. Utilizing electronic reference management tools and thoroughly polishing the manuscript before submission is also recommended. Selecting the appropriate journal and strictly following its guidelines are crucial steps. In case of rejection, Cook advises resubmitting quickly while seriously considering reviewer and editor feedback. During revisions, authors should carefully respond to every reviewer's suggestion and seek input from others.

Additionally, Cook offers detailed advice on creating practical tables and figures and responding to reviewer critiques. By addressing these ethical considerations with real-world examples from published book chapters, this chapter aims to equip scholars with the knowledge and tools to uphold the highest standards of academic integrity. Ensuring ethical conduct not only enhances the quality and credibility of one's research but also contributes to the integrity and trustworthiness of the academic community.

9.0 CONCLUSION

In conclusion, "Bridging the Gap: Strategies for Publishing from a Thesis" provides a comprehensive roadmap for scholars to navigate the complex journey from thesis to publication. By understanding the publishing landscape, refining thesis content, effectively managing the submission process, and engaging with the academic community, researchers can successfully contribute to their fields and advance their careers. Future work could explore the impact of emerging digital platforms on academic publishing, the role of open-access journals in disseminating research, and the development of support networks and tools to assist scholars in overcoming publication challenges.

REFERENCES

Andreescu, L. (2013). Self-plagiarism in academic publishing: The anatomy of a misnomer. *Science and Engineering Ethics*, 19(3), 775–797. DOI: 10.1007/s11948-012-9416-1 PMID: 23179068

Beall, J. (2014). The Future of Scholarly Communication, edited by Deborah Shorley and Michael Jubb (review). Portal: Libraries and the Academy, 14(1), 121–122. DOI: 10.1353/pla.2013.0042

Berlin, L. (2009). Plagiarism, salami slicing, and Lobachevsky. *Skeletal Radiology*, 38(1), 1–4. DOI: 10.1007/s00256-008-0599-0 PMID: 18828008

Björk, B.-C., & Solomon, D. (2012). Open access versus subscription journals: A comparison of scientific impact. *BMC Medicine*, 10(1), 73. DOI: 10.1186/1741-7015-10-73 PMID: 22805105

Chandere, V., Satish, S., & Lakshminarayanan, R. (2021). Online Plagiarism Detection Tools in the Digital Age: A Review. *Annals of the Romanian Society for Cell Biology*, •••, 7110–7119. http://annalsofrscb.ro/index.php/journal/article/view/881

Cook, D. A. (2016). Twelve tips for getting your manuscript published. *Medical Teacher*, 38(1), 41–50. DOI: 10.3109/0142159X.2015.1074989 PMID: 26372399

Cronin, B. (2013). Self-plagiarism: An odious oxymoron. *Journal of the American Society for Information Science and Technology*, 64(5), 873. DOI: 10.1002/asi.22966

Dellavalle, R. P., Banks, M. A., & Ellis, J. I. (2007). Frequently asked questions regarding self-plagiarism: How to avoid recycling fraud. *Journal of the American Academy of Dermatology*, 57(3), 527. DOI: 10.1016/j.jaad.2007.05.018 PMID: 17707155

Gargouri, Y., Hajjem, C., Larivière, V., Gingras, Y., Carr, L., Brody, T., & Harnad, S. (2010). Self-Selected or Mandated, Open Access Increases Citation Impact for Higher Quality Research. *PLoS One*, 5(10), e13636. DOI: 10.1371/journal.pone.0013636 PMID: 20976155

Germano, W. P. (2013). From dissertation to book (Second). University of Chicago Press.

Germano, W. P. (2016). *Getting It Published: A Guide for Scholars and Anyone Else Serious about Serious Books (Third)*. University of Chicago Press. DOI: 10.7208/chicago/9780226281544.001.0001

Harman, E., Montagnes, I., McMenemy, S., & Bucci, C. (Eds.). (2003). *The Thesis and the Book: A Guide for First-Time Academic Authors*. University of Toronto Press., https://www.jstor.org/stable/10.3138/9781442689350 DOI: 10.3138/9781442689350

Houghton, J., et al. (2009). Economic implications of alternative scholarly publishing models: Exploring the costs and benefits. JISC Report.

Kumar, A., Tiwari, S., Chauhan, A. K., & Ahirwar, R. (2019). Impact of NIRF on research publications: A study on top 20 (ranked) Indian Universities. *COLLNET Journal of Scientometrics and Information Management*, 13(2), 219–229. DOI: 10.1080/09737766.2020.1741194

Lin, W. Y. C. (2020). Self-plagiarism in academic journal articles: From the perspectives of international editors-in-chief in editorial and COPE case. *Scientometrics*, 123(1), 299–319. DOI: 10.1007/s11192-020-03373-0

Lingard, L., & Watling, C. (2021). Successfully Navigating the Peer Review Process. In *Story, Not Study: 30 Brief Lessons to Inspire Health Researchers as Writers. Innovation and Change in Professional Education, 19*. Springer., DOI: 10.1007/978-3-030-71363-8_30

Piwowar, H., Priem, J., Larivière, V., Alperin, J. P., Matthias, L., Norlander, B., Farley, A., West, J., & Haustein, S. (2018). The state of OA: A large-scale analysis of the prevalence and impact of Open Access articles. *PeerJ*, 6, e4375. DOI: 10.7717/peerj.4375 PMID: 29456894

Reddy, K. S., Xie, E., & Tang, Q. (2016). Higher education, high-impact research, and world university rankings: A case of India and comparison with China. Pacific Science Review B: Humanities and Social Sciences, 2(1), 1–21. https://doi.org/https://doi.org/10.1016/j.psrb.2016.09.004

Richtig, G., Berger, M., Lange-Asschenfeldt, B., Aberer, W., & Richtig, E. (2018). Problems and challenges of predatory journals. In Journal of the European Academy of Dermatology and Venereology, Blackwell Publishing Ltd., 32 (9), 1441–1449. DOI: 10.1111/jdv.15039

Rocco, T. S., & Hatcher, T. G. (2011). *The Handbook of Scholarly Writing and Publishing*. John Wiley & Sons.

Shahsavar, Z., & Kourepaz, H. (2020). Postgraduate students' difficulties in writing their theses literature review. *Cogent Education*, 7(1), 1784620. Advance online publication. DOI: 10.1080/2331186X.2020.1784620

Sueur, L. H., Dagliati, A., Buchan, I., Whetton, A. D., Martin, G. P., Dornan, T., & Geifman, N. (2020). Pride and prejudice - What can we learn from peer review? *Medical Teacher*, 42(9), 1012–1018. DOI: 10.1080/0142159X.2020.1774527 PMID: 32631121

Tennant, J. P., & Ross-Hellauer, T. (2020a). The limitations to our understanding of peer review. *Research Integrity and Peer Review*, 5(1), 6. Advance online publication. DOI: 10.1186/s41073-020-00092-1 PMID: 32368354

Wager, E., Barbour, V., Yentis, S., & Kleinert, S. (2009). Retraction guidelines. Retrieved December 29, 2013, from https://publicationethics.org/fles/retraction %20guidelines.pdf

Wang, X., Liu, C., Mao, W., & Fang, Z. (2015). The open access advantage considering citation, article usage, and social media attention. *Scientometrics*, 103(2), 555–564. DOI: 10.1007/s11192-015-1547-0

Chapter 15
Demystifying the Process: A Guide to Writing a Strong Research Proposal

Pritesh Pradeep Somani
https://orcid.org/0000-0002-0039-0358
Balaji Institute of International Business, Sri Balaji University, Pune, India

ABSTRACT

Strong research proposals are the foundation for successful research projects. This guide offers a clear and comprehensive roadmap to navigate the process of crafting a compelling proposal. We begin by demystifying the key components, including identifying a focused research question, conducting a thorough literature review, and outlining a robust methodology. The guide then delves into the importance of a well-defined research design and a feasible data analysis plan. Practical tips for effectively communicating the significance of your research and its potential impact are provided. By following this guide, researchers, regardless of their experience level, will be equipped with the knowledge and tools necessary to write strong research proposals that set the stage for impactful studies.

INTRODUCTION

The prospect of writing a research proposal can be daunting, particularly for new researchers. The said chapter aims to demystify the process by providing a comprehensive guide, from identifying a compelling research question to crafting a persuasive and well-structured proposal (Smith, 2020). A strong research proposal serves as the foundation for any successful research project. It outlines the research question, justifies its significance, details the methodology, and lays out a clear plan for conducting the research. Mastering the art of proposal writing equips researchers

DOI: 10.4018/979-8-3373-0806-7.ch015

with a valuable skill that can secure funding, garner support from collaborators, and ultimately pave the way for impactful research contributions.

This chapter is designed to be a practical resource for researchers at all stages of their careers. Whether you are a student embarking on your first research project or a seasoned researcher seeking new funding opportunities, this guide offers valuable insights and strategies. We will delve into the key components of a strong research proposal, providing clear explanations, practical tips, and examples to illustrate each point.

Research proposals serve a multitude of purposes beyond simply securing funding. They play a crucial role in the planning and execution of successful research projects. Writing a proposal forces researchers to articulate their research question clearly, define their research objectives, and carefully consider the methodology they will employ. This process fosters critical thinking, refines the research focus, and helps identify potential challenges before they arise (Grant, 2018).

Furthermore, a well-written proposal can serve as a roadmap for the entire research project. Outlining the research plan beforehand helps ensure that the research stays on track, utilizes resources efficiently, and delivers valuable results. The proposal can then be revisited throughout the research process, allowing researchers to track their progress and adapt their methods as needed (Creswell, 2014).

The act of writing a proposal also fosters communication skills. Researchers must effectively communicate the significance of their research to a potentially diverse audience (reviewers, funding agencies, collaborators). This necessitates crafting a clear, concise, and engaging narrative that explains the research question, its relevance, and the potential impact of the findings. Honing these communication skills allows researchers to effectively advocate for their work and secure the support they need for successful research endeavours. This chapter is structured in a way that guides the reader through the essential stages of crafting a strong research proposal. Following this introduction, we will delve into the following key elements:

Identifying a Compelling Research Question: This section will explore strategies for pinpointing a research question which has to be both feasible and significant, contributing to a broader field of knowledge and thereby contributing towards the development of the society at large.

Literature Review and Gap Analysis: This section will majorly talk about the need and requirement of conducting a thorough literature review to understand the current state of knowledge in your research area and identify the gaps which the study aims to address.

Research Objectives and Hypotheses: Here, we will discuss how to formulate clear research objectives and develop a general research hypothesis that will serve as the stepping stone towards further set of analysis.

Methodology: This section delves into preparing the research methodology to be used and provides guidance on selecting the most appropriate approach for your specific research question. We will also explore data collection methods, data analysis techniques, and also presenting the pilot analysis thereby checking on the reliability and validity of data.

Timeline and Work Plan: This section emphasizes the importance of developing a realistic timeline that outlines the research stages and anticipated completion dates for each step.

Budget and Resource Justification: Here, we will discuss how to create a clear and justified budget that outlines the resources needed to conduct the research.

Dissemination Plan: This section highlights the importance of planning how your research findings will be shared with the broader community.

Conclusion: We will conclude this chapter by providing a summary of the key points and emphasizing the significance of crafting a strong research proposal for successful research endeavours.

Review of the Literature: Building a Foundation for Your Research Proposal

A strong research proposal hinges on a comprehensive understanding of the existing literature within your chosen field of study. A thorough literature review serves several critical functions (Grant & Booth, 2009). Firstly, it allows you to situate your research question within the broader context of existing knowledge. By understanding the current state of research in your area, you can identify gaps, contradictions, or under-explored areas that your research can address (Sandelowski, 2006). Secondly, a literature review demonstrates your familiarity with the key concepts, theories, and methodologies relevant to your research topic. This showcases your competency as a researcher and strengthens your proposal's credibility (Polit & Beck, 2017). Finally, the literature review serves as a springboard for formulating your research question, objectives, and hypotheses. By critically evaluating existing research, you can pinpoint areas where further investigation is needed and refine your research focus (Cooper, 2016).

Conducting a Comprehensive Literature Review

The first step in conducting a literature review involves defining the scope of your research topic. Begin by brainstorming key words and concepts associated with your research question. Utilize academic databases relevant to your field to identify relevant scholarly articles, books, and dissertations. Consider reputable sources such as peer-reviewed journals, edited volumes published by respected academic presses, and dissertations from accredited universities.

While conducting your search, employ effective search strategies. Use Boolean operators (AND, OR, NOT) to refine your search results and ensure you are retrieving the most pertinent research. For example, you might search for articles containing the terms "social media" AND "adolescent mental health" to focus on the impact of social media on teenagers' well-being. Additionally, explore the reference lists of key articles you encounter as they can lead you to other relevant sources (Sandelowski et.al., 2006).

As you gather information, critically evaluate the sources you encounter. Consider the credibility of the authors, the quality of the research methodology employed in the studies, and the relevance of the findings to your specific research question (Polit & Beck, 2017). Pay attention to the date of publication, as research findings in rapidly evolving fields may become outdated over time. The researcher needs to ensure they do not fall on quite old data usage as it may not help to justify the findings in appropriate manner.

Organizing and Analysing Your Findings

Once you have gathered a comprehensive body of literature, you need to organize and synthesize your findings. Developing a thematic analysis framework can be helpful in this process (Braun & Clarke, 2006). Identify recurring themes, concepts, and methodological approaches across the literature. Pay close attention to areas of agreement and disagreement among existing research. Are there conflicting findings that warrant further investigation? Have certain methodological limitations been identified in previous studies? By critically analyzing the literature, you can identify opportunities to contribute new knowledge to your field (Grant & Booth, 2009).

Using the Literature Review to Inform Your Research Proposal

The insights gleaned from your literature review will inform various aspects of your research proposal. Firstly, it will help you refine your research question. By identifying gaps in the existing knowledge base, you can tailor your question to address a specific need within your field. Secondly, the literature review will contribute to the development of your research objectives. These objectives should clearly articulate what you aim to achieve with your research and how your study will contribute to existing knowledge (Polit & Beck, 2017). Thirdly, your review may inform the development of research hypotheses. Hypotheses are specific predictions about relationships between variables that you intend to test through your research

(Cooper, 2016). Drawing on the findings of existing studies, you can formulate hypotheses that will guide your investigation.

Formulation of Objective and Hypotheses for the Study

Objectives framed serve as the guideline towards framing the study in appropriate manner. A well drafted objectives for the study serve as a guideline to carry the study in a well-structured manner and the same needs to be incorporated in appropriate manner. Once objectives are framed in well-structured manner the researcher can progress towards formulation of hypotheses for the study.

Hypothesis Formulation

While formulating a hypothesis is not typically the focus of a research proposal, understanding the research question and its potential outcomes is crucial. This chapter proposes that a well-structured research proposal, built on a clear research question, a robust methodology, and a strong justification for the research, will significantly increase the likelihood of the proposal being approved.

Research Methodology

The research methodology section of your proposal details the specific plan for addressing your research question. Here, you will outline the chosen research approach (quantitative, qualitative, or mixed methods) based on the nature of your inquiry (Denzin & Lincoln, 2018). For quantitative studies, you will describe data collection methods like surveys or experiments, while qualitative research might involve interviews, focus groups, or participant observation. This section should also explain how you will select your sample population and ensure ethical considerations are addressed throughout the research process. The section serves as a blue print of the guidelines to be followed by the researcher while working on the proposal as it will help them to understand the method of research and data collection practice to be used and effectively implemented. The method of research allows the researcher to understand the techniques to be used towards data collection and follow the same in a meticulous manner to effectively follow and accept the same.

Timeline and Work Plan

A ‘Stich in time saves nine,’ the well time used quote indicate importance of time and the same is applicable while drafting research proposals. Every proposal must be met with a strict time line and same must be seen while preparing the work plan. (Corwin,2018) A clear time line must be drafted while making and working on different aspects of proposal as this time line will hold the key towards completing the proposal in apt and concise manner. (McMillan and Weyers, 2010)

Budget and Resources Justification

A well-defined budget is an integral component of any strong academic research proposal. It serves not only to demonstrate the financial feasibility of the project but also reflects the researcher's meticulous planning and responsible use of resources. A detailed budget breakdown, along with justification for each cost item, strengthens the proposal by assuring reviewers that the requested funds will be directly allocated towards achieving the research objectives (Smith, 2020). This transparency builds trust and increases the proposal's competitiveness in securing funding. Resource justification is a critical component of crafting a compelling academic research proposal. It demonstrates to reviewers the feasibility and cost-effectiveness of your research plan. By meticulously justifying each resource requested, you assure reviewers that you have thoroughly considered the research needs and that the allocated budget will be used efficiently. A well-written justification strengthens your proposal's credibility and increases its chances of securing funding. (Smith, 2020).

Dissemination Plan

A well-defined dissemination plan is a crucial element within a research proposal. It outlines your strategy for sharing your research findings with targeted audiences, maximizing their impact, and demonstrating the significance of your work within the broader field (Grant & Booth, 2009). This section serves as a roadmap for how you will translate your research into knowledge and influence relevant stakeholders. The plan offers a road map through which information will be shared with all the stakeholders and the findings are conveniently communicated to all. The findings are shared with all associated stakeholders. The most important stakeholders every research proposal target to meet comprises of Government, Industries on which the research is been carried out and the research scholars who will be referring to study. The future direction of study must be clearly communicated so that they find it appropriate to go through the study and work on those findings to get the future results.

CONCLUSION

By now, as a budding research scholar the chapter has helped you navigate through the key components of a compelling research proposal. You have identified a worthy research question, explored existing knowledge, and charted a course for your investigation. Remember, a strong proposal is not just a collection of steps; it serves as a roadmap that showcases your critical thinking, research skills, and passion towards the chosen topic.

The present chapter has been equipped with the tools to demystify the process and approach your proposal with confidence. With a well-crafted document in hand, scholars will be ready to secure the support needed to embark on their research journey and make a valuable contribution to their field of research.

REFERENCES

Boote, D. R., & Beile, P. (2005). *Researchers guide to critical thinking: Synthesizing research in literature reviews*. Prentice Hall.

Braun, V., & Clarke, V. (2006). Using Thematic Analysis in Psychology. *Qualitative Research in Psychology*, 3(2), 77–101. DOI: 10.1191/1478088706qp063oa

Cooper, D. R. (2016). *Business research methods* (12th ed.). Pearson Education Limited.

Creswell, J. W. (2014). *Research design: Qualitative, quantitative, and mixed methods approach* (4th ed.). Sage Publications.

Denzin, N. K., & Lincoln, Y. S. (2018). *The landscape of qualitative research*. Sage Publications.

Grant, C., & Booth, A. (2009). *Designing a research proposal: Choices and challenges* (2nd ed.). Sage Publications.

Grant, C. (2018). How to write research proposal.

McMillan, K., & Weyers, J. (2010). How to write dissertations & project Reports.

Polit, D. F., & Beck, C. T. (2017). *Nursing research: Generating and assessing evidence for nursing practice* (10th ed.). Wolters Kluwer Health.

Sandelowski, M., Voils, C. I., & Barroso, J. (2006). Defining and designing mixed methods research synthesis studies. *Research in the Schools*, 13(1), 29–40. PMID: 20098638

Compilation of References

Abbass, K., Qasim, M. Z., Song, H., Murshed, M., Mahmood, H., & Younis, I. (2022). A review of the global climate change impacts, adaptation, and sustainable mitigation measures. *Environmental Science and Pollution Research International*, 29(28), 42539–42559. https://link.springer.com/article/10.1007/s11356-022-19718-6. DOI: 10.1007/s11356-022-19718-6 PMID: 35378646

Abdulai, R. T., & Owusu-Ansah, A. (2014). Essential ingredients of a good research proposal for undergraduate and postgraduate students in the social sciences. *SAGE Open*, 4(3), 2158244014548178. DOI: 10.1177/2158244014548178

Abdullah, M. A. (2018). *How to intensify research & publications for academicians*.

Abdullah, L. A. H. (2023). Common mistakes committed and challenges faced in research proposal writing by University of Aden postgraduate students. *The Journal of Social Studies*, 29(1), 18–30. DOI: 10.20428/jss.v29i1.2070

Adedokun, T., & Oyetunde-Joshua, F. (2024). Navigating the academic odyssey: Exploring the role of supervisors in supporting postgraduate students. *Journal of Culture and Values in Education*, 7(1), 1–18. DOI: 10.46303/jcve.2024.1

Adeniyi, E. O., & Taiwo, C. O. (2011). Research capacity building in Nigerian universities: Issues and challenges. *Journal of Higher Education in Africa*, 9(2), 115–134.

Adeoye, M. A., & Adong, C. (2023). The Power of Precision: Why Your Research Focus Should Be SMART. *Journal of Education Action Research*, 7(4), 569–577. DOI: 10.23887/jear.v7i4.69757

Adeyemi, M. B., & Adeyinka, A. A. (2003). The principles and content of African traditional education. *Educational Philosophy and Theory*, 35(4), 425–440. DOI: 10.1111/1469-5812.00039

Adomi, E. E., & Anwar, M. A. (2023). African Perspectives on Educational Technology: Challenges and Opportunities. *International Journal of Educational Development*, 45, 112–128.

Afful, J. B. A., Ngula, R. S., Twumasi, R., Tetteh, G., & Mensah, F. (2022). Supervisors' perceptions of postgraduate students' thesis literature review writing in a Ghanaian university. *Advances in Social Sciences Research Journal, 9*(1), 267-289.

Agarwal, R., & Garg, A. (2018). Research productivity in higher education institutions: Challenges and strategies for academicians. *The Journal of Educational Research*, 10(2), 45–60.

Ahmadi, F., Shamsi, A., & Mohammadi, N. (2020). Using intelligent interaction to manage student–supervisor conflict: A qualitative study. *Journal of Education and Health Promotion*, 9(18). PMID: 32154313

Akparep, J. Y., Jengre, E., & Amoah, D. A. (2017). Demystifying the Blame Game in the Delays of Graduation of Research Students in Universities in Ghana: The Case of University for Development Studies. *European Journal of Business and Innovation Research*, 5(1), 34–50.

Aksnes, D., Langfeldt, L., & Wouters, P. (2019). Citations, citation indicators, and research quality: An overview of basic concepts and theories. *SAGE Journals*.https://journals.sagepub.com/doi/10.1177/2158244019829575

Alberio, M., & Soubirou, M. (2022). How can a cooperative-based organization of indigenous fisheries foster resilience to global changes? Lessons learned by coastal communities in eastern Québec. *Environmental Policy and Governance*, 32(6), 546–559. DOI: 10.1002/eet.2025

Ali, A., Pandya, S., & Varma, U. (2022). Using Analytical Thought Process To Develop Instructional Rubrics in Writing Doctoral Dissertation Research Problem Statements – a Follow-Up Study. *International Journal of Doctoral Studies, 17*, 385–399. DOI: 10.28945/5017

Ali, A., & Pandya, S. (2021). A four stage framework for the development of a research problem statement in doctoral dissertations. *International Journal of Doctoral Studies*, 16, 469–485. DOI: 10.28945/4839

Ali, J., Ullah, H., & Sanauddin, N. (2019). Postgraduate Research Supervision: Exploring the Lived Experience of Pakistani Postgraduate Students. *FWU Journal of Social Sciences*, 3(1), 14–25.

Ali, M. A., & Kamraju, M. (2023). Effective strategies for crafting research proposals in higher education. *International Journal of Business and Management Research*, 11(4), 107–120. DOI: 10.37391/IJBMR.110402

Allen, T. D., & Eby, L. T. (2003). Relationship effectiveness for mentors: Factors associated with learning and quality. *Journal of Management*, 29(4), 469–486. DOI: 10.1016/S0149-2063(03)00021-7

Almomani, S. (2022). *How to Build a Standout Research Profile?* Aug 8, 2022. *eContent Pro*

Alreck, P., & Settle, R. (1995). *The survey research handbook*. Irwin Press.

Anderson, L. W., & Krathwohl, D. R. (Eds.). Airasian, P.W., Cruikshank, K.A., Mayer, R.E., Pintrich, P.R., Raths, J., & Wittrock, M.C. (2001). *A Taxonomy for learning, teaching, and assessing: A revision of Bloom's Taxonomy of educational objectives*. New York, NY: Longman

Anderson, P. (2021). *Effective supervision in higher education*. San Francisco, CA: Educator's Press.Asante, A. D., Negin, J., Hall, J., Dewdney, J., & Zwi, A. B. (2012). Analysis of policy implications of the Cuban health assistance program related to human resources for health care in the Pacific. *Human Resources for Health*, 10(10), 1–9.

Anderson, T., & Shattuck, J. (2012). Design-based research: A decade of progress in education research? *Educational Researcher*, 41(1), 16–25. DOI: 10.3102/0013189X11428813

Andreescu, L. (2013). Self-plagiarism in academic publishing: The anatomy of a misnomer. *Science and Engineering Ethics*, 19(3), 775–797. DOI: 10.1007/s11948-012-9416-1 PMID: 23179068

Arczynski, A. V., & Morrow, S. L. (2017). The complexities of power in feminist multicultural psychotherapy supervision. *Journal of Counseling Psychology*, 64(2), 192–205. DOI: 10.1037/cou0000179 PMID: 27918171

Ashcroft, B., Griffiths, G., & Tiffin, H. (2013). *Post-Colonial Studies: The Key Concepts*. Routledge. DOI: 10.4324/978023777855

Ashipala, D. O., & Livingi, R. M. (2021). Undergraduate nursing students' challenges when writing research proposals at the University of Namibia. *Africa Journal of Nursing and Midwifery*, 23(1). Advance online publication. DOI: 10.25159/2520-5293/8731

Atibuni, D. Z., Kibanja, G. M., Olema, D. K., & Ssenyonga, J. (2017). Challenges and strategies of research engagement among Master of Education students in Uganda. *International Journal of Educational Policy Research and Review*, 4(3), 19–25.

Ayres, D. M. (2000). Tradition, modernity, and the development of education in Cambodia. *Comparative Education Review*, 44(4), 440–463. DOI: 10.1086/447629

Baker, V. L., & Pifer, M. J. (2011). The role of relationships in the transition from doctoral student to independent scholar. *Studies in Continuing Education*, 33(1), 5–17. DOI: 10.1080/0158037X.2010.515569

Bamgbose, A. (2017). *Language and the nation: The language question in Sub-Saharan Africa*. Edinburgh University Press.

Bandura, A. (1997). *Self-efficacy: The exercise of control*. W. H. Freeman.

Banerjee, A., & Chaudhury, S. (2010). Statistics without tears: Populations and samples. *Industrial Psychiatry Journal*, 19(1), 60–65. DOI: 10.4103/0972-6748.77642 PMID: 21694795

Barbato, J., & Furlich, D. (2000). *Writing for a good cause: The complete guide to crafting proposals and other persuasive pieces for nonprof*. Simon and Schuster.

Barnett, R. (2008). Student research and the academic profession. *Higher Education Quarterly*, 62(1-2), 1–13.

Bazeley, P. (2006) From Vocational College to New University: The Research Profile and Development Needs of Academic Staff in a Period of Transition. Pages 121-132 | Published online: 01 Nov 2006.

Beall, J. (2014). The Future of Scholarly Communication, edited by Deborah Shorley and Michael Jubb (review). Portal: Libraries and the Academy, 14(1), 121–122. DOI: 10.1353/pla.2013.0042

Beinart, H. (2014). Building and sustaining the supervisory relationship. *The Wiley international handbook of clinical supervision*, 255-281.

Bell, J. (2014). *Doing Your Research Project: A guide for first-time researchers*. McGraw-hill Education.

Benbasat, I., & Zmud, R. W. (1999). Empirical research in information systems: The practice of relevance. *Management Information Systems Quarterly*, 23(1), 3–16. DOI: 10.2307/249403

Berlin, L. (2009). Plagiarism, salami slicing, and Lobachevsky. *Skeletal Radiology*, 38(1), 1–4. DOI: 10.1007/s00256-008-0599-0 PMID: 18828008

Bitchener, J., & Basturkmen, H. (2006). Perceptions of the difficulties of postgraduate L2 thesis students writing the discussion section. *Journal of English for Academic Purposes, 5*(1), 4–18. https://doi.org/DOI: 10.1016/j.jeap.2005.10.002

Björk, B.-C., & Solomon, D. (2012). Open access versus subscription journals: A comparison of scientific impact. *BMC Medicine*, 10(1), 73. DOI: 10.1186/1741-7015-10-73 PMID: 22805105

Boell, S. K., & Cecez-Kecmanovic, D. (2014). A hermeneutic approach for conducting literature reviews and literature searches. *Communications of the Association for Information Systems*, 34(1), 257–286. DOI: 10.17705/1CAIS.03412

Boice, R. (1990). *Professors as writers: A self-help guide to productive writing*. New Forums Press.

Boote, D. R., & Beile, P. (2005). *Researchers guide to critical thinking: Synthesizing research in literature reviews*. Prentice Hall.

Boshoff, N., & De Jong, S. P. (2015). Building research capacity at a developing university: The role of a research proposal writing course. *South African Journal of Science*, 111(11/12), 1–9.

Boud, D., & Lee, A. (2005). 'Peer learning' as pedagogic discourse for research education. *Studies in Higher Education*, 30(5), 501–516. DOI: 10.1080/03075070500249138

Bowen, G. A. (2009). Document analysis as a qualitative research method. *Qualitative Research Journal*, 9(2), 27–40. DOI: 10.3316/QRJ0902027

Bowling, A. (2002). *Research methods in health: investigating health and health services*. Open University Press.

Braun, V., & Clarke, V. (2006). Using Thematic Analysis in Psychology. *Qualitative Research in Psychology*, 3(2), 77–101. DOI: 10.1191/1478088706qp063oa

Brink, H., van der Walt, C., & van Rensburg, G. (2018). *Fundamentals of research methodology for healthcare professionals*. Juta & Co.

Brock-Utne, B. (2016). *Language of instruction and student performance: New insights from research in Africa and elsewhere*. Nordiska Afrikainstitutet.

Brown, A. (2020). *Foundations of higher education: Policy and practice*. University Press.

Brown, A., & Green, B. (2020). *The Role of Academic Authors in Research Publishing*. Academic Press.

Brown, J., & Johnson, M. (2017). Journal editorials: Insights into academic publishing strategies. *Journal of Academic Publishing*, 45(2), 112–130.

Brown, S. (2019). Methodological preferences in research area selection: A comparative analysis. *Research Methodologies Journal*, 22(3), 345–367.

Brown, T. D., Patel, V. B., & Jenkins, K. E. (2023). *Quantitative research methods for business students*. Pearson Education.

Bryman, A., & Bell, E. (2020). *Business research methods*. Oxford University Press.

Bunting, I., Cloete, N., & Van Schalkwyk, F. (2014). *An empirical overview of eight flagship universities in Africa: 2001-209*. African Minds.

Busza, J. (2004). Participatory research in constrained settings: Sharing challenges from Cambodia. *Action Research*, 2(2), 191–208. DOI: 10.1177/1476750304043730

Campbell, T. A., & Campbell, D. E. (2007). Outcomes of mentoring at-risk college students: Gender and ethnic matching effects. *Mentoring & Tutoring*, 15(2), 135–148. DOI: 10.1080/13611260601086287

Cardilini, A. P., Risely, A., & Richardson, M. F. (2021). Supervising the PhD: Identifying common mismatches in expectations between candidate and supervisor to improve research training outcomes. *Higher Education Research & Development*, •••, 1–15.

Carless, D. (2020). Longitudinal perspectives on students' experiences feedback: A need for teacher-student partnership. *Higher Education Research & Development*, 39(3), 425–498. DOI: 10.1080/07294360.2019.1684455

Carton, D. J. (2023, May 4). *Thing 8: Building Professional Relationships in Research Supervision*. Retrieved from May 4, 2023: https://www.23thingsinternational.com/post/thing-9-building-professional-relationships-in-research-supervision#:~:text=the%20fundamental%20touchstone%20for%20all,course%20of%20their%20research%20programm

Castrillón, J. C. A. (2020). Raise and formulate a research problem: An exercise of reasoning. *Revista Lasallista de Investigacion*, 17(1), 301–313. DOI: 10.22507/rli.v17n1a4

Cekiso, M., Tshotsho, B., Masha, R., & Saziwa, T. (2019). Supervision Experiences of Postgraduate Research Students at One South African Higher Education Institution. *South African Journal of Higher Education, 33(3), 8-25., 33*(3), 8-25. DOI: 10.20853/33-3-2913

Celikates, R. (2019). Critical Theory and the Unfinished Project of Mediating Theory and Practice. In Gordon, P. E., Hammer, E., & Honneth, A. (Eds.), *The Routledge Companion to the Frankfurt School* (pp. 206–220). Routledge.

Chambers, M. M. (1960). Selection, Definition, and Delimitation of a Doctoral Research Problem. *Phi Delta Kappan*, 42(2), 71–73. https://www.jstor.org/stable/20342507

Chamnan, S., & Ford, N. (2018). Higher education in Cambodia: Perspectives of students and faculty. *Journal of Asian Public Policy*, 11(2), 199–213.

Chandere, V., Satish, S., & Lakshminarayanan, R. (2021). Online Plagiarism Detection Tools in the Digital Age: A Review. *Annals of the Romanian Society for Cell Biology*, •••, 7110–7119. http://annalsofrscb.ro/index.php/journal/article/view/881

Chanty. (2023, June 8). *Implementing the Transactional Model of Communication for Better Collaboration*. Retrieved from Chanty: https://www.chanty.com/blog/transactional-model-of-communication/

Charmaz, K. (2006). *Constructing Grounded Theory: a practical guide through qualitative analysis*. Sage.

Chea, C. (2020). Enhancing research culture in Cambodian universities. *Asia Pacific Education Review*, 21(4), 623–635.

Chen, C. Y., Sok, P., & Sok, K. (2007). Benchmarking potential factors leading to education quality: A study of Cambodian higher education. *Quality Assurance in Education*, 15(2), 128–148. DOI: 10.1108/09684880710748901

Chen, D. H., & Dahlman, C. J. (2005). *The knowledge economy, the KAM methodology and World Bank operations*. World Bank Institute.

Chilisa, B. (2017). Decolonizing Transdisciplinary Research Approaches: An African Perspective for Enhancing Knowledge Integration in Sustainability Science. *Sustainability Science*, 12(1), 811–823.

Chilisa, B. (2017). *Indigenous research methodologies*. Sage Publications.

Chin, P., Em, S., & Phann, S. (2023). Enhancing research knowledge and productivity through professorship titles in Cambodia: Striking a balance. *Jurnal As-Salam*, 7(2), 133–143. DOI: 10.37249/assalam.v7i2.653

Chugh, R., Macht, S., & Harreveld, B. (2022). Supervisory feedback to postgraduate research students: A literature review, Assessment & Evaluation in. *Assessment & Evaluation in Higher Education*, 47(5), 683–697. DOI: 10.1080/02602938.2021.1955241

Clark, E. (2020). *Writing successful research proposals*. Academic Insights.

Clayton, T. (1998). Building the new Cambodia: Educational destruction and construction under the Khmer Rouge, 1975–1979. *History of Education Quarterly*, 38(1), 1–16. DOI: 10.2307/369662

Clayton, T. (2005). Re-orientations in moral education in Cambodia since 1975. *Journal of Moral Education*, 34(4), 505–517. DOI: 10.1080/03057240500410236

Cloete, N., Bailey, T., & Maassen, P. (2015). *Knowledge production and contradictory functions in African higher education*. African Minds. DOI: 10.47622/978-1-920677-85-5

Cluett, E. R., & Bluff, R. (2002). *Principles and practice of midwifery research*. Churchill Livingstone.

Cook, D. A. (2016). Twelve tips for getting your manuscript published. *Medical Teacher*, 38(1), 41–50. DOI: 10.3109/0142159X.2015.1074989 PMID: 26372399

Cooksey, R., & McDonald, G. (2019). Surviving and Thriving in Postgraduate Research. In *Surviving and Thriving in Postgraduate Research*. DOI: 10.1007/978-981-13-7747-1

Cooper, H. (2011). *Research synthesis and meta-analysis: A step-by-step approach*. Sage Publications.

Cresswell, J. W. (2002). *Research design: qualitative, quantitative and mixed methods approaches*. Sage Publications.

Creswell, J. W. (2014). *Research design: Qualitative, quantitative, and mixed methods approach* (4th ed.). Sage Publications.

Creswell, J. W., & Creswell, J. D. (2017). *Research design: Qualitative, quantitative, and mixed methods approaches* (5th ed.). SAGE Publications.

Creswell, J. W., & Creswell, J. D. (2018). *Research Design: Quantitative, Qualitative, Mixed Methods*. Sage Publications.

Creswell, J. W., & Plano Clark, V. L. (2020). *Designing and conducting mixed methods research*. Sage Publications.

Crisp, G., & Cruz, I. (2009). Mentoring college students: A critical review of the literature between 1990 and 2007. *Research in Higher Education*, 50(6), 525–545. DOI: 10.1007/s11162-009-9130-2

Cronin, B. (2013). Self-plagiarism: An odious oxymoron. *Journal of the American Society for Information Science and Technology*, 64(5), 873. DOI: 10.1002/asi.22966

Crossman, P. (2004). Knowledge society and higher education: The case of South Africa. *Comparative Education*, 40(4), 525–539.

Cutillas, A., Benolirao, E., Camasura, J., Golbin, R.Jr, Yamagishi, K., & Ocampo, L. (2023). Does Mentoring Directly Improve Students' Research Skills? Examining the Role of Information Literacy and Competency Development. *Education Sciences*, 13(7), 1–21. DOI: 10.3390/educsci13070694

Dar, M. A., & Dar, H. M. (2019). contribution of effective communication in educational settings: A critical review. *International Journal of Research in all Subjects in Multi Languages, 7(8) 13-20.*, *7*(20), 13-20.

Davies, M. B. (2007). *Doing a successful research project: using qualitative or quantitative methods*. Palgrave Macmillan.

Davis, C. (2023). *Exploring Research Areas: A Guide for Scholars*. Springer.

Davis, L. (2021). *Advanced studies: A guide to postgraduate education*. Scholarly Press.

Deci, E. L., & Ryan, R. M. (2008). Facilitating optimal motivation and psychological well-being across life's domains. *Canadian Psychology*, 49(1), 14–23. DOI: 10.1037/0708-5591.49.1.14

Dei, G. J. S. (2000). Rethinking the role of indigenous knowledges in the academy. *International Journal of Inclusive Education*, 4(2), 111–132. DOI: 10.1080/136031100284849

Dei, G. J. S. (2018). Reframing the Dialogue on Indigenous Knowledge Systems: Implications for Education Policy and Practice. *Journal of Education Policy*, 33(4), 437–454.

Delamont, S., Atkinson, P., & Parry, O. (2000). *The doctoral experience: Success and failure in graduate school*. Routledge.

Dellavalle, R. P., Banks, M. A., & Ellis, J. I. (2007). Frequently asked questions regarding self-plagiarism: How to avoid recycling fraud. *Journal of the American Academy of Dermatology*, 57(3), 527. DOI: 10.1016/j.jaad.2007.05.018 PMID: 17707155

DeNisi, A. S. (1994). Is relevant research irrelevant?: On evaluating the contribution of research to management practice. *Journal of Managerial Issues*, 6(2), 145.

Denzin, N. K., & Lincoln, Y. S. (2018). *The landscape of qualitative research*. Sage Publications.

Deuchar, R. (2008). Facilitator, director or critical friend?: Contradiction and congruence in doctoral supervision styles. *Teaching in Higher Education*, 13(4), 489–500. DOI: 10.1080/13562510802193905

Dissanayake, D. M. N. S. W. (2013). Research, Research Gap and the Research Problem. *Munich Personal RePEc Archive*, 47519, 1–5.

Drabble, S. J., O'Cathain, A., Thomas, K. J., Rudolph, A., & Hewison, J. (2014). Describing qualitative research undertaken with randomized controlled trials in grant proposals: A documentary analysis. *BMC Medical Research Methodology*, 14(1), 1–11. DOI: 10.1186/1471-2288-14-24 PMID: 24533771

Dredge, D. (2015). Does relevance matter in academic policy research? *Journal of Policy Research in Tourism, Leisure & Events*, 7(2), 173–177. DOI: 10.1080/19407963.2014.990661

Drew, C. (2023, July 22). *Home: All 8 Models of Communication, Explained!* Retrieved May 06, 2024, from Helpful Professor.com: https://helpfulprofessor.com/communication-models/

Dugartsyrenova, V. A., & Sardegna, V. G. (2022). Enhancing genre instruction on research proposal introductions with an online academic writing tutor. *Journal of Second Language Writing*, 58, 100908. DOI: 10.1016/j.jslw.2022.100908

Dutton, G. H. (1999). Introduction and problem statement. *A Hierarchical Coordinate System for Geoprocessing and Cartography*, 1-18.

Dy, S. S. (2004). Strategies and policies for basic education in Cambodia: Historical perspectives. *International Education Journal*, 5(1), 90–97.

Eam, P. (2018). Current state, key challenges and ways forward for Cambodian research capacities: A review of four studies. *Cambodia Education Review*, 2(1), 108–143.

Eam, P., Keo, B., Leng, P., Song, S., & Khieng, S. (2021). Correlates of STEM major choice: A quantitative look at Cambodian university freshmen. *Research in Science & Technological Education*, 39(2), 206–224. DOI: 10.1080/02635143.2019.1682987

Easterby-Smith, M., Jaspersen, L. J., Thorpe, R., & Valizade, D. (2021). Management and business research. *Sage (Atlanta, Ga.).*

Eccles, J. S., & Wigfield, A. (2002). Motivational beliefs, values, and goals. *Annual Review of Psychology*, 53(1), 109–132. DOI: 10.1146/annurev.psych.53.100901.135153 PMID: 11752481

Edelson, D. C. (2002). Design research: What we learn when we engage in design. *Journal of the Learning Sciences*, 11(1), 105–121. DOI: 10.1207/S15327809JLS1101_4

Ehrich, L. C., Hansford, B., & Tennent, L. (2004). Formal mentoring programs in education and other professions: A review of the literature. *Educational Administration Quarterly*, 40(4), 518–540. DOI: 10.1177/0013161X04267118

Ellis, T. J., & Levy, Y. (2009). Towards a Guide for Novice Researchers on Research Methodology: Review and Proposed Methods. *Issues in Informing Science & Information Technology, 6*.

Em, S. (2023a, August 6). *The issue of title acquisition through proxy writing in Cambodian academia*. Cambodianess.

Em, S. (2023b, December 27). *Concerns of professor nominations and self-nominated professors in Cambodia*. English Academic Essay (EAE).

Em, S. (2024). Groundbreaking issue of the Cambodian Journal of Educational and Social Sciences (CJESS). [CJESS]. *Cambodian Journal of Educational and Social Sciences*, 1(1), 1–25. DOI: 10.69496/cjess.v1i1.4

Em, S., Khan, S., & Nun, N. (2022). Education system in Cambodia: A brief review from the prehistoric period to the present, and an education strategic plan for the future. *FIRE: Forum for International Research in Education*, 7(3), 141–164. DOI: 10.32865/fire202273320

Englund, R. L., & Bucero, A. (2006). *Project sponsorship: Achieving management commitment for project success*. John Wiley & Sons.

Evans, C., & Green, C. (2007). Why did I call you a genius? Or whose ethics is it anyway? Moving beyond a dualistic approach to debates about ethical issues in supervision. *International Journal of Qualitative Studies in Education : QSE*, 20(6), 723–729.

Evans, C., Kandiko Howson, C., Forsythe, A., & Edwards, C. (2021). What constitutes high-quality higher education pedagogical research? *Assessment & Evaluation in Higher Education*, 46(4), 525–546. DOI: 10.1080/02602938.2020.1790500

Fandino, W. (2019, August). Formulating a good research question: Pearls and pitfalls. *Indian Journal of Anaesthesia*, 63(8), 611–616. DOI: 10.4103/ija.IJA_198_19 PMID: 31462805

Faryadi, Q. (2018). PhD thesis writing process: A systematic approach--How to write your introduction. *Online Submission, 9*, 2534-2545.

Fauth, G. C. (2008). Women in Educational Administration: A Research Profile. *The Educational Forum*, •••, 49.

Field, A. (2018). *Discovering statistics using IBM SPSS Statistics* (5th ed.). SAGE Publications.

Fink, A. (2020). *Conducting research literature reviews: From the Internet to paper*. Sage Publications.

Firth, A., & Martens, E. (2008). Transforming supervisors? A critique of post-liberal approaches to research supervision. *Teaching in Higher Education*, 13(3), 279–289. DOI: 10.1080/13562510802045303

Flick, U. (2022). *The SAGE handbook of qualitative research design*. Sage. DOI: 10.4135/9781529770278

Fraenkel, J. R., Wallen, N. E., & Hyun, H. H. (2012). *How to design and evaluate research in education* (8th ed.). McGraw-Hill.

Freire, P. (1970). Pedagogy of the Oppressed. *Continuum : an Interdisciplinary Journal on Continuity of Care*.

Frumkin, P. (2016). *Strategic giving: The art and science of philanthropy*. University of Chicago Press.

Fry, G., & Attawell, K. (2020). Evaluating the impact of a research proposal development program: Experiences from the African continent. *Research Evaluation*, 29(1), 78–90.

Gall, M. D., Gall, J. P., & Borg, W. R. (2020). *Educational research: An introduction*. Pearson Education.

Garcia, D., & Gluesing, J. C. (2013). Qualitative research methods in international organizational change research. *Journal of Organizational Change Management*, 26(2), 423–444. DOI: 10.1108/09534811311328416

Garcia, M., & Smith, J. (2017). Integrating ecological resilience with urban design for climate change mitigation. *Environmental Science & Urban Planning Review*, 25(2), 112–130.

Gardner, S. K., Mendoza, P., & Austin, A. E. (2023). ON BECOMING A SCHOLAR: Socialization and Development in Doctoral Education. In *On Becoming a Scholar: Socialization and Development in Doctoral Education*. Taylor \& Francis. DOI: 10.4324/9781003446187

Gargouri, Y., Hajjem, C., Larivière, V., Gingras, Y., Carr, L., Brody, T., & Harnad, S. (2010). Self-Selected or Mandated, Open Access Increases Citation Impact for Higher Quality Research. *PLoS One*, 5(10), e13636. DOI: 10.1371/journal.pone.0013636 PMID: 20976155

Gazzola, N., De Stefano, J., Theriault, A., & Audet, C. (2014). Positive experiences of doctoral-level supervisors-in-training conducting group-format supervision: A qualitative investigation. *British Journal of Guidance & Counselling*, 42(1), 26–42. DOI: 10.1080/03069885.2013.799263

Germano, W. P. (2013). From dissertation to book (Second). University of Chicago Press.

Germano, W. P. (2016). *Getting It Published: A Guide for Scholars and Anyone Else Serious about Serious Books (Third)*. University of Chicago Press. DOI: 10.7208/chicago/9780226281544.001.0001

Ghosh, U. (2022). Role of communication in education. *Journal for Research Trends and Innovation*, 7(6), 326–329.

Giroux, H. A. (2018). Critical Theory and Educational Practice. *Educational Theory*, 68(2), 145-158.Fanon, F. (2008). Black Skin, White Masks. Grove Press.

Gleasure, R. (2013). What is a "Wicked Problem" for IS Research? *The 2nd International SIG Prag Workshop on "IT Artefact Design & Workpractice Improvement," Conklin 2008*, 1–12. http://www.vits.org/adwi2013/RGleasure-ADWI2013.pdf %0Awww.vits.org/adwi2013/RGleasure-ADWI2013.pdf

Gonzalez, D., Abdel Sattar, R., Budhathoki, R., Carrard, N., Chase, R. P., Crawford, J., Halcrow, G., Kozole, T., MacArthur, J., Nicoletti, C., Toeur, V., Basnet, M. P., Chhetri, A., Gurung, H., Yadav, A., Vourchnea, P., & Willetts, J. (2022). A partnership approach to the design and use of a quantitative measure: Co-producing and piloting the WASH gender equality measure in Cambodia and Nepal. *Development Studies Research*, 9(1), 142–158. DOI: 10.1080/21665095.2022.2073248

Gordon, R. A., & Howell, J. E. (1959). Higher Education for Business. In *Higher Education for Business*. Columbia University Press., DOI: 10.7312/gord91408

Govender, N. (2023). Govender, N. (2023). Attributes and Relationships in the Supervisor-Supervisee Partnership: An autoethnographic Study. In Samuel, M. A., & Mariaye, H. (Eds.), *Transforming Postgraduate in Africa. In M. Samuel, & H. Mariaye*. Transforming Postgraduate in Africa.

Gramsci, A. (1971). *Selections from the Prison Notebooks*. International Publishers.

Grant, C. (2018). How to write research proposal.

Grant, B. (2003). Mapping the pleasures and risks of supervision. *Discourse (Abingdon)*, 24(2), 175–190. DOI: 10.1080/01596300303042

Grant, C., & Booth, A. (2009). *Designing a research proposal: Choices and challenges* (2nd ed.). Sage Publications.

Gray, D. E. (2010). Towards the lifelong skills and business development of coaches: An integrated model of supervision and mentoring. *Coaching (Abingdon, UK)*, 3(1), 60–72. DOI: 10.1080/17521880903559705

Gray, D. E. (2017). *Doing Research in the Real World*. Sage Publications.

Gray, L., & Smith, T. (2021). Theoretical frameworks in document analysis: A comparative study. *Journal of Document Analysis*, 28(5), 621–643.

Greenhalgh, T. (2019). *How to read a paper: The basics of evidence-based medicine*. Wiley- Blackwell.

Greenhow, C., & Galvin, S. (2020). Teaching with social media: Evidence-based strategies for making remote higher education less remote. *Information and Learning Science*, 121(7/8), 513–524. DOI: 10.1108/ILS-04-2020-0138

Green, R.. (2018). Research policy documents and their impact on academic research agendas. *Policy Studies Journal: the Journal of the Policy Studies Organization*, 36(4), 567–589.

Griffiths, V., Thompson, S. & Hrynie, L. (2010). Developing a research profile: mentoring and support for teacher educators. *Professional Development in Education*. (30), Issue 1-2.

Grove, J., Ripke, S., Als, T. D., Mattheisen, M., Walters, R. K., Won, H., Pallesen, J., Agerbo, E., Andreassen, O. A., Anney, R., Awashti, S., Belliveau, R., Bettella, F., Buxbaum, J. D., Bybjerg-Grauholm, J., Bækvad-Hansen, M., Cerrato, F., Chambert, K., Christensen, J. H., & Børglum, A. D.ChicagoGrove. (2019). Identification of common genetic risk variants for autism spectrum disorder. *Nature Genetics*, 51(3), 431–444. DOI: 10.1038/s41588-019-0344-8 PMID: 30804558

Gruwell, C. (2022). *Critical thinking in academic research*. Open Textbook Library.

Gupta, A. K., Sharma, S.K. and Kumar, R. (2020). "The Impact of Assumptions on Research Proposals: A Systematic Review. Journal: Journal of Research Methodology. DOI: DOI: 10.1080/16138776.2020.1722349

Gurr, G. M. (2001). Negotiating the 'rackety bridge'—A dynamic model for aligning supervisory style with research student development. *Higher Education Research & Development*, 20(1), 81–92. DOI: 10.1080/07924360120043882

Guthrie, J., Parker, L. D., & Gray, R. (2004). Requirements and understandings for publishing academic research: an insider view. Elsevier., 3/50026-8.DOI: 10.1016/B978-008043972-

Habermas, J. (1984). The Theory of Communicative Action, Volume 1: Reason and the Rationalization of Society. Beacon Press.

Hackman, J. R., & Oldham, G. R. (1976). Motivation through the design of work: Test of a theory. *Organizational Behavior and Human Performance*, 16(2), 250–279. DOI: 10.1016/0030-5073(76)90016-7

Hair, J. F., Black, W. C., Babin, B. J., & Anderson, R. E. (2019). *Multivariate data analysis* (8th ed.). Cengage Learning.

Ha, J. F., & Longnecker, N. (2010). Nurse-patient communication: A review. *The Ochsner Journal*, 10(1), 38–43. PMID: 21603354

Halse, C., & Malfroy, J. (2010). Retheorizing doctoral supervision as professional work. *Studies in Higher Education*, 35(1), 79–92. DOI: 10.1080/03075070902906798

Hamid, M. O., Hanna, B., Gannaway, D., & Nguyen, T. T. T. (2024). The ethics and epistemology of researching higher degree by research supervision: An encounter with institutional ethics review. *International Journal of Research & Method in Education*, •••, 1–14. DOI: 10.1080/1743727X.2023.2299973

Hancock, D. R., Algozzine, B., & Lim, J. H. (2021). Doing case study research: A practical guide for beginning researchers.

Harman, E., Montagnes, I., McMenemy, S., & Bucci, C. (Eds.). (2003). *The Thesis and the Book: A Guide for First-Time Academic Authors*. University of Toronto Press., https://www.jstor.org/stable/10.3138/9781442689350 DOI: 10.3138/9781442689350

Harris, R. (2015). The Impact of research on development policy and practice: This much we know. In Chib, A., May, J., & Barrantes, R. (Eds.), *Impact of Information Society Research in the Global South*. Springer., DOI: 10.1007/978-981-287-381-1_2

Harris, S. K., Smith, J. R., & Johnson, K. (2023). *Research methods in education: An introduction to qualitative and quantitative research methods*. Routledge.

Hart, C. (2018). *Doing a literature review: Releasing the research imagination*. Sage Publications.

Hart, J., Hakim, J., Kaur, R., Jeremy, R., Coorey, G., Kalman, E., Jenkin, R., & Bowen, D. (2022). Research supervisors' views of barriers and enablers for research projects undertaken by medical students; a mixed methods evaluation of a post-graduate medical degree research project program. *BMC Medical Education*, 22(1), 370. DOI: 10.1186/s12909-022-03429-0 PMID: 35562832

Hartley, J., & Jory, S. (2015). Lifting the veil on the viva: The experience of Psychology PhD candidates in the UK. *Psychology Teaching Review*, 9(2), 79–90.

Harvard Graduate School of Education. (2020). Writing a Research Proposal. https://cepr.harvard.edu/proposal-writing-tips

Harzing, A. W. (2010). *The publish or perish book: Your guide to effective and responsible citation analysis*. Tarma Software Research.

Hayes, S., Jopling, M., Hayes, D., Westwood, A., Tuckett, A., & Barnett, R. (2021). Raising regional academic voices (alongside data) in Higher Education (HE) debate. *Postdigital Science and Education, 3*, 242-260. https://link.springer.com/article/10.1007/s42438-020-00131-6

Heath, M., & Tynan, C. (2010). Crafting a research proposal. *The Marketing Review*, 10(2), 147–168. DOI: 10.1362/146934710X505753

Heath, T. (2002). A quantitative analysis of PhD students' views of supervision. *Higher Education Research & Development*, 21(1), 41–53. DOI: 10.1080/07294360220124648

Helldén, D., Sok, S., Chea, T., Nordenstedt, H., Kuruvilla, S., Alvesson, H. M., & Alfvén, T. (2023). Sustainable development goals and multisectoral collaborations for child health in Cambodia: A qualitative interview study with key child health stakeholders. *BMJ Open*, 13(11), 1–9. DOI: 10.1136/bmjopen-2023-073853 PMID: 37989366

Heng, K. (2022, February 21). *Cambodian academics' conceptions of research and the research-teaching nexus: A qualitative analysis*. Cambodian Education Forum.

Heng, K., & Sol, K. (2022). Education: Key to making Cambodia great again. *Cambodia Development Center, 4*(3), 1-18.

Heng, K., Sol, K., Kaing, S., & Ros, V. (2020). *Introduction: The need to promote a strong research and publication culture in Cambodia*. Cambodian Education Forum.

Heng, K., Sol, K., Kaing, S., & Ros, V. (2021). *Introduction: Engaging and supporting Cambodian youth to write for publication in English*. Cambodian Education Forum.

Heng, K. (2024). Challenges and developments in university research in Cambodia: A case study of two universities. *Higher Education*, 87(6), 1593–1613. https://link.springer.com/article/10.1007/s10734-023-01080-2. DOI: 10.1007/s10734-023-01080-2

Heng, K., Hamid, M. O., & Khan, A. (2022). Academics' conceptions of research and the research-teaching nexus: Insights from Cambodia. *International Journal of Educational Development*, 90, 102569. DOI: 10.1016/j.ijedudev.2022.102569

Heng, K., Hamid, M. O., & Khan, A. (2023). Research engagement of academics in the Global South: The case of Cambodian academics. *Globalisation, Societies and Education*, 21(3), 322–337. DOI: 10.1080/14767724.2022.2040355

Heng, K., & Heng, K. (2023). The implementation of the professorial ranking system in Cambodia: Opportunities and challenges. *Cambodian Journal of Educational Research*, 3(2), 62–77. DOI: 10.62037/cjer.2023.03.02.04

Heng, K., & Sol, K. (2021). Academic research in Cambodia: Progress, challenges, and ways forward. *Cambodian Journal of Educational Research*, 1(2), 6–23. DOI: 10.62037/cjer.2021.01.02.02

Heng, K., Sol, K., & Em, S. (2023). COVID-19 and digital transformation of Cambodian higher education: opportunities, challenges, and the way forward. In Ordóñez de Pablos, P., Zhang, X., & Almunawar, M. N. (Eds.), *Handbook of research on education institutions, skills, and jobs in the digital era* (pp. 307–327). IGI Global., DOI: 10.4018/978-1-6684-5914-0.ch018

Heng, M. (2022). Cambodia's gender policy: An analysis of women and girls' education strategies under the Neary Rattanak IV Strategic Plan (2014-2018). *Cambodian Journal of Educational Research*, 2(1), 37–61. DOI: 10.62037/cjer.2022.02.01.03

Henson, K. T. (2011). *Writing for academic publication*. Pearson.

Henze, R. (2010). Cultural diversity and its impact on global university administration: The case of leadership and staff development. *International Journal of Educational Management*, 24(6), 445–455.

Hernandez, M. (2010). Promoting inclusion and diversity in the academic workplace: A pilot intervention. *Equality, Diversity and Inclusion*, 29(3), 321–334.

Herwix, A., & Haj-Bolouri, A. (2021). Revisiting the Problem of the Problem - An Ontology and Framework for Problem Assessment in IS Research. *ECIS 2021 Proceedings, June*, 154. https://aisel.aisnet.org/ecis2021_rp/154

Herzberg, F., Mausner, B., & Snyderman, B. B. (1959). *The motivation to work* (2nd ed.). John Wiley & Sons.

Hewson, D., & Carroll, M. (2016). *Reflective practice in supervision: companion volume to the reflective supervision toolkit*. Moshpit Publishing.

Higgs, P. (2016). African philosophy and the decolonisation of education in Africa: Some critical reflections. *Educational Philosophy and Theory*, 48(4), 336–346.

Hizriani, N., Khairatunnisa, K., Nor, H., & Handrianto, C. (2022). Online and offline students' presentation in thesis proposal: Challenges and solution. *Randwick International of Social Science Journal, 3(3)*, 549-560.

Hockey, J. (1996). Strategies and tactics in the supervision of UK social science PhD students. *International Journal of Qualitative Studies in Education : QSE*, 9(4), 481–500. DOI: 10.1080/0951839960090409

Horkheimer, M., & Adorno, T. W. (2002). *Dialectic of Enlightenment: Philosophical Fragments*. Stanford University Press.

Houghton, J., et al. (2009). Economic implications of alternative scholarly publishing models: Exploring the costs and benefits. JISC Report.

Howard, P. J., Gorzycki, M., Desa, G., & Allen, D. D. (2018). Academic reading: Comparing students' and faculty perceptions of its value, practice, and pedagogy. *Journal of College Reading and Learning*, 48(3), 189–209. DOI: 10.1080/10790195.2018.1472942

Huang, R. (2012). International experience and graduate employability: Perceptions of Chinese international students in the UK. *Journal of Studies in International Education*, 16(3), 355–368.

Hum, C., & Heng, K. (2024). English in Cambodia. In Moody, A. J. (Ed.), *The Oxford Handbook of Southeast Asian Englishes* (pp. 186–201). Oxford University Press., DOI: 10.1093/oxfordhb/9780192855282.013.8

Indrawati, S. M., & Kuncoro, A. (2021). Improving competitiveness through vocational and higher education: Indonesia's vision for human capital development in 2019–2024. *Bulletin of Indonesian Economic Studies*, 57(1), 29–59. DOI: 10.1080/00074918.2021.1909692

Iqbal, J. (2007). Learning from a doctoral research project: Structure and content of a research proposal. *Electronic Journal of Business Research Methods*, 5(1), 11–20.

Ireland assignments. (2020, April 23). *Home: blog/dissertation writing guide/ Methods Of Communicating Effectively With The Ph.D. Dissertation Supervisor*. Retrieved from Ireland Assignment Help.com: https://www.irelandassignmenthelp .com/blogs/methods-communicating-effectively-phd-supervisor/

Ives, G., & Rowley, G. (2005). Supervisor selection or allocation and continuity of supervision: PhD. students' progress and outcomes. *Studies in Higher Education*, 30(5), 535–555. DOI: 10.1080/03075070500249161

J. Ellis, T., & Levy, Y. (2008). Framework of Problem-Based Research: A Guide for Novice Researchers on the Development of a Research-Worthy Problem. *Informing Science: The International Journal of an Emerging Transdiscipline, 11*(January), 017–033. DOI: 10.28945/438

Jadad, A. R., & Enkin, M. (2007). *Randomized Controlled Trials: questions, answers and musings*. Blackwell Publishing Ltd. DOI: 10.1002/9780470691922

Jalongo, M. R. (2023). Choosing a Topic: Focus and Title. In *Scholarly Writing: Publishing Manuscripts That Are Read, Downloaded, and Cited* (pp. 91–114). Springer International Publishing. DOI: 10.1007/978-3-031-39516-1_5

Jan Mendoza, C. (2020). *Home: Academia: Communication and Communication Models*. Retrieved from Academia.edu website: https://www.academia.edu/ 44018864/Communication_and_Communication_sssss

Janeiro, M. G., & González, M. G. (2015). Developing Students' Autonomy And Self-Regulation Through A Co-Teaching Research Methods Experience. *Contemporary Issues in Education Research*, 8(1), 25–32. DOI: 10.19030/cier.v8i1.9059

Japheth, N., Ssentamu, P., & Wambua, B. (2023). Strategies Used for Effective Research Supervision in the Completion of Postgraduate Studies in Selected Universities of Uganda. *Journal of the National Council for Higher Education Vol. 10, Issue 2, 2023 1, 10*(2).

Jenkins, K. E. H., Stephens, J. C., Reames, T. G., & Hernández, D. (2020). Towards impactful energy justice research: Transforming the power of academic engagement. *Energy Research and Social Science, 67*(October 2019), 101510. DOI: 10.1016/j. erss.2020.101510

Johnson, B. W., Rose, G., & Schlosser, L. Z. (2007). Student–faculty mentoring: Theoretical and methodological issues. The Blackwell handbook of mentoring: A multiple perspectives approach, 49-68.

Johnson-Bailey, J., & Cervero, R. M. (2004). Mentoring in black and white: The intricacies of cross-cultural mentoring. *Mentoring & Tutoring*, 12(1), 7–21. DOI: 10.1080/1361126042000183075

Johnson, E. (2018). Exploring novel therapeutic targets in Alzheimer's disease: A neuroscientific approach. *Journal of Neuroscience Research*, 42(3), 311–328.

Johnson, E. (2019). *Research Methodologies in Academic Inquiry*. Oxford University Press.

Johnson, E., & Brown, S. (2019). Targeting niche areas in Alzheimer's research: From discovery to clinical implications. *Neuroscience Advances*, 18(1), 89–107.

Johnson, R. B., & Christensen, L. B. (2020). *Educational research: Quantitative, qualitative, and mixed approaches*. Pearson Education.

Johnson, S. L., Fitzsimmons, K. J., & Dingwall, R. (2014). The role of niche research in scientific advancement. *Research Policy*, 43(2), 432–445.

Johnson, W. B. (2015). *On being a mentor: A guide for higher education faculty*. Routledge. DOI: 10.4324/9781315669120

Joireman, J., & Van Lange, P. A. (2015). The context of research in the behavioral sciences: Facing and embracing complexity. *The American Psychologist*, 70(3), 213–221. DOI: 10.1037/a0038999

Jones, A.. (2020). Empirical approaches in selecting research areas: A cross-disciplinary analysis. *Research Perspectives*, 18(1), 89–107.

Jones, R. (2022). *Overcoming challenges in higher education*. Boston, MA: Education Publishers.Komba, S. C. (2016). Challenges of writing theses and dissertations among postgraduate students in Tanzanian higher learning institutions. *International Journal of Research Studies in Education*, 5(3), 71–80.

Jones, R.. (2021). *Crafting Scholarly Articles: A Comprehensive Roadmap*. Routledge.

Kafu, P. A. (2023). Research proposal and thesis/dissertation writing course in modern higher education, an emerging issue in sustainable development in the modern world. *African Journal of Education, Science and Technology, 7*(3), 442-449.

Kam, B. H. (1997). Style and quality in research supervision: The supervisor dependency factor. *Higher Education*, 34(1), 81–103. DOI: 10.1023/A:1002946922952

Katikireddi, S. V., & Reilly, J. (2017). Characteristics of good supervision: A multi-perspective qualitative exploration of the Masters in Public Health dissertation. *Journal of Public Health (United Kingdom), 39(3), pp.625 632. doi:, 39*(3), 625-632.DOI: 10.1093/pubmed/fdw

Kaya, H. O., & Seleti, Y. N. (2013). African indigenous knowledge systems and relevance of higher education in South Africa. *International Education Journal: Comparative Perspectives*, 12(1), 30–44.

Keane, M. (2016). Supervising PhD students: A practical guide and toolkit. *Innovations in Education and Teaching International*, 53(4), 496–498.

Keet, A. (2018). Engaging Critical Pedagogy in South Africa: Decolonization, Education, and the Politics of Knowledge. *Education as Change*, 22(1), 1–16. DOI: 10.17159/1947-9417/2017/2741

Kelly, A., & Kelly, M. (2018). *The Research Process: A Practical Guide*. Sage Publications.

Khan Academy. (n.d.). What is a SMART goal? Retrieved from https://www.khanacademy.org/test-prep/mcat/behavioral-sciences/smart-goals/a/what-is-a-smart-goal

Khan, M. A. W. (2023). The art of writing a "Title" for an original research paper. *Bangladesh Journal of Medical Education*, 14(1), 70–77. DOI: 10.3329/bjme.v14i1.63685

Khatri, N., Ojha, A. K., Budhwar, P., Srinivasan, V., & Varma, A. (2012). Management research in India: Current state and future directions. *IIMB Management Review*, 24(2), 104–115. DOI: 10.1016/j.iimb.2012.05.002

Khupe, C., & Keane, M. (2017). Towards an African education research methodology: Decolonising new knowledge. *Educational Research for Social Change*, 6(1), 25–37. DOI: 10.17159/2221-4070/2017/v6i1a3

Kiernan, B. (2002). Introduction: Conflict in Cambodia, 1945-2002. *Critical Asian Studies*, 34(4), 483–495. DOI: 10.1080/1467271022000035893

Kieser, A., & Leiner, L. (2009). Why the rigour-relevance gap in management research is unbridgeable. *Journal of Management Studies*, 46(3), 516–533. DOI: 10.1111/j.1467-6486.2009.00831.x

Killian, K. D. (2013). Differences making a difference: Cross-cultural interactions in supervisory relationships. In *Integrating gender and culture in family therapy training* (pp. 61–103). Routledge.

Kim, H., Lee, J. and Park, S. (2020). "Assumptions in Research Proposals: A Study of Graduate Students' Perceptions. Journal: Journal of Educational Psychology. DOI: DOI: 10.1037/edu0000366

Kim, H., Lee, Y., & Kim, J. (2021). *Research methods for business students.* McGraw-Hill Education.

Kim, S., Sok, P., & Eng, S. (2020). Aligning research with national development goals in Cambodia. *Higher Education Policy*, 33(2), 237–256.

Kincheloe, J. L., McLaren, P., & Steinberg, S. R. (2011). Critical pedagogy and qualitative research: Moving to the bricolage. In *The SAGE Handbook of Qualitative Research* (4th ed., pp. 163–178). Sage Publications.

King, K. P. (2013). *Successful Academic Writing: Five Roadblocks to Overcome.* Wiley., DOI: 10.1002/nha.20019

Kivunja, C. (2016). How to write an effective research proposal for higher degree research in higher education: Lessons from practice. *International Journal of Higher Education*, 5(2), 163–172. DOI: 10.5430/ijhe.v5n2p163

Kong, M. L. F. (2017). On teaching methods: The personal experiences of teachers of English. *RELC Journal*, 48(2), 197–209. DOI: 10.1177/0033688216661251

Koskela, L. (2011). Fifty years of irrelevance: The wild goose chase of management science. *Lean Construction Journal*, 2011(Special Issue), 1–11.

Koskela, L. (2017). Why is management research irrelevant? *Construction Management and Economics*, 35(1–2), 4–23. DOI: 10.1080/01446193.2016.1272759

Kothari, C. R. (2004). *Research methodology: Methods and techniques* (2nd ed.). New Age International.

Kovach, M. (2018). *Indigenous Methodologies: Characteristics, Conversations, and Contexts*. University of Toronto Press.

Krueger, R. A., & Casey, M. A. (2015). Focus group interviewing. In Hatry, H. P., Newcomer, K. E., & Wholey, J. S. (Eds.), *Handbook of practical program evaluation* (pp. 506–534). DOI: 10.1002/9781119171386.ch20

Kuhn, T. S. (2012). *The Structure of Scientific Revolutions*. University of Chicago Press. DOI: 10.7208/chicago/9780226458144.001.0001

Kumar, A., Patel, V. B., & Brown, T. D. (2024). A systematic review of the literature on [topic]. *Journal of Research in Education*, 22(1), 1–15. PMID: 35694371

Kumar, A., Tiwari, S., Chauhan, A. K., & Ahirwar, R. (2019). Impact of NIRF on research publications: A study on top 20 (ranked) Indian Universities. *COLLNET Journal of Scientometrics and Information Management*, 13(2), 219–229. DOI: 10.1080/09737766.2020.1741194

Kumar, R., Kumar, S., & Singh, R. K. (2020). *Research methodology: An introduction*. PHI Learning.

Lai, H. (2024). Selection of a Topic of a Research Proposal of Emerging Academic Writers in a Blended Linked EAP Course: An Interactional Ethnographic Perspective. *TESOL Quarterly*, tesq.3316. DOI: 10.1002/tesq.3316

Lave, J., & Wenger, E. (1991). *Situated learning: Legitimate peripheral participation*. Cambridge University Press. DOI: 10.1017/CBO9780511815355

Lawrence A, Lewis L, Hofmeyr GJ, Dowswell T, Styles C (2009) Maternal positions and mobility during first stage labour. Cochrane Database Syst Rev (2) Issue 2. Art. No. CD003934. https://doi.org/DOI: 10.1002/14651858.CD003934.pub2

Le Ha, P., & Barnawi, O. Z.John Wiley & Sons. (2015). Where English, neoliberalism, desire and internationalization are alive and kicking: Higher education in Saudi Arabia today. *Language and Education*, 29(6), 545–565. DOI: 10.1080/09500782.2015.1059436

Lebeau, Y., & Mills, D. (2008). From 'Crisis' to 'Transformation'? Shifting Orthodoxies of African Higher Education Policy and Research. *African Studies Review*, 51(1), 1–22.

Lee, A. (2018). How can we develop supervisors for the modern doctorate?', Studies in Higher Education. *Taylor & Francis, 33*(5), 878-890.

Lee, A. (2008). How are doctoral students supervised? Concepts of doctoral research supervision. *Studies in Higher Education*, 33(3), 267–281. DOI: 10.1080/03075070802049202

Lee, A. (2012). *Successful research supervision: Advising students doing research*. Routledge. DOI: 10.4324/9780203816844

Lee, S. (2018). *Identifying Gaps in Literature: Strategies and Applications*. Sage Publications.

Lee, S., & Bozeman, B. (2005). The impact of research collaboration on scientific productivity. *Social Studies of Science*, 35(5), 673–702. DOI: 10.1177/0306312705052359

Lee, Y., Kim, H., & Lee, J. (2023). The effects of social media on mental health among adolescents: A systematic review. *Journal of Psychology and Health*, 32(2), 147–162. PMID: 34328044

Leon-Roa, C., Zuñiga-Collazos, A., Castillo, H. S. V., Guarin, H. P., Franco, C. M. G., Gómez, D. C. R., & Acosta, E. B. G. (2024). Valorization of research results for knowledge and technology transfer in public higher education institutions. *Journal of Open Innovation*, 10(1), 1–10. DOI: 10.1016/j.joitmc.2024.100245

Leslie, L. Z. (2010). *Communication research methods in postmodern culture*. Routledge.

Lessing, A. C., & Schulze, S. (2002). Postgraduate supervision and academic support: Students' perceptions. *South African Journal of Higher Education*, 16(2), 139–149. DOI: 10.4314/sajhe.v16i2.25253

Lessing, A. C., & Schulze, S. (2002). Postgraduate supervision and academic support: Students' perceptions. *The Journal of Higher Education*, 16(2), 139–149. DOI: 10.4314/sajhe.v16i2.2525

Leta, J., Glanzel, W., & Thijs, B. (2006). Science in Brazil. Part 2: Sectoral and institutional research profiles. *Scientometrics*, 67(1), 1–2. DOI: 10.1007/s11192-006-0051-y

Levac, D., Colquhoun, H., & O'Brien, K. K. (2010). Scoping studies: Advancing the methodology. *Implementation Science : IS*, 5(1), 1–9. DOI: 10.1186/1748-5908-5-69 PMID: 20854677

Li, A. W. (2023). Using Peerceptiv to support AI-based online writing assessment across the disciplines. *Assessing Writing*, 57, 100746. Advance online publication. DOI: 10.1016/j.asw.2023.100746

Li, A. W., & Hebert, M. (2024). Unpacking an online peer-mediated and self-reflective revision process in second-language persuasive writing. *Reading and Writing*, 37(6), 1545–1573. DOI: 10.1007/s11145-023-10466-8

Liao, Y., Deschamps, F., Loures, E. D. F. R., & Ramos, L. F. P. (2017). Past, present and future of Industry 4.0-a systematic literature review and research agenda proposal. *International Journal of Production Research*, 55(12), 3609–3629. DOI: 10.1080/00207543.2017.1308576

Li, J., & Pilz, M. (2023). International transfer of vocational education and training: A literature review. *Journal of Vocational Education and Training*, 75(2), 185–218. DOI: 10.1080/13636820.2020.1847566

Li, M., Chen, X., & Wang, Y. (2021). *Introduction to statistics for social sciences*. Cengage Learning.

Lingard, L., & Watling, C. (2021). Successfully Navigating the Peer Review Process. In *Story, Not Study: 30 Brief Lessons to Inspire Health Researchers as Writers. Innovation and Change in Professional Education, 19*. Springer., DOI: 10.1007/978-3-030-71363-8_30

Lin, W. Y. C. (2020). Self-plagiarism in academic journal articles: From the perspectives of international editors-in-chief in editorial and COPE case. *Scientometrics*, 123(1), 299–319. DOI: 10.1007/s11192-020-03373-0

Liu, Z., Liu, S. and Zhang, J. (2021). Assumptions in Research Proposals: A Review of the Literature. Journal of Educational Evaluation and Research. DOI: DOI: 10.1037/edr0000354

Locke, E. A., & Latham, G. P. (2002). Building a practically useful theory of goal setting and task motivation. *The American Psychologist*, 57(9), 705–717. DOI: 10.1037/0003-066X.57.9.705 PMID: 12237980

Lund, T. (2022). Research Problems and Hypotheses in Empirical Research. *Scandinavian Journal of Educational Research*, 66(7), 1183–1193. DOI: 10.1080/00313831.2021.1982765

Lyon, R. E. C., & Potkar, K. A. (2011). The supervisory relationship. In *Counselor supervision* (pp. 41–78). Routledge.

Maharani, I. D., & Santoso, D. R. (2023). Enhancing Verb Writing Skills in Eighth-Grade Students with Flashcards: A Classroom Action Research in Indonesia. *Indonesian Journal of Education Methods Development*, 18(3). Advance online publication. DOI: 10.21070/ijemd.v21i3.777

Mahlangu, V. (2021). Exploring Challenges of Supervising Postgraduate Students in Open Distance Learning in Higher Education Settings. In Books, B. C. (Ed.), *New Challenges to Education: Lessons from Around the World*. Bulgarian Comparative Education Society.

Mainhard, T., van der Rijst, R. M., van Tartwijk, J., & Wubbels, T. (2009). A model for the supervisor–doctoral student relationship. *Higher Education*, 58(3), 359–373. DOI: 10.1007/s10734-009-9199-8

Mainstone, F., & Wonnacott, J. (2021). The integrated supervision model: The 4× 4× 4. In *The Routledge International Handbook of Social Work Supervision* (pp. 321–333). Routledge.

Makina, A. (2024). A Critical Reflection of the Feedback Practices in the Postgraduate Supervision in an ODL Institution. *Reflections on Postgraduate Supervision and Academic Development in an African Distance Education Environment*, 49.;Mbogo, R. W., & Wambua, J. M. (2020). Supervision Challenges and Delays in Completion of PhD Programmes in Public and Private Universities: Experiences of Supervisors and Graduate Students in Kenya. *European Journal of Education Studies*, 6(7), 261–279. DOI: 10.5281/zenodo.3653866

Mali, Y. C. G. (2023). EFL Students 'Challenges in Writing Research Proposals. *LLT Journal: A Journal on Language and Language Teaching, 26*(1), 272-289.

Mamdani, M. (2018). *Citizen and subject: Contemporary Africa and the legacy of late colonialism.* Princeton University Press.

Mamdani, M. (2018). Decolonizing Universities. *Social Dynamics*, 44(2), 181–192.

Management & Business Excellence Consultancy. (n.d.). *Business Problem Identification.* Retrieved January 2, 2023, from https://www.mnbec.com/business-problem-identification/

Manathunga, C. (2007). Supervision as mentoring: The role of power and boundary crossing. *Studies in Continuing Education*, 29(2), 207–221. DOI: 10.1080/01580370701424650

Manathunga, C. (2012). Supervisors watching supervisors: The deconstructive possibilities and tensions of team supervision. *Australian Universities Review*, 54(1), 29–37.

Manchez, J. M., Rodriguez, M. A. and F. Garcia. F. (2020)."Assumptions in Research Proposals: A Critical Examination of the Concept. Journal: Journal of Educational Theory. Doi: DOI: 10.1111/1468-5965.12345

Mapolisa, T., & Mafa, O. (2012). Challenges being experienced by undergraduate students in conducting research in open and distance learning. *International Journal of Asian Social Science*, 2(10), 1672–1684.

Marchisotti, G. G., & Farias Filho, J. R. (2022). Application of a Multi-method to Identify a Research Problem. *Revista de Administracao Contemporanea, 26*(6), 1–17. https://doi.org/DOI: 10.1590/1982-7849rac2022210020.en

Marcuse, H. (2015). *One-Dimensional Man: Studies in the Ideology of Advanced Industrial Society*. Beacon Press.

Marginson, S. (2010). Higher education in the global knowledge economy. *Procedia: Social and Behavioral Sciences*, 2(5), 6962–6980. DOI: 10.1016/j.sbspro.2010.05.049

Marginson, S. (2011). Higher education and public good. *Higher Education Quarterly*, 65(4), 411–433. DOI: 10.1111/j.1468-2273.2011.00496.x

Maslow, A. H. (1943). A theory of human motivation. *Psychological Review*, 50(4), 370–396. DOI: 10.1037/h0054346

Mawere, J. & Ndwamato, T.W. 2023. Exploring the Impact of Colonialism on South Africa with Focus on the Venda Community: Examining the Role of Traditional Knowledge Systems in Achieving Restorative Justice In book: Indigenous People - Traditional Practices and Modern Development [Working Title].

Maxwell, D. (2016). The research lifecycle as a strategic roadmap. *Journal of Library Administration*, 56(2), 111–123. DOI: 10.1080/01930826.2015.1105041

May, T., & Perry, B. (2022). *Social research: Issues, methods and process*. McGraw-Hill Education.

Mbembe, A. (2016). Decolonizing the university: New directions. *Arts and Humanities in Higher Education*, 15(1), 29–45. DOI: 10.1177/1474022215618513

McCallin, A., & Nayar, S. (2012). Postgraduate research supervision: A critical review of current practice. *Teaching in Higher Education*, 17(1), 63–74. DOI: 10.1080/13562517.2011.590979

McMillan, K., & Weyers, J. (2010). How to write dissertations & project Reports.

Mertens, D. M. (2009). *Transformative Research and Evaluation*. Guilford Press.

Mertens, D. M., Cram, F., & Chilisa, B. (2016). *Indigenous Pathways into Social Research: Voices of a New Generation*. Left Coast Press. DOI: 10.4324/9781315426693

Mertens, D. M., Sullivan, W. M., & McPhee-Jones, K. J. (2020). *Research methods in education: An introduction*. Routledge.

Merton, R. K. (1973). *The Sociology of Science: Theoretical and Empirical Investigations*. University of Chicago Press.

Meuter, R., Gallois, C., Segalowitz, N., Ryder, A., & Hocking, J. (2015). Overcoming language barriers in healthcare: A protocol for investigating safe and effective communication when patients or clinicians use a second language. *BMC Health Services Research*, 15(371), 1–5. DOI: 10.1186/s12913-015-1024-8 PMID: 26357948

Meyer, K. E. (2006). Asian management research needs more self-confidence. *Asia Pacific Journal of Management*, 23(2), 119–137. DOI: 10.1007/s10490-006-7160-2

Mgqwashu, E. M. (2016). On developing academic literacy through decolonised curricula. *Journal of Education*, 67, 141–158.

Mildeová, S. (2013). Research problem description and definition: From mental map to connection circle. *Journal on Efficiency and Responsibility in Education and Science*, 6(4), 328–335. DOI: 10.7160/eriesj.2013.060409

Miller, A. N., Taylor, S. G., & Bedeian, A. G. (2011). Publish or perish: Academic life as management faculty lives it. *Career Development International*, 16(5), 422–445. DOI: 10.1108/13620431111167751

Miller, P., & Williams, D. (2017). *Effective Dissemination Strategies in Academic Publishing*. Cambridge University Press.

Miller, S. (2022). *Undergraduate studies: An introduction*. Student Press.

Ministry of Education, Youth and Sport [MoEYS]. (2014). *Education strategic plan 2014-2018.*

Ministry of Education, Youth and Sport [MoEYS]. (2024b). *Guidelines for professorship choosing, evaluation, and provision in the education sector.*

Ministry of Education., Youth and Sport [MoEYS]. (2024a). *Education Congress 2024: Reviewing what has been done in 2022-2023 and directions for 2023-2024.*

Minnaert, L. (2015). Does relevance matter in academic policy research? A comment on Dredge. *Journal of Policy Research in Tourism, Leisure & Events*, 7(2), 190–194. DOI: 10.1080/19407963.2014.990663

Mkandawire, T. (2005). *African Intellectuals: Rethinking Politics, Language, Gender, and Development*. Zed Books. DOI: 10.5040/9781350218147

Mlambo, N. (2018). The Lingering Legacy of Colonial Education in Southern Africa. *Education and Society*, 36(2), 212–230.

Mncina, T. C., Letsie, H., Nkhi, S. E., & Mofana, M. (2024). Effective Communication in Postgraduate Supervision: Shaping Experiences and Overcoming Challenges. *Interdisciplinary Journal of Education Research*, 6(6), 1–21. DOI: 10.38140/ijer-2024.vol6.13

Molaba, M., Gololo, S. S., & Seeletse, S. M. (2024). Optimizing postgraduate supervision using multilevel modeling techniques in a constrained higher education environment. *International Journal of Research in Business and Social Science (2147-4478), 13*(2), 375-381.

Montgomery, C. (2010). *Understanding the international student experience*. Palgrave Macmillan. DOI: 10.1007/978-0-230-36500-1

Morgan, D. L., & Spencer, L. M. (2024). The use of mixed-methods research in education: A critical review. *Journal of Educational Research Review*, 21(1), 1–20.

Morris, V., & Laipple, J. (2015). How to approach supervising your international postgraduate student. *Innovations in Education and Teaching International*, 52(2), 137–147.

Motshoane, P., & McKenna, S. (2021). Crossing the border from candidate to supervisor: The need for appropriate development. *Teaching in Higher Education*, 26(3), 387–403. DOI: 10.1080/13562517.2021.1900814

Murphy, N., Bain, J. D., & Conrad, L. (2007). Orientations to research higher degree supervision. *Higher Education*, 53(2), 209–234. DOI: 10.1007/s10734-005-5608-9

Murray, R. (2012). *How to write a thesis*. Buckingham, UK: Open University Press.

Muthukrishna, M., & Henrich, J. (2019). A problem in theory. *Nature Human Behaviour*, 3(3), 221–229. DOI: 10.1038/s41562-018-0522-1 PMID: 30953018

Mwambari, D., Ali, F. A., & Barak, C. (2022). The impact of open access on knowledge production, consumption, and dissemination in Kenya's higher education system. *Third World Quarterly*, 43(6), 1408–1424. DOI: 10.1080/01436597.2022.2056010

Nasution, M. K. M., Onrizal,, , & Aulia, I. (2019). Design of the research problem statement. *Journal of Physics: Conference Series*, 1235(1), 012115. Advance online publication. DOI: 10.1088/1742-6596/1235/1/012115

National Science Foundation. (2020). The Role of Literature Reviews in Research Proposals. Retrieved from https://www.nsf.gov/bias/literature-review.cfm

National Science Foundation. (n.d.). Writing the Research Proposal: A Guide. Retrieved from https://www.nsf.gov/bias/writing_guide.cfm

Ndlovu-Gatsheni, S. J. (2018). *Epistemic Freedom in Africa: Deprovincialization and Decolonization*. Routledge. DOI: 10.4324/9780429492204

Ndofirepi, T., & Chibaya, W. T. (2022). Decolonizing the Curriculum: African Perspectives on Inclusive Education. *Journal of Education for Sustainable Development*, 36(2), 245–261.

Neau, V. (2003). The teaching of foreign languages in Cambodia: A historical perspective. *Language, Culture and Curriculum*, 16(3), 253–268. DOI: 10.1080/07908310308666673

Nehru Pasoloran Pongsapan, M. P., Allo, M. D. G., & Rumpa, L. D. (2023). *A Guide to Research Proposal*. Deepublish.

Nelson, C., & Bérubé, M. (2020). Introduction: A report from the front. In M. Bérubé & C. Nelson (Eds.), *Higher education under fire* (pp. 1-32). Routledge. DOI: 10.4324/9781003061489-1

Nettles, M. T., & Millett, C. M. (2006). *Three magic letters: Getting to Ph.D.* Johns Hopkins University Press. DOI: 10.56021/9780801882326

Ngulube, P. (2018). Overcoming the Western Dominance in Research: Critical Insights from an African Perspective. *African Journal of Library, Archives* & Mac-Donald, C. (2012). Understanding Participatory Action Research: A Qualitative Research Methodology Option. Canadian Journal of Action Research, 13(2), 34-50. *Information Science*, 28(2), 123–136.

Nhem, D. (2023). So, is there a teaching-research nexus in Cambodia? Evidence from EFL students at one public university. *Research in Post-Compulsory Education*, 28(1), 48–69. DOI: 10.1080/13596748.2023.2166692

Nolen, A., & Talbert, T. (2011). Qualitative assertions as prescriptive statements. *Educational Psychology Review*, 23(2), 263–271. DOI: 10.1007/s10648-011-9159-6

Nsamenang, A. B., & Tchombe, T. M. S. (Eds.). (2011). *Handbook of African Educational Theories and Practices: A Generative Teacher Education Curriculum.* Human Development Resource Centre.

Ntabe, A. C. (2020). *Rethinking the practice of community engagement in health research: The case of the tenofovir trials in Cambodia and Cameroon* (Master's thesis). https://core.ac.uk/download/pdf/304374342.pdf

Nyamnjoh, F. B. (2018). Rethinking African Higher Education: Beyond Colonial Paradigms. *Journal of Higher Education in Africa*, 16(1-2), 1–21.

Nyoni, P. (2018). Socio-Economic Inequalities and Education in Africa: A Critical Analysis. *African Educational Review*, 14(3), 276–294.

Oanda, I. O., & Jowi, J. O. (2013). University expansion and the challenges to social development in Kenya: Dilemmas and pitfalls. *Journal of Higher Education in Africa*, 11(1-2), 49–71.

Odora Hoppers, C. A. (2002). *Indigenous knowledge and the integration of knowledge systems: Towards a conceptual and methodological framework.* New Africa Books.

Odora Hoppers, C. A. (2015). Cognitive justice and integration without duress: The future of development education - perspectives from the South. *International Journal of Development Education and Global Learning*, 7(2), 89–106. DOI: 10.18546/IJDEGL.07.2.08

Odora Hoppers, C. A. (2015). Thinking the Unthinkable: From the Universalisation of Knowledge to Decolonisation of the University in Africa. *South African Journal of Higher Education*, 29(5), 199–217.

Odularu, O. I., & Akande, J. O. (2024). Reflection of experiences with academic supervisors, supervisees, and issues of power. *South African Journal of Higher Education*, 38(2), 254–273. DOI: 10.20853/38-2-5706

Ogunleye, J. K., & Afolabi, F. M. (2021). Empowering African Teachers Through Reflective Practices: Insights for Educational Reform. *Journal of African Educational Research*, 18(3), 78–94.

Okeke, C. (2018). Education and Marginalization in Africa: A Critical Examination. *Journal of African Studies*, 19(2), 159–175.

Olesen, J. L., & Stoecklin, A. (2018). Teaching research proposal writing to social science students: A collaboration model. *Journal of Academic Writing*, 8(1), 1–16.

Oyewole, O. (2010). Africa's public service delivery & performance review. African Higher Education in the 21st Century: Quality. *Relevance and Leadership in a Time of Global Change*, 3(4), 65–89.

Padmadewi, N. N., Artini, L. P., Ratminingsih, N. M., Suhardiana, I. P. A., Zamzam, A., & Juniarta, P. A. K. (2023). Designing Project-Based Learning in research proposal writing: Its effect, problems, and scaffolding utilized. *Studies in English Language and Education*, 10(2), 841–862. DOI: 10.24815/siele.v10i2.27408

Paglis, L. L., Green, S. G., & Bauer, T. N. (2006). Does adviser mentoring add value? A longitudinal study of mentoring and doctoral student outcomes. *Research in Higher Education*, 47(4), 451–476. DOI: 10.1007/s11162-005-9003-2

Panda, A., & Gupta, R. K. (2014). Making academic research more relevant: A few suggestions. *IIMB Management Review*, 26(3), 156–169. DOI: 10.1016/j.iimb.2014.07.008

Pardede, P. (2018). Identifying and Formulating the Research Problem. *Research in ELT, 1*(October), 1–13. https://www.researchgate.net/publication/329179630_Identifying_and_Formulating_the_Research_Problem

Parker-Jenkins, M. (2018). Mind the gap: Developing the roles, expectations, and boundaries in the doctoral supervisor–supervisee relationship. *Studies in Higher Education*, 43(1), 57–71. DOI: 10.1080/03075079.2016.1153622

Patel, R., Patel, D., & Shah, P. (2021). Improving student writing skills through scaffolding strategies: A randomized controlled trial. *Journal of Educational Psychology*, 113(2), 341–353.

Pawar, P. B., Verma, R., Daniel, C. O., & Sayyad, L. (2023). *Foundation of research methodology: A comprehensive guide*. Red Unicorn Publication.

Pearson, M., & Brew, A. (2002). Research training and supervision development. *Studies in Higher Education*, 27(2), 135–150. DOI: 10.1080/03075070220119986c

Perry, C. (1998). A structured approach for presenting theses. [AMJ]. *Australasian Marketing Journal*, 6(1), 63–85. DOI: 10.1016/S1441-3582(98)70240-X

Petraki, E., & Khat, K. (2022). Challenges and constraints in the design of an ESP course in Cambodia: Implications for higher education institutions. *Asia Pacific Journal of Education*, 42(2), 260–275. DOI: 10.1080/02188791.2020.1798738

Peycam, P. (2020). An integrated capacity-building program. In Peycam, P. (Ed.), *Cultural renewal in Cambodia* (pp. 115–150). Brill., DOI: 10.1163/9789004437357_009

Pietersen, C. (2014). Content issues in students' research proposals. *Mediterranean Journal of Social Sciences*, 5(20), 1533–1541.

Pilbeam, C., & Denyer, D. (2009). Lone scholar or community member? The role of student networks in doctoral education in a UK management school. *Studies in Higher Education*, 34(3), 301–318. DOI: 10.1080/03075070802597077

Piwowar, H., Priem, J., Larivière, V., Alperin, J. P., Matthias, L., Norlander, B., Farley, A., West, J., & Haustein, S. (2018). The state of OA: A large-scale analysis of the prevalence and impact of Open Access articles. *PeerJ*, 6, e4375. DOI: 10.7717/peerj.4375 PMID: 29456894

Polgar, S. D., & Thomas, S. A. (2008). *Introduction to Research in the Health Sciences*. Churchill Livingston.

Polit, D. F., & Beck, C. T. (2017). *Nursing research: Generating and assessing evidence for nursing practice* (10th ed.). Wolters Kluwer Health., DOI: 10.1016/j.iccn.2015.01.005

Prah, K. K. (2018). Language and Cultural Identity in Post-Colonial Africa. *International Journal of African Renaissance Studies*, 13(1), 45–59.

Prihandoko, L. A., Anggawirya, A., & Rahman, F. (2021). Students' Perceptions Towards Autonomous Learners Concept in Academic Writing Classes: Sequential Mixed-Method. *Advances in Social Science, Education and Humanities Research*, 603, 487–491.

Prilutski, M. A. (2010). A brief look at effective health communication strategies in Ghana. *The Elon Journal of Undergraduate Research in Communications*, 1(2), 51.

Ragins, B. R., & Kram, K. E. (2007). The roots and meaning of mentoring. The handbook of mentoring at work: Theory, research, and practice, 3-15.

Rajan Varadarajan, P. (2003). Musings on Relevance and Rigor of Scholarly Research in Marketing. *Journal of Technological Advancements*, 31(4), 368–376. DOI: 10.1177/0092070303258240

Rany, S., Zain, A. N. M., & Jamil, H. (2012). Cambodia's higher education development in historical perspectives (1863-2012). *International Journal of Learning and Development*, 2(2), 224–241. DOI: 10.5296/ijld.v2i2.1670

Ravitch, S. M., & Riggan, M. (2016). *Reason & Rigor: How Conceptual Frameworks Guide Research*. Sage Publications.

Reddy, K. S., Xie, E., & Tang, Q. (2016). Higher education, high-impact research, and world university rankings: A case of India and comparison with China. Pacific Science Review B: Humanities and Social Sciences, 2(1), 1–21. https://doi.org/https://doi.org/10.1016/j.psrb.2016.09.004

Reiter-Palmon, R., & Robinson, E. J. (2009). Problem Identification and Construction: What Do We Know, What Is the Future? *Psychology of Aesthetics, Creativity, and the Arts*, 3(1), 43–47. DOI: 10.1037/a0014629

Richtig, G., Berger, M., Lange-Asschenfeldt, B., Aberer, W., & Richtig, E. (2018). Problems and challenges of predatory journals. In Journal of the European Academy of Dermatology and Venereology, Blackwell Publishing Ltd., 32 (9), 1441–1449. DOI: 10.1111/jdv.15039

Riley, T. (2007). Developing effective research proposals. *Qualitative Research Journal*, 7(2), 84–88. DOI: 10.3316/QRJ0702084

Roberts, A. (2000). Mentoring revisited: A phenomenological reading of the literature. *Mentoring & Tutoring*, 8(2), 145–170. DOI: 10.1080/713685524

Robinson, E. (2018). Editorial perspectives on interdisciplinary collaboration in research area selection. *Journal of Interdisciplinary Studies*, 30(2), 201–218.

Robson, C. (2024). *Real-world research*. John Wiley & Sons.

Rocco, T. S., & Hatcher, T. G. (2011). *The Handbook of Scholarly Writing and Publishing*. John Wiley & Sons.

Rosser, P., & Soler, S. (2024). Reflective Practice and Frame Theory in Educational History: A Communicative Narrative Research Proposal. *Indonesian Journal of Education and Social Sciences*, 3(2), 151–16. DOI: 10.56916/ijess.v3i2.724

Rudolph, H. R., & Peluchette, J. V. (2011). Significance Of Usefulness: A Congruency Model Of Relevant Research Criteria. [JABR]. *Journal of Applied Business Research*, 8(3), 83. DOI: 10.19030/jabr.v8i3.6148

Rusdiana, A. (2024). Educational Supervision and Evaluation: Semester II Academic Year 2023/2024.

Ryan, J., & Viete, R. (2009). Respectful interactions: Learning with international students in the English-speaking academy. *Teaching in Higher Education*, 14(3), 303–314. DOI: 10.1080/13562510902898866

Rynes, S. L., Mc Natt, D. B., & Bretz, R. D. (1999). Academic research inside organizations: Inputs, processes, and outcomes. *Personnel Psychology*, 52(4), 869–898. DOI: 10.1111/j.1744-6570.1999.tb00183.x

Saeed, M. A., & Qunayeer, A. I. S. H., & AL-Jaberi, M. A. (2021, April). Exploring Supervisory Feedback Formulation on Academic Writing of Research Proposals and Postgraduates' Responses to Feedback: A Case Study. *Sage publications Open access*, 1-15. doi:, P., Muntean, C. H., Sheridan, F., Thornbury, E., & Pathak, P. (2024, March). Research Supervision Framework: A Student's Experience. In *2024 IEEE World Engineering Education Conference (EDUNINE)* (pp. 1-5). IEEE.DOI: 10.1177/ Stynes

Saldaña, J. (2015). *The coding manual for qualitative researchers* (3rd ed.). SAGE Publications.

Salmi, J. (2009). *The challenge of establishing world-class universities*. World Bank Publications. DOI: 10.1596/978-0-8213-7865-6

Sanak, S. D., Nalley, H. M., & Suek, L. A. (2023). An analysis of the introductory part of Research Proposals written by the Students of the English study Program. *Sparkle Journal of Language. Education and Culture*, 2(1), 52–65.

Sandberg, E., Oghazi, P., Chirumalla, K., & Patel, P. C. (2022). Interactive research framework in logistics and supply chain management: Bridging the academic research and practitioner gap. *Technological Forecasting and Social Change, 178*(September 2021), 121563. DOI: 10.1016/j.techfore.2022.121563

Sandelowski, M., Voils, C. I., & Barroso, J. (2006). Defining and designing mixed methods research synthesis studies. *Research in the Schools*, 13(1), 29–40. PMID: 20098638

Santos, J., Fernandes, L., & Silva, C. (2024). A mixed-methods study on the impact of gamification on student engagement in online learning environments. *Journal of Business Research*, 133(1), 15–25.

Santos, S. J., & Reigadas, E. T. (2004). Understanding the student-faculty mentoring process: Its effects on at-risk university students. *Journal of College Student Retention*, 6(3), 337–357. DOI: 10.2190/KGVC-7218-DPER-RMBC

Saunders, A. M., & Lewis, P. (2019). *Research Methods for Business Students* (8th ed.). Pearson.

Savitha, K., & Venkatachalam, J. (2024). Crafting and validating the essence of professional identity in higher secondary educators: *Educational Administration. Theory into Practice*, 30(5), 10565–10572. DOI: 10.53555/kuey.v30i5.3801

Sawyerr, A. (2004). African universities and the challenge of research capacity development. *Journal of Higher Education in Africa*, 2(1), 211–240. DOI: 10.57054/jhea.v2i1.1687

Schwartz-Shea, P., & Yanow, D. (2012). *Interpretive Research Design: Concepts and Processes*. Routledge.

Scoffham, S., & Rawlinson, S. (2022). *Sustainability education: A classroom guide*. Bloomsbury Publishing. DOI: 10.5040/9781350262119

Scott, P. (1998). Massification, internationalization, and globalization. In Scott, P. (Ed.), *The globalization of higher education* (pp. 108–129). Open University Press.

Semali, L., & Kincheloe, J. L. (1999). *What is indigenous knowledge?: Voices from the academy*. Falmer Press.

Seng, C., Carlon, M. K. J., & Cross, J. S. (2020). Research self-efficacy of Cambodian undergraduate students at province-based universities. *The International Journal of Social Education*, 9(2), 155–190. DOI: 10.17583/rise.2020.4267

Sergeant, S. (2021). *Working together, learning together: Towards universal design for research*. Gompel and Svacina.

Shahsavar, Z., & Kourepaz, H. (2020). Postgraduate students' difficulties in writing their theses literature review. *Cogent Education*, 7(1), 1784620. Advance online publication. DOI: 10.1080/2331186X.2020.1784620

Shannon, A. G. (1995). Research degree supervision: "More mentor than master.". *Australian Universities Review*, 38(2), 12–15.

Shatimwene, G. P., & Ashipala, D. O. (2023). Nursing educators' experiences and expectations of research production at a Faculty of Health Sciences: A qualitative study. *Africa Journal of Nursing and Midwifery*, 25(2). Advance online publication. DOI: 10.25159/2520-5293/14153

Shinton Consulting. (2013). *Guide to building a research profile.* Sussex Researcher Careers. https://sussexresearchercareers.org/wp-content/uploads/2017/08/guide-to-building-a-research-profile.pdf

Shrivastava, P. (1987). Rigor and practical usefulness of research in strategic management. *Strategic Management Journal*, 8(1), 77–92. DOI: 10.1002/smj.4250080107

Sidor, J. (2015). Debate Over Rigor and Relevance in Scientific Study of Management. *Management and Business Administration.Central Europe*, 23(3), 32–46. DOI: 10.7206/mba.ce.2084-3356.149

Siler, K., & Frenken, K. (2020). The pricing of open access journals: Diverse niches and sources of value in academic publishing. *Quantitative Science Studies*, 1(1), 28–59. DOI: 10.1162/qss_a_00016

Silverman, D. (2005). *Doing Qualitative Research: A Practical Handbook.* Sage.

Simelane, N., Klopper, R., & Lubbe, S. (2018). Hands-on postgraduate research proposal design. *Alternation (Durban)*, 25(1), 295–315. DOI: 10.29086/2519-5476/2018/v25n1a13

Singh, M. K., Singh, S.K. and Singh, R.K. (2021). "The Role of Assumptions in Research Proposals: A Critical Analysis. Journal: Journal of Research Methodology. DOI: DOI: 10.1080/16138776.2021.1901423

Slife, B. D., Wright, C. D., & Yanchar, S. C. (2016). Using operational definitions in research: A best-practices approach. *Journal of Mind and Behavior*, •••, 119–139.

Smith, J. (2021). *The role of academics in modern education.* Academic Press.

Smith, J. (2022). *Crafting Publishable Articles: Best Practices for Academic Authors*. Wiley.

Smith, J.. (2020). Collaborative strategies for urban sustainability and climate resilience. *Journal of Urban Planning and Development*, 36(4), 567–589.

Smith, J., Brown, T. D., & Johnson, K. (2022). *Research design: Qualitative and quantitative approaches*. Sage Publications.

Smith, L. T. (2010). *Decolonizing methodologies: Research and indigenous peoples.* Zed Books.

Smith, L. T. (2012). *Decolonizing Methodologies: Research and Indigenous Peoples*. Zed Books.

Smith, P. (2019). Niche exploration in academic research: Strategies and outcomes. *Academic Research Quarterly*, 42(3), 311–328.

Snyder, H. (2019). Literature review as a research methodology: An overview and guidelines. *Journal of Business Research*, 104, 333–339. DOI: 10.1016/j.jbusres.2019.07.039

Sok, S., & Bunry, R. (2021). Internationalization of higher education in Cambodia: Toward an agenda for higher education development. *International Journal of Comparative Education and Development*, 23(3), 193–211. DOI: 10.1108/IJCED-08-2020-0049

Sol, K., & Heng, K. (2022). The Cambodian Education Forum at two years old: Keeping the momentum going. *Cambodian Journal of Educational Research*, 2(1), 1–7. DOI: 10.62037/cjer.2022.02.01.01

Sol, K., & Heng, K. (2023). The Cambodian Journal of Educational Research's 3rd anniversary: Reflecting on its development. *Cambodian Journal of Educational Research*, 2(2), 1–6. DOI: 10.62037/cjer.2023.03.02.01

Sosin, K. (2019). *Productive writing: How to write more efficiently and effectively*. Academic Press.

Soulard, E. C., Walker, J., & Petrakis, R. E. (2020). Implementation of a surface water extent model in Cambodia using cloud-based remote sensing. *Remote Sensing (Basel)*, 12(6), 1–19. DOI: 10.3390/rs12060984

Sovansophal, K., & Shimizu, K. (2020). A review on STEM enrollment in higher education of Cambodia: Current status, issues, and implications of initiatives. *Journal of International Development and Cooperation*, 26(1), 123–134. DOI: 10.15027/48743

Sreekumar, D. (2023). What Are Research Objectives and How to Write Them? https://researcher.life/blog/article/what-are-research-objectives-how-to-write-them-with-examples

Ssegawa, J. K., & Rwelamila, P. D. (2009). The research skill factor as a cause for high postgraduate attrition rate. *Journal of Engineering. Design and Technology*, 7(3), 293–322. DOI: 10.1108/17260530910998703

Stanovich, K. E. (1999). *Who is rational? Studies of individual differences in reasoning*. Psychology Press. DOI: 10.4324/9781410603432

Starkey, K., & Madan, P. (2001). Bridging the Relevance Gap: Aligning Stakeholders in the Future of Management Research. *British Journal of Management*, 12(SPEC. ISS.). Advance online publication. DOI: 10.1111/1467-8551.12.s1.2

Steinbrecher, W. J., Booth, W. J., & Mullen, C. K. (2016). *The Craft of Research.* University of Chicago Press.

Stremersch, S., Gonzalez, J., Valenti, A., & Villanueva, J. (2023). The value of context-specific studies for marketing. *Journal of the Academy of Marketing Science*, 51(1), 50–65. DOI: 10.1007/s11747-022-00872-9

Strydom, P. (2011). *Contemporary critical theory and methodology.* Routledge. DOI: 10.4324/9780203875568

Sueur, L. H., Dagliati, A., Buchan, I., Whetton, A. D., Martin, G. P., Dornan, T., & Geifman, N. (2020). Pride and prejudice - What can we learn from peer review? *Medical Teacher*, 42(9), 1012–1018. DOI: 10.1080/0142159X.2020.1774527 PMID: 32631121

Sun, P., Doh, J. P., Rajwani, T., & Siegel, D. (2021). Navigating cross-border institutional complexity: A review and assessment of multinational nonmarket strategy research. *Journal of International Business Studies*, 52(9), 1818–1853. https://link.springer.com/article/10.1057/s41267-021-00438-x. DOI: 10.1057/s41267-021-00438-x PMID: 34176993

Suri, H. (2011). Purposeful sampling in qualitative research synthesis. *Qualitative Research Journal*, 11(2), 63–75. DOI: 10.3316/QRJ1102063

Tavakol, M., & O'Brien, D. (2023). The importance of crafting a good introduction to scholarly research: Strategies for creating an effective and impactful opening statement. *International Journal of Medical Education*, 14, 84–87. DOI: 10.5116/ijme.6499.82af PMID: 37384945

Taylor, L. (2020). Aligning research agendas with policy objectives: A case study analysis. *Public Policy Research*, 25(2), 176–193.

Taylor, M. (2023). *Research methodologies in social sciences.* Research Publishing.

Taylor, S., & Beasley, N. (2015). *A handbook for doctoral supervision.* Routledge.

Teferra, D. (2004). Higher education in Sub-Saharan Africa. International Handbook of Higher Education, 1, 557-569.

Teferra, D. (2000). Brain circulation: Unparalleled opportunities, underlying challenges, and outmoded presumptions. *International Journal of African Higher Education*, 3(1), 1–16.

Teferra, D., & Altbach, P. G. (2004). African higher education: Challenges for the 21st century. *Higher Education*, 47(1), 21–50. DOI: 10.1023/B:HIGH.0000009822.49980.30

Teferra, D., & Altbach, P. G. (2018). African Higher Education: Challenges for the 21st Century. *Industry and Higher Education*, 45, 15–18.

Tennant, J. P., & Ross-Hellauer, T. (2020a). The limitations to our understanding of peer review. *Research Integrity and Peer Review*, 5(1), 6. Advance online publication. DOI: 10.1186/s41073-020-00092-1 PMID: 32368354

Thomas, H. (2009). Business Schools and Management Research: A UK Perspective. *Journal of Management Development*, 28(8), 660–667. DOI: 10.1108/02621710910985432

Thomas, K. W., & Tymon, W. G. (1982). Necessary Properties of Relevant Research: Lessons from Recent Criticisms of the Organizational Sciences. *Academy of Management Review*, 7(3), 345–352. DOI: 10.2307/257326

Thompson, D. (2020). Policy implications of digital learning platforms in K-12 education. *Educational Policy Review*, 25(2), 176–193.

Thompson, D., & Johnson, M. (2021). Digital learning and student engagement: Evidence from policy-driven research. *The Journal of Educational Research*, 35(5), 673–702.

Tinto, V. (1975). Dropout from higher education: A theoretical synthesis of recent research. *Review of Educational Research*, 45(1), 89–125. DOI: 10.3102/00346543045001089

Tinto, V. (1993). *Leaving college: Rethinking the causes and cures of student attrition*. University of Chicago Press.

Tomaselli, K. G. (2017). Research proposal development: A framework for researchers in developing countries. *South African Review of Sociology*, 48(3), 79–95.

Tracy, S. J. (2019). *Qualitative research methods: Collecting evidence, crafting analysis, communicating impact*. John Wiley & Sons.

Trafford, V., & Leshem, S. (2008). *Stepping stones to achieving your doctorate: By focusing on your viva from the start*. McGraw-Hill Education.

TrinityCollege. (2024, March 8). *Current Students>Postgraduate Research Students>Conflict IN Supervision*. Retrieved from Trinity College Dublin: https://www.tcd.ie/graduatestudies/students/research/conflict-resolution/

Trochim, W. M. K., & Donnelly, J. P. (2007). *The research methods knowledge base* (3rd ed.). Thomson.

Trudell, B. (2009). Local-language Literacy and Sustainable Development in Africa. *International Journal of Educational Development*, 29(1), 73–79. DOI: 10.1016/j.ijedudev.2008.07.002

Tsui, M. S., O'Donoghue, K., & Ng, A. K. (2014). Culturally competent and diversity-sensitive clinical supervision. *The Wiley international handbook of clinical supervision*, 238-254.

Tsui, A. S., & McKiernan, P. (2022). Understanding Scientific Freedom and Scientific Responsibility in Business and Management Research. *Journal of Management Studies*, 59(6), 1604–1627. DOI: 10.1111/joms.12816

Tuck, E. (2018). Participatory Action Research as Public Science. *Educational Policy*, 32(4), 620–651.

Tutar, H., Şahin, M., & Sarkhanov, T. (2024). Problem areas of determining the sample size in qualitative research: A model proposal. *Qualitative Research Journal*, 24(3), 315–336. DOI: 10.1108/QRJ-06-2023-0099

UC. (2020). *cumming.ucalgary.ca.proskills.resources*. Retrieved from University of Calgary: https://cumming.ucalgary.ca/sites/default/files/teams/13/proskills/resources/resolving_conflicts_in_the_graduate_student-supervisor_relationship.pdf

Ullah, M., Shahid, A., Din, I. U., Roman, M., Assam, M., Fayaz, M., Ghadi, Y., & Aljuaid, H. (2022). Analyzing Interdisciplinary Research Using Co-Authorship Networks. *Complexity*, 2022(1), 1–13. DOI: 10.1155/2022/2524491

UNESCO. (2019). *Higher education in Cambodia: Challenges and perspectives*.

University of Birmingham. (n.d.) *Five ways to raise your research profile*. University of Birmingham. https://intranet.birmingham.ac.uk/as/libraryservices/library/research/5-ways-improve-research-impact.aspx

University of California. (2020). Writing a Research Proposal. San Francisco (UCSF) School of Medicine. https://www.ctsi.ucsf.edu/funding-support/research-proposal-guidelines

University of California. Berkeley. (2019). Writing a Research Proposal: A Step-by-Step Guide. Retrieved from https://www.ucberkeley.edu/units/osp/research/propose/writing-research-proposal

Ursić, L., Baldacchino, G., Bašić, Ž., Sainz, A. B., Buljan, I., Hampel, M., Kružić, I., Majić, M., Marušić, A., Thetiot, F., Tokalić, R., & Markić, L. V. (2022). Factors influencing interdisciplinary research and industry-academia collaborations at six European universities: A qualitative study. *Sustainability (Basel)*, 14(15), 9306. DOI: 10.3390/su14159306

Van Aken, J. E. (2004). Management Research Based on the Paradigm of the Design Sciences: The Quest for Field-Tested and Grounded Technological Rules. *Journal of Management Studies*, 41(2), 219–246. DOI: 10.1111/j.1467-6486.2004.00430.x

Van Hoof, H., Estrella, M., Eljuri, M., & Torres, L. (2013). Ecuador's system of higher education in times of change. *Journal of Hispanic Higher Education*, 12(4), 345–355. DOI: 10.1177/1538192713495060

van Riel, A., & Snyder, H. (2024). Enhancing the impact of literature reviews: guidelines for making meaningful contributions. *Spanish Journal of Marketing-ESIC*.

Vehviläinen, S. (2009). Problems in the research problem: Critical feedback and resistance in academic supervision. *Scandinavian Journal of Educational Research*, 53(2), 185–201. DOI: 10.1080/00313830902757592

Veit, D. R., Lacerda, D. P., Camargo, L. F. R., Kipper, L. M., & Dresch, A. (2017). Towards Mode 2 knowledge production: Analysis and proposal of a framework for research in business processes. *Business Process Management Journal*, 23(2), 293–328. DOI: 10.1108/BPMJ-03-2016-0045

Veles, N., & Danaher, P. A. (2024). Transformative research collaboration as third space and creative understanding: Learnings from higher education research and doctoral supervision. *Research Papers in Education*, 39(1), 50–66. DOI: 10.1080/02671522.2022.2089212

Vereijken, M., van der Rijst, R., van Driel, J., & Dekker, F. (2018). Factors of effective postgraduate student-supervisor relationships at selected universities in South Africa. [JAE]. *Journal of African Education*, 2(2), 135–155.

Visser-Wijnveen, G.J., Van Driel, J.H., Van der Rijst, R.M., Verloop, N. & Visser, A. (2010). The ideal research-teaching nexus in the eyes of academics: building profiles. Pages 195-210 | Received 24 Mar 2009, Accepted 29 Nov 2009, Published online: 02 Mar 2010

Waal, T., Verbruggen, A., & Wright, T. (2010). University research for sustainable development: Definitions and characteristics explored. *Journal of Cleaner Production*, 18(7), 629–636. DOI: 10.1016/j.jclepro.2009.09.017

Wager, E., Barbour, V., Yentis, S., & Kleinert, S. (2009). Retraction guidelines. Retrieved December 29, 2013, from https://publicationethics.org/fles/retraction%20guidelines.pdf

Waghid, Y. (2004). African philosophy of education: Implications for teaching and learning. *South African Journal of Higher Education*, 18(3), 56–64.

Wang, T., & Li, L. Y. (2008). Understanding international postgraduate research students' challenges and pedagogical needs in thesis writing. *International Journal of Pedagogies and Learning, 4*(3), 88–96. http://dx.doi.org/DOI: 10.5172/ijpl.4.3.8

Wang, J., & Waltman, L. (2016). Measuring scientific impact beyond academia: An assessment of existing impact metrics and proposed improvements. *PLoS One*, 11(8), e0167653.

Wang, X., Liu, C., Mao, W., & Fang, Z. (2015). The open access advantage considering citation, article usage, and social media attention. *Scientometrics*, 103(2), 555–564. DOI: 10.1007/s11192-015-1547-0

Wanyama, S. B., & Eyamu, S. (2021). Perceived organizational support, graduate research supervision, and research completion rate. *Employee Relations*, 43(6), 1414–1430. DOI: 10.1108/ER-05-2020-0205

Wasti, S. A., & Robert, C. (2003). Out of Touch? an Evaluation of the Correspondence Between Academic and Practitioner Concerns in IHRM. *Advances in International Management*, 15, 207–239. DOI: 10.1016/S0747-7929(03)15010-X

Watkins, D. C. (2022). Secondary data in mixed methods research. *Sage (Atlanta, Ga.)*.

Webster, J. (2024). Building A Research Profile: Why And How? Published Jan 24, 2024.

Webster, M. (2024). *Defining Strategizing: Concepts and Applications*. Harvard Business Review Press.

Wenger, E. (1998). *Communities of practice: Learning, meaning, and identity.* Cambridge University Press. DOI: 10.1017/CBO9780511803932

White, D. (2023). *Student life and learning: A comprehensive guide*. University Publications.

White, G., & Black, D. (2016). Ethical considerations in research area selection: Insights from journal editorials. *Journal of Ethics in Academic Research*, 15(4), 445–462.

Williams, K. (2021). Research proposal writing for social work students: A guide to success. *Social Work Education*, 40(4), 462–476.

Williams, M., Wiggins, R., Vogt, W. P., & Vogt, P. R. (2022). Beginning quantitative research. *Sage (Atlanta, Ga.)*.

Wilson, K. (2019). *The life of a researcher: Ethics and practices*. Academic Books.

Wisker, G. (2012). *The good supervisor: Supervising postgraduate and undergraduate research for doctoral theses and dissertations*. Macmillan International Higher Education. DOI: 10.1007/978-1-137-02423-7

World Bank. (2010). *Financing higher education in Africa*. World Bank.

World Bank. (2018). *Cambodia's quest for knowledge: Investing in research and innovation*.

Wu, S., Oubibi, M., & Bao, K. (2024). How supervisors affect students' academic gains and research ability: An investigation through a qualitative study. *Heliyon*, 10(10), e31079. DOI: 10.1016/j.heliyon.2024.e31079 PMID: 38779017

Yilmaz, K. (2013). Comparison of quantitative and qualitative research traditions: Epistemological, theoretical, and methodological differences. *European Journal of Education*, 48(2), 311–325. DOI: 10.1111/ejed.12014

Yin, R. K. (2018). *Case study research and applications: Design and methods* (6th ed.). SAGE Publications.

Zeleza, P. T. (2016). The study of Africa: The challenges and opportunities of a new global intellectual order. *African Studies Review*, 59(1), 7–35.

Zeleza, P. T. (2018). The Struggles for African Knowledge: Beyond the Western Paradigm. *African Studies Review*, 61(2), 1–19.

Zikmund, W. G., Babin, B. J., Carr, J. C., & Griffin, M. (2013). Business Research Methods. In *Business Research Methods* (9th ed.). Cengage India Private Limited.

Zina, O. (2021). The essential guide to doing your research project. *Sage (Atlanta, Ga.)*.

Zuber-Skerritt, O., & Fletcher, M. (2007). The quality of an action research thesis in the social sciences. *Quality Assurance in Education, 15(4)*, 413-436.

Zuber-Skerritt, O., & Knight, N. (1986). Problem definition and thesis writing - Workshops for the postgraduate student. *Higher Education, 15*(1–2), 89–103. DOI: 10.1007/BF00138094

Żuk, P., & Szulecki, K. (2020). Unpacking the right-populist threat to climate action: Poland's pro-governmental media on energy transition and climate change. *Energy Research & Social Science*, 66, 101485. DOI: 10.1016/j.erss.2020.101485

About the Contributors

Andile Alfred Mdikana is currently employed as Senior Lecturer at the University of South Africa in the Department of Inclusive Education since November 2012 to date. He is currently mentoring colleagues who have PhD's on article writing for publication in peer reviewed journals. During his tenure at UNISA, he has developed modules on UNDERSTANDING INCLUSIVE EDUCATION AND ADVOCACY FOR PEOPLE WITH DISABILITIES. His academic citizenship includes the co-ordinatorship of the RESEARCH and MENTORSHIP committees and has been a member of the Higher Degrees in the Department of Inclusive Education. Furthermore, he has represented the department at college level in the Research and Higher Degrees committees. In addition to that he is also the primary lecturer and coordinator of the module called 'Addressing barriers to learning' and 'Advocacy for people with Disabilities'. Dr. A.A. Mdikana is a Registered Educational Psychologist since 2008. Prior to this appointment, he worked for Gauteng Department of Education as the Deputy Chief Education Specialist for Inclusion and Special Schools. It is worth noting that he spent four years running a private practice. He obtained his BEd (Guidance & Counselling) from the University of the Western Cape, his MEd (Specialized Education) from the University of Stellenbosch, his MEd (Educational Psychology) from the University of Johannesburg and his DEd (Inclusive Education) from the University of South Africa. He worked as a Psychology and Guidance Methodology lecturer at Wits University. He has worked as a Psychologist/Student Counsellor at the Universities of Johannesburg and Fort Hare. His main activities included counselling and psychotherapy, psychological assessment, workshop design (conceptualization, implementation and evaluation). At Fort Hare University he, also co-ordinated student development programs in the residences and conceptualized and implemented Gender Based Violence Wellness Programs and he also on a part-time basis at UFH worked as a Psychology 3 Lecturer teaching Psychotherapy/therapeutic interventions. In 2006 he served as one of the presiding officers for the 2006 SRC elections and in the same year he served as the

Hall Co-ordinator of the Esselen Residence at the University of the Witwatersrand. He has held several positions such as HIV/AIDS Consultant/Training Officer, School Psychologist, Life-Skills Facilitator, ABET Educator and ABET Curriculum Advisor, Special and High School Educator and Acting Principal. He has been the external examiner for the Universities of Pretoria, Stellenbosch, Johannesburg (UJ), Wits University, University of Zambia and the University of KwaZulu-Natal

***

Fillemon Shekutamba Albanus is a graduate of Bachelor of Nursing Science (Clinical) (Honours) degree in Nursing Science at the University of Namibia, School of Nursing and Public Health, General Nursing department in the Faculty of Health Sciences and Veterinary Medicine (FHSVM) at the University of Namibia, Namibia. He actually has a greater interest of partaking in all the heath care research fields. His research interests would specifically focus, in Primary health Care, Health care services, Nursing education, and Mental health.

Daniel Opotamutale Ashipala (MN Sc) is a senior lecturer and Head of Department for the School of nursing and Public Health (SoNPH), General Nursing Science department in the Faculty of Health Sciences and Veterinary Medicine (FHSVM) at the University of Namibia, Namibia. His research interests include Primary health Care, Health care services and Nursing education, and Mental health. He has published a quite number of peer reviewed articles and book chapters on the delivery of teaching and learning in higher education studies. He is a doctoral candidate at Stellenbosch University in Cape Town, South Africa. His doctoral work focuses on developing a framework to facilitate the implementation of a task shifting approach for nurses in the Namibia PHC services.

Alok Behera is an Assistant Professor at Unitedworld Institute of Technology, Karnavati University, Gandhinagar, Gujrat. Before this, he held a position as a postdoctoral researcher at reputed research institutions like Indian Institute of Science (IISc), Bengaluru, and Indian Institute of Technology (IIT), Hyderabad. Dr. Behera obtained his Ph.D. from the Department of Metallurgical and Materials Engineering at Visvesvaraya National Institute of Technology (VNIT), Nagpur, India. His research expertise lies in advanced composites, sustainable manufacturing, and mechanics of polymer matrix composites. Dr. Behera has authored over 17 research papers, including 7 in Q1 SCI journals.

Doris Chasokela is a senior lecturer at the National University of Science and Technology (NUST) in Zimbabwe. She holds a Doctorate in Education, a Master of

Engineering in Electrical Systems Control and Information Technology; a Bachelor of Technical Education in Electrical and Electronic Engineering; a Diploma in Technical Vocational Education. She has supervised several undergraduate projects and postgraduate dissertations. She has eleven years of experience in higher education. Her interests are higher education, engineering education, artificial intelligence, control and e-learning.

Sereyrath Em is a Cambodian government teacher of English with a higher education degree, a visiting lecturer at the National University of Cheasim Kamchaymear (NUCK) and some other universities, and an Editor at the Cambodian Journal of Educational Research (CJER). He is also the founder and the Editor-in-Chief of the Cambodian Journal of Educational and Social Sciences (CJESS). Prior to this, he was a secondary school teacher of English and Khmer languages. In 2017, he graduated with a Master of Arts in Teaching English to Speakers of Other Languages (MA in TESOL) from Human Resource University (HRU), and in 2019, he graduated with a Master of Education in Educational Administration from the National Institute of Education (NIE) with the support from Cambodia International Education Support Foundation (CIESF), and he earned his MBA from the National Institute of Entrepreneurship and Innovation (NIEI) in 2023. Then in late 2023, he earned his Ph.D. in Educational Science from Khemarak University, Phnom Penh, Cambodia. From 2020 until now, he has studied and become a Ph.D. candidate in Educational Administration at the University of Cambodia too. He was also trained at the Regional Language Center (RELC) two times between early 2020 and early 2021 concerning teaching listening, speaking, reading, and writing. He also attended a seminar on Green Energy Development for Cambodia in Beijing, China in September 2024. His research interests include English language teaching, teaching methodology, educational management and science, educational leadership, learning and teaching motivation, and learning and teaching challenges. Currently, he is also a Ph.D. candidate specializing in Cooperative Education at the Institute of Social Technology, Suranaree University of Technology, Thailand.

G. Kumar holds a B.E. in Electronics and Communication (SRM Engineering College, 2001), an MBA (Bharathidhasan University, 2007), and a Ph.D. in Management (SRM Institute of Science and Technology, 2009). His doctoral research focused on mobile banking adoption in Chennai. With 14+ years of teaching experience at SRM's Faculty of Management, he has published 70+ articles in reputable journals. Additionally, he served as a programmer (2003-2009) with system administration and documentation responsibilities. His areas of expertise include business analysis, analytics, programming (Python, R), data structures, storytelling, statistics, big data, SQL, and machine learning. He has held the role of

Domain Coordinator-Systems (2022-2023) and currently mentors and leads classes for management students

Sharad Gupta is a seasoned professional with over 20 years of experience in both industry and academia, having worked nationally and internationally. He holds an MBA, dual MCom, an MA in Education, and a diploma from Harvard Business School, underscoring his strong academic background and business acumen. Mr. Gupta has contributed significantly across various industries including media, software development, finance, and education. His keen interest in innovation and entrepreneurial activities has led to numerous successful ventures and initiatives. An accomplished author, Mr. Gupta has co-authored several notable books such as "Innovative Pedagogy: A Beginner's Guide to Modern Teaching Strategies" with Dr. Manjari Sharma, and "WealthWise: Lessons from India's Warren Buffett." His work reflects a blend of practical experience and scholarly insight. Mr. Gupta is also a co-inventor of a patented 'IoT Based Fertilizer Spraying Drone,' demonstrating his commitment to leveraging technology for practical solutions. His extensive experience and innovative approach make him a distinguished figure in both business and education.

Buratin Khampirat received the B.Ed., M.Ed., and Ph.D. degrees from Chulalongkorn University, Thailand. She was a Visiting Professor with the Universidad Nacional de Educación a Distancia (UNED), in 2019. She is currently an associate professor of educational research at the Suranaree University of Technology, Thailand. Her research interests include the assessment of efficiency and effectiveness in HEIs, employability, career and student development, and assessment of competencies using SEM and multilevel modeling. She received a number of national and international awards/honors, most notably the best refereed research paper award from WACE, in 2017 and 2018. She is also a Co-Convenor of the European Educational Research Association (EERA) in Network 11-Educational Effectiveness and Quality Assurance.

Kaushik Kumar, holds a Ph.D. in Engineering from Jadavpur University, India, an MBA in Marketing Management from Indira Gandhi National Open University, India and a Bachelor of Technology from Regional Engineering College (Now National Institute of Technology), Warangal, India. For 11 years, he worked in a manufacturing unit of Global repute. He is currently working as a Professor in the Department of Mechanical Engineering, Birla Institute of Technology, Mesra, Ranchi, India. He has 23 Years of Teaching and Research Experience. His research interests include Composites, Optimization, Non-Conventional Machining, CAD / CAM, Rapid Prototyping and Quality Management Systems towards product development for societal and industrial usage and has received 35+ patents for them most of which are donated to Start-ups and Industrial fraternity through Jharkhand Small Industries Association (JSIA). He has published 55+ Books, 95+ Book Chapters and 200+ Research Papers in peer-reviewed reputed national and international journals. Kaushik has also served as Editor–in–Chief, Series Editor, Guest Editor, Editor, Editorial Board Member and Reviewers for International and National Journals. He has been felicitated with many awards and honours including Distinguished Alumnus Award for Professional Excellence 2023 under Academic and Research from National Institute of Technology, Warangal, India. He has also received Sponsored Research and Consultancy Projects of more than 1 Crore from Govt. of India and abroad. Kaushik has delivered expert lectures as Keynote Speaker at International & National Conferences, Resource Person at Various Workshops, FDP's and Short-Term Courses. He has guided many students of Doctoral, Masters and Under Graduate programmes of his home and other institutions in India and abroad. He also has served as Reviewer and Examiner of Doctoral and Masters dissertation for institutes in India and abroad.

Lwazi Sibanda is a professor and holds a PhD in Education, Master of Education in Educational Management, Bachelor of Education in Educational Administration, Planning and Policy Studies, and Diploma in Education.

Mohsen Mahmoudi-Dehaki is an Assistant Professor of TEFL at the Islamic Azad University of Najafabad, Iran. With over a decade of experience teaching undergraduate and postgraduate students, he has established himself as a respected educator and scholar. His research has been widely published in national and international journals, solidifying his reputation as a prominent voice in the field. Dr. Mahmoudi-Dehaki serves as a peer reviewer for several esteemed scientific journals, including RELP, ensuring the highest standards of academic rigor and integrity. His academic achievements include winning the prestigious first place in the English Writing Award in Iran, a distinction earned among all university students from state universities and institutes of higher education. His paper "COVID-19 Lingo or Covidipedia" has garnered significant attention, with over 70,000 visits, making it the most visited paper in Iranian scientific journals. This remarkable achievement underscores his significant impact on the academic community. Dr. Mahmoudi-Dehaki's research and teaching interests span a diverse range of novel branches in applied linguistics, including psycho/cognitive/neurocognitive/forensic/jury/sociocultural/ecological, and clinical linguistics. He is also well-versed in educational technology, educational psychology, and psychometrics, demonstrating his versatility and breadth of expertise in the interdisciplinary fields of education and linguistics.

Agnes Maradze is a lecturer at the Lupane State University in Zimbabwe. She holds an MSc in Procurement and Supply Chain Management (Bindura University of Science); and a BCom Hons. Degree in Purchasing and Supply Chain Management (Lupane State University); a Higher Diploma in Business Administration and Management (ABMA, UK), and a Diploma in Tertiary and Higher Education (National University of Science and Technology). She has supervised several undergraduates and postgraduate students in her field. Her research interests are Public Procurement Governance; Supply Chain Innovation; Integrated Logistics systems and applications and Quality in Higher Education Institutions.

Elangovan N is currently a Professor in the School of Business and Management and PhD –Coordinator for Management programme at Christ University, Bengaluru, India. He is also the Associate Director of Research and Development Cell of the University. Before that he served as the Campus Director of National Institute of Fashion Technology, Kannur, Kerala. His educational background spans Engineering, Science and Management with Bachelor degrees with a BE in Mechanical Engineering and BA in Psychology, Post Graduate degrees of MBA Marketing and MSc Psychology, and a MPhil in Psychology. He holds a Doctorate in Management Science specialising in Information systems from Anna University, Chennai. His interest of research is in Strategic Information System Area. Before moving to academics, his experience spans in the textile area where he managed his family business of screen printing for almost 16 years. He has completed a research project funded by Indo-Korean Science and Technology centre (IKST) established by Korea Institute of Science and Technology (KIST), South Korea focused on Social Network Services and Social Medial Platform/ Technology. He has published in many Scopus indexed Journals.

Nasim Nasr-Esfahani, PhD, is an Assistant Professor of TEFL at the Islamic Azad University of Isfahan (Khorasgan), Iran. She earned her PhD in TEFL at the Islamic Azad University of Isfahan, Khorasgan. She has over two decades of experience teaching undergraduate and postgraduate students in Iran. Her publications have appeared in leading international journals, such as the Australian Journal of Teacher Education. She serves as a reviewer and article editor of RELP Journal. Her areas of research and teaching interests are dynamic and evolving and span language education, generative artificial intelligence (GenAI) in education, critical thinking skills, oral and written communication skills, student engagement, advancing equity in education, brain-based teaching and learning, neurolinguistics, psycholinguistics as well as teaching and research methodologies.

Peter Khwezi Nyewe is a distinguished registered psychologist and an education lecturer at the Cape Peninsula University of Technology (CPUT). With extensive experience in both psychology and education, Dr. Nyewe has dedicated his career to advancing mental health awareness and educational practices. He holds a PhD in Psychology, which has equipped him with a profound understanding of psychological theories and their application in educational settings. His academic and professional journey reflects a commitment to integrating psychological principles into teaching methodologies to enhance student learning and well-being. At CPUT, Dr. Nyewe is known for his innovative teaching approaches and his research contributions to the field of educational psychology. He has published numerous articles in peer-reviewed journals, focusing on topics such as mental health in education, inclusive education practices, and teacher-student dynamics. His work not only enriches the academic community but also has practical implications for improving educational outcomes. Through his lectures, workshops, and research, Dr. Nyewe continues to inspire and mentor future educators, emphasizing the importance of psychological well-being in achieving academic success.

R. Krishna Kumari holds a diverse array of qualifications including M.Sc., M.Phil., M.Ed., TNSET, and a Ph.D. With over 5 years of teaching experience, she has contributed significantly to the academic field. Dr. Krishna Kumari has authored a total of 18 publications, all indexed in SCOPUS, with 11 being published post-Ph.D. Additionally, she has authored 3 books/book chapters and holds 1 patent published and 1 granted. She serves as a reviewer in 4 SCOPUS indexed journals and 2 SCOPUS and ESCI indexed journals. Dr. Krishna Kumari has been a resource person for 3 Faculty Development Programs (FDPs), presented at 2 conferences/symposiums, and chaired 2 conferences. She has also attended 3 workshops and participated in 10 FDPs. Furthermore, Dr. Krishna Kumari is an active member of 10 professional memberships, showcasing her commitment to professional development and academic excellence.

R.Vijay Raja is an Assistant Professor in College of Management at SRM Institute of Science and Technology. He has more than 12 years of academic experience and has 3.5 years of Industry experience. He completed his Doctor of Philosophy (Ph.D) in the area of "Electronic Recruitment" in Human Resource Management in the Department of Management studies, University of Madras. He completed his Master of Business Administration (MBA) in the year 2010 from London Metropolitan University, United Kingdom and had an opportunity of working as a lecturer in Business Studies, London Victoria College, UK. He holds an undergraduate degree in B.E (Electronics and Communication). As an academician, he expertise in Human Resources and Marketing and taught various subjects like Organizational Behaviour, Human Resource Management, Manpower Planning and Selection, Marketing Management, Consumer Behaviour etc. He also coordinated with TATA Consultancy Services (TCS) for the preparation of Learning and Curriculum Based Framework (LOCF) for B.Tech, Computer Science in Business Systems. At present, he is the program Coordinator for MBA in Hospital Management. He has published various research articles in Scopus Indexed and other peer reviewed journals.

Chikesh Ranjan, a highly accomplished Mechanical Engineer, PhD in Mechanical Engineering at the Department of Mechanical Engineering, Birla Institute of Technology Mesra, Ranchi, Jharkhand. He holds a Master of Engineering (M.E) degree in Mechanical Engineering with a specialization in Design of Mechanical Equipment from the same institution. Chikesh's academic journey began with a Bachelor of Engineering (B.E) in Mechanical Engineering from MIT, Aurangabad, Maharashtra, where he graduated with First Class with distinction. His innovative contributions have been recognized through the publication of 15 journal papers, 10 authored books, and 4 book chapters. He has organized 10 workshops and 18 events, and coordinated 13 events. Additionally, he has participated in 31 training/FDP programs and attended 34 national/international workshops, seminars, and webinars. Chikesh has been granted 10 design patents and 6 foreign patents, demonstrating his dedication to advancing mechanical design and technology.

Salony Sahu completed her Master of Commerce at Gangadhar Meher University, Sambalpur, Odisha, in 2021. Her master's thesis on the impact of the COVID-19 pandemic on the stock prices of leading Indian pharmaceutical companies received high praise from the faculty. She also holds a Bachelor of Commerce with honors in Accountancy from the same institution. Salony has presented her research at various national seminars and is proficient in accounting, bookkeeping, and Microsoft Office, alongside strong team management and leadership skills. Her achievements include a national scholarship and several academic and artistic awards. In her free time, she enjoys painting, sustainable art and craft, and cooking.

Manjari Sharma is an Associate Professor at CHRIST (Deemed to be University), Pune Lavasa Campus. With a Ph.D., an MBA in Finance & Marketing, M.Com., and B.Com., she specializes in Managerial Economics, Financial Management, Securities Analysis, Portfolio Management, and Operations Management. With over 15 years of teaching experience across Asia, Dr. Sharma is known for her innovative teaching methods and extensive contributions to academia. She has authored several books, including “11 Ways to Earn Passive Income: Earn Your Financial Independence” and “Innovative Pedagogy: A Beginner's

Guide to Modern Teaching Strategies." Her research includes a Ph.D. thesis on the price behavior of securities in the Indian capital market, and she co-convened the 'International Conference on Finance and Business Analytics (ICFBA 2021).' Dr. Sharma also holds a patent for an 'IoT Based Fertilizer Spraying Drone.' Her dynamic approach to education and practical insights make her a prominent figure in business and commerce education.

V. Mohana Sundari, with over eleven years of academic experience in Business Management, strongly believes in the democratic and proactive system of teaching and management, where each student is encouraged to participate and contribute to the learning process. With an impressive research portfolio, she has published several papers in well-reputed journals indexed in SCOPUS, showcasing a deep understanding of the subject matter and a passion for research. She also demonstrates a commitment to lifelong learning, having continuously upgraded her skills and knowledge through various certifications from NPTEL, ATAL, and other FDPs, workshops, and trainings. She has served as a resource person for numerous events and conferences, and has also been a member of the Board of Studies for other institutions.

Index

E

F

H

I

K

L

M

N

O

P

R

S

T

U

W

www.ingramcontent.com/pod-product-compliance
Lightning Source LLC
LaVergne TN
LVHW080610300926
849162LV00005B/1104